AF358808

Federated and Transfer Learning Applications

Federated and Transfer Learning Applications

Editors

George Drosatos
Pavlos S. Efraimidis
Avi Arampatzis

Basel • Beijing • Wuhan • Barcelona • Belgrade • Novi Sad • Cluj • Manchester

Editors

George Drosatos
Institute for Language and
Speech Processing, Athena
Research Centre
Xanthi
Greece

Pavlos S. Efraimidis
Department of Electrical and
Computer Engineering,
Democritus University of
Thrace
Xanthi
Greece

Avi Arampatzis
Department of Electrical and
Computer Engineering,
Democritus University of
Thrace
Xanthi
Greece

Editorial Office
MDPI
St. Alban-Anlage 66
4052 Basel, Switzerland

This is a reprint of articles from the Special Issue published online in the open access journal *Applied Sciences* (ISSN 2076-3417) (available at: https://www.mdpi.com/journal/applsci/special_issues/Federated_Transfer_Learning).

For citation purposes, cite each article independently as indicated on the article page online and as indicated below:

Lastname, A.A.; Lastname, B.B. Article Title. *Journal Name* **Year**, *Volume Number*, Page Range.

ISBN 978-3-7258-0075-9 (Hbk)
ISBN 978-3-7258-0076-6 (PDF)
doi.org/10.3390/books978-3-7258-0076-6

© 2024 by the authors. Articles in this book are Open Access and distributed under the Creative Commons Attribution (CC BY) license. The book as a whole is distributed by MDPI under the terms and conditions of the Creative Commons Attribution-NonCommercial-NoDerivs (CC BY-NC-ND) license.

Contents

About the Editors

George Drosatos

George Drosatos has been a researcher with the position title "Privacy Technologies in Content Analysis and Retrieval" since July 2020 at the Institute for Language and Speech Processing, Athena Research Center in Xanthi (Greece). Before that and after receiving his PhD in Privacy-Enhancing Technologies, he was an Associate Researcher for more than 7 years, participating in National and European research projects. During this period, he also received a Fellowship as a PostDoc Researcher, researching the "assessment of news reliability in social networks of influence". Dr. Drosatos has an interdisciplinary background (engineering, computer science, and biomedical science), and his research interests focus mainly on privacy-enhancing technologies, information security, and biomedical science. Last but not least, he was elected in March 2023 as a Secretary General to the European Alliance for Medical and Biological Engineering and Science (EAMBES).

Pavlos S. Efraimidis

Pavlos S. Efraimidis is a Professor of Computer Science at the Department of Electrical and Computer Engineering of the Democritus University of Thrace (Greece). He received his PhD in Informatics in 2000 from the University of Patras and the diploma of Computer Engineering and Informatics (CEID) from the same university in 1995. He has served as a director of the Division of Software and Application Development and as a director of the Programming and Data Processing Lab. His main work is on algorithms, and his current research interests are in the fields of design and analysis of algorithms, graph theory and network analysis, federated machine learning, algorithmic game theory, and algorithmic aspects of privacy. His past industrial experience includes working as a computer engineer in the high-performance computing field (Parsytec Computer GmbH, Aachen, Germany) and the financial sector (ASYK—Athens Stock Exchange, Athens, Greece).

Avi Arampatzis

Avi Arampatzis is an Associate Professor of Computer Science at the Department of Electrical and Computer Engineering of the Democritus University of Thrace (Greece). He received his PhD in Informatics (Information Retrieval and Information Systems group) from the Computing Science Department, University of Nijmegen, the Netherlands, on June 21st, 2001. After that, he was a Postdoctoral Fellow at the Department of Information and Computing Sciences, Center for Geometry, Imaging and Virtual Environments, Utrecht University, the Netherlands (2003–2005) and at the Department of Media Studies, Archives and Information Studies group, University of Amsterdam, the Netherlands (2006–2009). Prof. Dr. Arampatzis has a lot of experience (more than 26 years) in information retrieval, including teaching related courses, and around 100 publications in the field.

Editorial

Federated and Transfer Learning Applications

George Drosatos [1,*], Pavlos S. Efraimidis [1,2] and Avi Arampatzis [2]

[1] Institute for Language and Speech Processing, Athena Research Center, 67100 Xanthi, Greece; pefraimi@ee.duth.gr

[2] Department of Electrical and Computer Engineering, Democritus University of Thrace, 67100 Xanthi, Greece

* Correspondence: gdrosato@athenarc.gr; Tel.: +30-25410-78787 (ext. 322)

1. Introduction

The classic example of machine learning is based on isolated learning—a single model for each task using a single dataset. Most deep learning methods require a significant amount of labelled data, preventing their applicability in many areas where there is a shortage. In these cases, the ability of models to leverage information from unlabelled data or data that are not publicly available (for privacy and security reasons) can offer a remarkable alternative. Transfer learning and federated learning are such alternative approaches that have emerged in recent years. More precisely, transfer learning (TL) is defined as a set of methods that leverage data from additional fields or tasks to train a model with greater generalizability and usually use a smaller amount of labelled data (via fine-tuning) to make them more specific for dedicated tasks. Accordingly, federated learning (FL) is a learning model that seeks to address the problem of data management and privacy through joint training with this data, without the need to transfer the data to a central entity. Figure 1 illustrates the comparison of federated and transfer learning applications with traditional machine learning approaches.

Citation: Drosatos, G.; Efraimidis, P.S.; Arampatzis, A. Federated and Transfer Learning Applications. *Appl. Sci.* **2023**, *13*, 11722. https://doi.org/10.3390/app132111722

Received: 20 October 2023

Accepted: 24 October 2023

Published: 26 October 2023

Copyright: © 2023 by the authors. Licensee MDPI, Basel, Switzerland. This article is an open access article distributed under the terms and conditions of the Creative Commons Attribution (CC BY) license (https://creativecommons.org/licenses/by/4.0/).

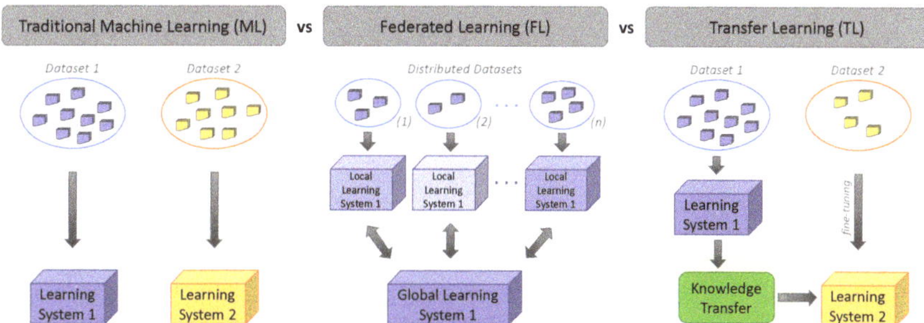

Figure 1. Comparison of federated and transfer learning applications with traditional machine learning.

With this in mind, the present Special Issue of Applied Sciences on "Federated and Transfer Learning Applications" provides an overview of the latest developments in this field. Twenty-four papers were submitted to this Special Issue, and eleven papers [1–11] were accepted (i.e., a 45.8% acceptance rate). The presented papers explore innovative trends of federated learning approaches that enable technological breakthroughs in high-impact areas such as aggregation algorithms, effective training, cluster analysis, incentive mechanisms, influence study of unreliable participants and security/privacy issues, as well as innovative breakthroughs in transfer learning such as Arabic handwriting recognition, literature-based drug–drug interaction, anomaly detection, and chat-based social engineering attack recognition.

2. Federated Learning Approaches

The federated learning approaches published in this Special Issue include aggregation algorithms, layer regularization for effective training, cluster analysis, blockchain-based incentive mechanisms, influence study of unreliable participants, and security/privacy issues.

FL aggregation algorithms: Xu et al. [1] proposed an FL framework, called FedLA (Federated Lazy Aggregation), which reduces aggregation frequency to obtain high-quality gradients and improve robustness to non-independent and identically distributed (non-IID) training data. This framework is particularly used to deal with data heterogeneity. Based on FedAvg, FedLA allows more devices, which come from consecutive rounds, to be trained sequentially, without needing additional information about the devices. Due to sequential training, it performs well on both homogeneous and heterogeneous data, simply by discovering enough samples to aggregate them in time. Regarding the aggregation timing, this work proposes weighted deviation (WDR) as the change rate of the models to monitor sampling. Furthermore, the authors propose a FedLA with a Cross-Device Momentum (CDM) mechanism, called FedLAM, which alleviates the instability of the vertical optimization, helps the local models escape from the local optimum, and improves the performance of the global model in FL. Compared to benchmark algorithms (e.g., FedAvg, FedProx, and FedDyn), FedLAM performs best in most scenarios, achieving a 2.4–3.7% increase in accuracy compared to FedAvg on three image classification datasets.

Li et al. [10] presented a novel federated aggregation algorithm, named FedGSA, which relies on gradient similarity for wireless traffic prediction. FedGSA addresses critical aspects in the optimization of cellular networks, including load balancing, congestion control, and the promotion of value-added services. This method incorporates several key elements: (i) utilizing a global data-sharing strategy to overcome the challenge of data heterogeneity in multi-client collaborative training within federated learning, (ii) implementing a sliding window approach to construct dual-channel training data, enhancing the model's capacity for feature learning, and (iii) proposing a two-layer aggregation scheme based on gradient similarity to enhance the generalization capability of the global model. This entails the generation of personalized models through a comparison of gradient similarity for each client model, followed by aggregation at the central server to obtain the global model. In the experiments conducted using two authentic traffic datasets, FedGSA outperformed the commonly used Federated Averaging (FedAvg) algorithm, delivering superior prediction results while preserving the privacy of client traffic data.

Layer regularization for effective training: Federated learning is a widely used method for training neural networks on distributed data; however, its main limitation is the performance degradation that occurs when data are heterogeneously distributed. To address this limitation, Son et al. [3] proposed *FedCKA*, an alternative approach based on a more up-to-date understanding of neural networks, showing that only certain important layers in a neural network require regularization for effective training. In addition, the authors show that Centred Kernel Alignment (CKA) is the most appropriate when comparing representational similarity between layers of neural networks, in contrast with previous studies that used the L2-distance (FedProx) or cosine similarity (MOON). Experimental results reveal that FedCKA outperforms previous state-of-the-art methods in various deep learning tasks, while also improving efficiency and scalability.

Federated clustering framework: Stallmann and Wilbik [5] presented for first time a federated clustering framework that addresses three key challenges: (i) determining the optimal number of global clusters within a federated dataset, (ii) creating a data partition using a federated fuzzy c-means algorithm, and (iii) validating the quality of the clustering using a federated fuzzy Davies–Bouldin index. This framework is thoroughly evaluated through numerical experiments conducted on both artificial and real-world datasets. The findings indicate that, in most instances, the results achieved by the federated clustering framework are in line with those of the non-federated counterpart. Additionally, the authors incorporate an alternative federated fuzzy c-means formulation into the proposed framework and observe that this approach exhibits greater reliability when dealing with non-IID data, while performing equally well in the IID scenario.

Blockchain-based incentive mechanism: In response to the issues associated with centralization and the absence of incentives in conventional federated learning, Liu et al. [8] introduced a decentralized federated learning method for edge computing environments based on blockchain technology, called BD-FL. This approach integrates all edge servers into the blockchain network, where the edge server nodes that acquire bookkeeping rights collectively aggregate the global model. This addresses the centralization problem inherent in federated learning, which is prone to single points of failure. To encourage active participation of local devices in federated learning model training, an incentive mechanism is introduced. To minimize the cost of model training, BD-FL employs a preference-based stable matching algorithm which associates local devices with suitable edge servers, effectively reducing communication overhead. Furthermore, the authors propose the utilization of a reputation-based practical Byzantine fault tolerance (R-PBFT) algorithm to enhance the consensus process for global model training within the blockchain. The experimental results demonstrate that BD-FL achieves a notable reduction in model training time of up to 34.9% compared to baseline federated learning methods. Additionally, the incorporation of the R-PBFT algorithm enhances the training efficiency of BD-FL by 12.2%.

Influence of unreliable participants and selective aggregation: Esteves et al. [9] introduced a multi-mobile Android-based implementation of a federated learning system and investigated how the presence of unreliable participants and selective aggregation influence the feasibility of deploying mobile federated learning solutions in real-world scenarios. Furthermore, the authors illustrated that employing a more sophisticated aggregation method, such as a weighted average instead of a simple arithmetic average, can lead to substantial enhancements in the overall performance of a federated learning solution. In the context of ongoing challenges in the field of federated learning, this research argues that imposing eligibility criteria can prove beneficial to the context of federated learning, particularly when dealing with unreliable participants.

Survey on privacy and security in FL: Gosselin et al. [2] presented a comprehensive survey addressing various privacy and security issues related to federated learning (FL). Toward this end, the authors introduce an overview of the FL applications, network topology, and aggregation methods, and then present and discuss the existing FL-based studies in terms of security and privacy protection techniques aimed at mitigating FL vulnerabilities. The findings show that the current major security threats are poisoning, backdoor, and Generative Adversarial Network (GAN) attacks, while inference-based attacks are the most critical for FL privacy. Finally, this study concludes with FL open issues and provides future research directions on the topic.

3. Transfer Learning Approaches

The transfer learning approaches published in this Special Issue include Arabic handwriting recognition, literature-based drug–drug interaction, anomaly detection, and chat-based social engineering attack recognition.

Arabic handwriting recognition: Albattah and Albahli [4] proposed an approach for Arabic handwriting recognition by leveraging the advantages of machine learning for classification and deep learning for feature extraction, resulting in the development of hybrid models. The most exceptional performance among the stand-alone deep learning models, trained on both datasets (Arabic MNIST and Arabic Character), was achieved by the transfer learning model applied to the MNIST dataset, with an accuracy of 99.67%. In contrast, the hybrid models performed strongly when using the MNIST dataset, consistently achieving accuracies exceeding 90% across all hybrid models. It is important to note that the results of the hybrid models using the Arabic character dataset were particularly poor, suggesting a possible problem with the dataset itself.

Drug–drug interaction based on the biomedical literature: Zaikis et al. [6] introduced a Knowledge Graph schema integrated into a Graph Neural Network-based architecture designed for the classification of Drug–Drug Interactions (DDIs) within the biomedical literature. Specifically, they present an innovative Knowledge Graph (KG)-based approach that leverages a unique graph structure, combined with a Transformer-based Language Model (LM) and Graph Neural Networks (GNNs), to effectively classify DDIs found in the biomedical literature. The KG is meticulously constructed to incorporate the knowledge present in the DDI Extraction 2013 benchmark dataset without relying on additional external information sources. Each drug pair is classified based on the context in which it appears within a sentence, making use of transfer knowledge through semantic representations derived from domain-specific BioBERT weights, which act as the initial KG states. The proposed approach is evaluated on the DDI classification task using the same dataset, and it achieves an F1-score of 79.14% across the four positive classes, surpassing the current state-of-the-art methods.

Anomaly detection for latency in 5G networks: Han et al. [7] proposed a novel approach for detecting anomalies and issuing early warnings with application to abnormal driving scenarios involving Automated Guided Vehicles (AGVs) within the private 5G networks of China Telecom. This approach efficiently identifies high-latency cases through their proposed ConvAE-Latency model. Building on this, the authors introduce the LstmAE-TL model, which leverages the characteristics of Long Short-Term Memory (LSTM) to provide abnormal early warnings at 15 min intervals. Furthermore, they employ transfer learning to address the challenge of achieving convergence in the training process for abnormal early warning detection, particularly when dealing with a limited sample size. Experimental results demonstrate that both models excel in anomaly detection and prediction, delivering significantly improved performance compared to existing research in this domain.

Chat-based social engineering attack recognition: Tsinganos et al. [11] employed the terminology and methodologies of dialogue systems to simulate human-to-human dialogues within the realm of chat-based social engineering (CSE) attacks. The ability to accurately discern the genuine intentions of an interlocutor is a crucial element in establishing an effective real-time defence mechanism against CSE attacks. The authors introduce 'in-context dialogue acts' that reveal an interlocutor's intent and the specific information they aim to convey. This, in turn, enables the real-time identification of CSE attacks. This research presents dialogue acts tailored to the CSE domain, constructed with a meticulously designed ontology, and establishes an annotated corpus that employs these dialogue acts as classification labels. In addition, they introduce SG-CSE BERT, a BERT-based model that adheres to the schema-guided approach, for zero-shot CSE attack scenarios in dialogue-state tracking. The preliminary evaluation results indicate a satisfactory level of performance.

Author Contributions: Conceptualization, G.D., P.S.E. and A.A.; writing—original draft preparation, G.D.; writing—review and editing, P.S.E. and A.A. All authors have read and agreed to the published version of the manuscript.

Acknowledgments: We would like to thank the authors of the eleven papers, the reviewers, and the editorial team of Applied Sciences for their valuable contributions to this special issue.

Conflicts of Interest: The authors declare no conflict of interest.

References

1. Xu, G.; Kong, D.L.; Chen, X.B.; Liu, X. Lazy Aggregation for Heterogeneous Federated Learning. *Appl. Sci.* **2022**, *12*, 8515. [CrossRef]
2. Gosselin, R.; Vieu, L.; Loukil, F.; Benoit, A. Privacy and Security in Federated Learning: A Survey. *Appl. Sci.* **2022**, *12*, 9901. [CrossRef]
3. Son, H.M.; Kim, M.H.; Chung, T.M. Comparisons Where It Matters: Using Layer-Wise Regularization to Improve Federated Learning on Heterogeneous Data. *Appl. Sci.* **2022**, *12*, 9943. [CrossRef]
4. Albattah, W.; Albahli, S. Intelligent Arabic Handwriting Recognition Using Different Standalone and Hybrid CNN Architectures. *Appl. Sci.* **2022**, *12*, 10155. [CrossRef]

5. Stallmann, M.; Wilbik, A. On a Framework for Federated Cluster Analysis. *Appl. Sci.* **2022**, *12*, 10455. [CrossRef]

6. Zaikis, D.; Karalka, C.; Vlahavas, I. A Message Passing Approach to Biomedical Relation Classification for Drug–Drug Interactions. *Appl. Sci.* **2022**, *12*, 10987. [CrossRef]

7. Han, J.; Liu, T.; Ma, J.; Zhou, Y.; Zeng, X.; Xu, Y. Anomaly Detection and Early Warning Model for Latency in Private 5G Networks. *Appl. Sci.* **2022**, *12*, 12472. [CrossRef]

8. Liu, S.; Wang, X.; Hui, L.; Wu, W. Blockchain-Based Decentralized Federated Learning Method in Edge Computing Environment. *Appl. Sci.* **2023**, *13*, 1677. [CrossRef]

9. Esteves, L.; Portugal, D.; Peixoto, P.; Falcao, G. Towards Mobile Federated Learning with Unreliable Participants and Selective Aggregation. *Appl. Sci.* **2023**, *13*, 3135. [CrossRef]

10. Li, L.; Zhao, Y.; Wang, J.; Zhang, C. Wireless Traffic Prediction Based on a Gradient Similarity Federated Aggregation Algorithm. *Appl. Sci.* **2023**, *13*, 4036. [CrossRef]

11. Tsinganos, N.; Fouliras, P.; Mavridis, I. Leveraging Dialogue State Tracking for Zero-Shot Chat-Based Social Engineering Attack Recognition. *Appl. Sci.* **2023**, *13*, 5110. [CrossRef]

Disclaimer/Publisher's Note: The statements, opinions and data contained in all publications are solely those of the individual author(s) and contributor(s) and not of MDPI and/or the editor(s). MDPI and/or the editor(s) disclaim responsibility for any injury to people or property resulting from any ideas, methods, instructions or products referred to in the content.

Article

Lazy Aggregation for Heterogeneous Federated Learning

Gang Xu [1], De-Lun Kong [1,*], Xiu-Bo Chen [2,*] and Xin Liu [3,*]

[1] School of Information Science and Technology, North China University of Technology, Beijing 100144, China
[2] Information Security Center, State Key Laboratory of Networking and Switching Technology,
Beijing University of Posts and Telecommunications, Beijing 100876, China
[3] School of Information Engineering, Inner Mongolia University of Science and Technology,
Baotou 014010, China
* Correspondence: delunkong96@gmail.com (D.-L.K.); flyover100@163.com (X.-B.C.); lx2001.lx@163.com (X.L.)

Abstract: Federated learning (FL) is a distributed neural network training paradigm with privacy protection. With the premise of ensuring that local data isn't leaked, multi-device cooperation trains the model and improves its normalization. Unlike centralized training, FL is susceptible to heterogeneous data, biased gradient estimations hinder convergence of the global model, and traditional sampling techniques cannot apply FL due to privacy constraints. Therefore, this paper proposes a novel FL framework, federated lazy aggregation (FedLA), which reduces aggregation frequency to obtain high-quality gradients and improve robustness in non-IID. To judge the aggregating timings, the change rate of the models' weight divergence (WDR) is introduced to FL. Furthermore, the collected gradients also facilitate FL walking out of the saddle point without extra communications. The cross-device momentum (CDM) mechanism could significantly improve the upper limit performance of the global model in non-IID. We evaluate the performance of several popular algorithms, including FedLA and FedLA with momentum (FedLAM). The results show that FedLAM achieves the best performance in most scenarios and the performance of the global model can also be improved in IID scenarios.

Keywords: federated learning; heterogeneous data; lazy aggregation; cross-device momentum

Citation: Xu, G.; Kong, D.-L.; Chen, X.-B.; Liu, X. Lazy Aggregation for Heterogeneous Federated Learning. *Appl. Sci.* **2022**, *12*, 8515. https://doi.org/10.3390/app12178515

Academic Editors: George Drosatos, Avi Arampatzis and Pavlos S. Efraimidis

Received: 30 July 2022
Accepted: 20 August 2022
Published: 25 August 2022

Publisher's Note: MDPI stays neutral with regard to jurisdictional claims in published maps and institutional affiliations.

Copyright: © 2022 by the authors. Licensee MDPI, Basel, Switzerland. This article is an open access article distributed under the terms and conditions of the Creative Commons Attribution (CC BY) license (https://creativecommons.org/licenses/by/4.0/).

1. Introduction

With the development of information technology, the amount of data generated by human beings is unprecedented. To mine the potential value of data, a series of emerging information technologies have been promoted (for example, Internet of Things, Big Data, and Data Mining). Nevertheless, it is tough for industry competition, monopoly, and data protection laws to achieve data sharing. Thus, it is necessary to discuss how to break the "data silos" to achieve secure data sharing. In 2016, Google proposed a privacy-constrained distributed neural network training paradigm named FL. Different from centralized training, FL aggregates the models' weights of devices to integrate data knowledge of all parties without disclosing local data and improves the performance of the model on unknown data. FL realizes the efficient and safe sharing of data knowledge among multiple parties. FL has been widely explored and applied in data-sensitive fields such as medical treatment, personalized recommendation, and monetary risk assessment.

Unfortunately, there are some challenges in FL [1,2]. First of all, the open distributed architecture requires network transmission for synchronous training. In practice, the local training delay is much smaller than network transmission delay, especially in the case of large scale and big model. The network quality directly determines training efficiency. Secondly, the central server cannot access local data and data statistics of the participants, which restricts data preprocessing operations (e.g., data cleaning, deduplication, and enhancement). In addition, different distributions and data sparsity greatly weaken the final effect and efficiency, and even fail to converge, which has become a dark cloud over FL.

In order to resolve the issues of communication and data heterogeneity in FL, researchers have carried out a series of explorations. McManhan et al. [3] proposed the vanilla FL framework federated averaging (FedAvg), whose optimization focuses on increasing local epochs to reduce the synchronization frequency, and just some devices are chosen in each round. In Independent and Identically Distributed (IID), FedAvg per- forms very well and takes the first step of FL from theory to practice. However, FedAvg not only slows down the convergence speed but also greatly reduces the performance of the global model in non-IID. Li et al. [4] analyze theoretically convergence of FedAvg for the strongly convex and smooth problem, different data distributions expand the weight difference, and more epochs further exacerbate gradient offsets. Duan et al. [5] proved that it could also affect FL performance in the case of global unbalanced dataset, and they proposed to use a global shared and balanced dataset to alleviate gradient offsets caused by different distributions. As sharing proportion increased, it performs better, which maybe cause data leakage unfortunately. Zeng et al. [6] proposed a novel FL framework FedGSP, which counteracts perturbations for heterogeneous data by sequential execution. Sequential training is more robust to heterogeneous data than parallel training; however, the efficiency of sequential training is lower. For this reason, the author proposes the inter-cluster grouping (IGP), which is achieved inter-group homogeneity and intra-group heterogeneity. Grouping requires device characteristics, which has security risks.

In this paper, we propose a novel method for FL, which is shown in Figure 1. Federated lazy aggregating (FedLA) is used to address data heterogeneity. On the basis of FedAvg, FedLA allows more devices, which are from successional rounds, to train sequentially. FedLA doesn't need extra informations about devices. As paper [6] said, finding out heterogeneous device groups is an NP-hard problem. Due to sequential training performs well under both homogeneous and heterogeneous data, just finding out sufficient sampling to aggregate in time. As for the aggregation timing, this paper proposes the change rate of the models' weight divergence (WDR) to monitor sampling. Furthermore, we propose FedLA with Cross-Device Momentum (CDM) mechanism, FedLA with Momentum (FedLAM), which relieves vertical optimization jitters and helps local models jump out of local optima and release the performance of the global model in FL. In the end, we introduce several benchmark algorithms for comparison. The result shows that FedLAM achieves excellent performance in most scenarios. Compared with FedAvg, FedLAM promotes test accuracy by 3.1% on Mnist and is even higher by 7.3% in class 1. It has a comprehensive increase of 3.7% on Cifar10 and also has a 2.4% improvement on Emnist.

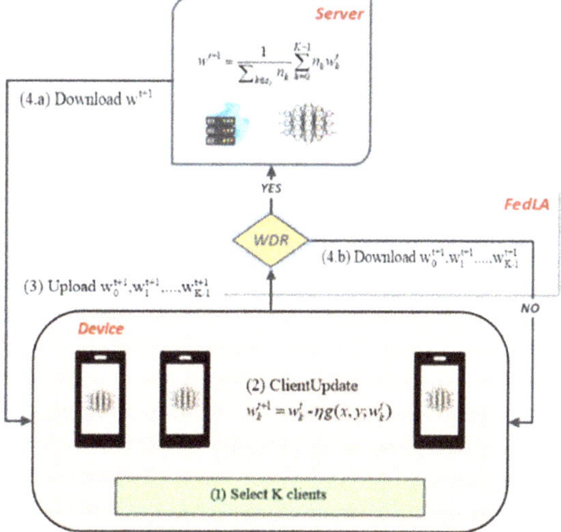

Figure 1. Federated lazy aggregating framework.

2. Related Work

At present, there are three main solutions to handle the FL problem in non-IID, including (1) adaptive optimization, (2) model personalization, and (3) data-based optimization.

2.1. Adaptive Optimization

FedAvg excels in IID but is not ideal in non-IID. To this end, there is a series of works to modify FedAvg to adapt to non-IID scenarios. Li et al. [7] eased the extreme point oscillations by dynamically reducing the local learning rate; moreover, a proximal term is added to local loss to limit the gradient offset. In the presence of straggler devices, aggregation stability is greatly improved. In general, it is similar to FedAvg. Shamir et al. [8] used multi-party distributed optimization to coordinate all parties to find the global optimal solution. However, FL cannot fully participate in each round. Karimireddy et al. [9] proposed a control state scheme, Scaffold which uses device-state and server-state control variables to coordinate the differences in training objectives of all parties, which greatly improves the convergence speed and upper bound of the global model, but its transmission overhead is twice as much as FedAvg. Durmus et al. [10] proposed a dynamic regularization method, FedDyn, by dynamically modifying local objectives to make them asymptotically consistent with the global objective and avoids transferring extra data. It should be noted that the method is very sensitive to learning rate.

2.2. Model Personalization

Just as its name implies, this kind of scheme allows device models to be personalized, and each model may be completely different. Although the personalized models are different, they still contain some global shared knowledge. Finn et al. [11] proposed a federated meta-learning framework, MAML which pretrains a meta-model using auxiliary dataset, and then fine-tunes the meta-model with local data. Fallah et al. proposed scheme [12] is to extract the feature extraction layers of pretraining model and fine-tune output layers to implement personalization. Arivazhagan et al. [13] proposed a personalized scheme which splits the model into two parts: base layer and personalized layer, and only upload weights of the base layer for aggregation in each round. It is personalized learning with the lower requirement for datasets. While the personalized model is excellent locally and not often satisfactory. Sattler et al. [14] proposed a fresh multitask framework, clustered federated learning (CFL), which introduced a recursive dual-partition mechanism based on FedAvg to separate heterogeneous devices and combine homogeneous devices to improve generalization and reduce gradient conflicts, but it is computationally inefficiency. Ghosh et al. [15] put forward an iterative federated clustering learning algorithm IFCA similar to EM. First, several group models are randomly initialized, and the group models are distributed to selected devices in each round, and selected devices choose the optimal weight base on local data. Finally, the server aggregates the gradients by group. The FedSem scheme of Xie et al. [16] is based on l_2 distance between device and group model weights. The above two schemes can improve grouping efficiency. The former needs to transmit more redundancy models, while the latter requires complex model initialization. Duan et al. [17] proposed a pre-grouping scheme, FedGroup, which uses decomposed cosine similarity instead of Euler distance to avoid the "Dimensional Curse" and improve the efficiency and stability of the cluster algorithms. Although CFL establishes the connections of devices to a certain extent, each group is independent. In reality, the relationships of devices are complex and diverse, and hard grouping will cut off connections between devices. Therefore, Li et al. [18] proposed a federated learning with soft clustering algorithm, FLSC based on IFCA. The devices of each group can overlap, and the updated weights can be shared by multiple groups. Its experiment shows that group models of FLSC have better performance than IFCA.

2.3. Data-Based Optimization

Data sharing [4,19,20] is a simple and effective scheme. FedShare [19] uses a global shared and balanced dataset G to initialize the global model, distributes the global model and partial G in each round, and devices use the distributed and local data to train jointly. Experiments show that only 5% of shared dataset makes the global model improve by 30% on Cifar10. It is difficult to obtain dataset such as this in practice. Data augmentation [21,22] can expand dataset through random transformation, knowledge transfer etc., which can effectively alleviate the impact of data sparsity. However, these methods need to acquire statistical informations, which will undoubtedly break the privacy constraint. Knowledge distillation [23] is a kind of promising solution for FL in non-IID, which transfers the knowledge of teacher model to student models with the help of auxiliary dataset. Lin et al. [24] used the ensemble distillation method to fuse multiple local models (teacher) to generate the next global model (student) and extract the knowledge of local models to ensure privacy, but the efficiency of knowledge distillation is relatively low.

In summary, these works have made improvements for heterogeneous federated learning from the data to model and system architecture. To mitigate the difference in data distribution and deal with the conflict gradient are the main goals. Our work focuses on the former, and continuously sampling multiple devices is to make sampling closer to global data distribution. Data sparsity is also very common on the device side, which makes the local model fall into local optimization. To this end, CDM is introduced. More importantly, the two-point optimization does not introduce extra communication overhead and FL privacy is effectively guaranteed, even reducing the server communication load.

3. Methods

3.1. Problem

FedAvg is the prototype of a series of FL algorithms whose theoretical basis is the federated optimization [25] problem, which is essentially a distributed optimization problem with privacy constraints. The distributed optimization goal of FL is:

$$\min_{w} \left\{ F(w) = \sum_{k=1}^{N} a_k L_k(w) \right\} \tag{1}$$

where N is the number of total devices and p_k is the global weight of the k-th device $a_k \geq 0$, $\sum_{k=1}^{N} a_k = 1$. Generally, $a_k = n_k / \sum_{k=1}^{N} n_k$; n_k is the number of the k-th device samples. In general, $L_k(w)$ is widely used cross-entropy loss as the loss function of the classification task:

$$L_k(w) = \sum_{i=1}^{C} p_k(y = i) E_{x|y=i}[\log f_i(x, w)] \tag{2}$$

The local optimization goal is to find the weight w_k with the minimum loss, and FL algorithms look for global optimal weight w with the minimum total loss. To cope with the complex network environments, FedAvg only selects a subset S_t of all devices $\{N\}$ in each round. Then, they execute training in parallel. Finally, the server aggregates recycled updates to renew the global model weight w^{t+1} whose transfer expression is:

$$w^{t+1} = \frac{1}{\sum_{k \in S_T} n_k} w_k^{t+1} \tag{3}$$

When the local distributions of devices are consistent with global distribution (IID), there are few distinctions between partial and full training in each round, while the amount of computation and communication is greatly reduced. In reality, local data are sparse and distributed inconsistently, and gradient conflicts are prone to occur. The effect of gradient aggregation is poor [4], and it is easy to form a low-quality model.

3.2. FedLA

When the sampling is sufficient and IID, the gradients are unbiased estimates, and training in parallel can accelerate convergence of the global model. Otherwise, the gradient directions are divergent or even conflicting, which impacts aggregation [26]. It is known from the literature [6] that sequential SGD is more robust to heterogeneous data. An intuitive idea is to make devices train one by one. However, this way is inefficient, especially for large-scale FL. Moreover, sequential training has higher requirements for network connectivity, and too long sequential training would cause knowledge forgetting phenomenon. On the one hand, the efficiency of parallel training is higher in IID. On the other hand, sequential training is non-negligible robustness. We combine the two methods to form a mutual and allow homogeneous devices to train parallelly and heterogeneous devices to train sequentially. In FL settings, the server cannot obtain data distributions on the device-side, grouping by data distribution becomes an NP-Hard like knapsack problem. Therefore, Zeng et al. [6] proposed an IGP approximation to solve the problem, but the operation of acquiring device characteristics may cause data leakage. We consider that it isn't necessary to judge whether devices are homogeneous or not, and more attention should be paid to how to sample fully. That is to say, the gradients of only one round of devices are often biased in non-IID, which makes aggregation unsatisfactory. Reducing the frequency of the server's aggregation maybe a better choice.

The accuracy of gradient depends on whether data sampling is sufficient. In non-IID settings, data distribution is inconsistent with global distribution on the device- side, which is likely to cause gradient bias. The cross-device sequential training can sample more samples and gain high-quality gradients. Too long sequential training also leads to knowledge forgetting and is low efficiency. It is necessary to aggregate and cut off sequential chains periodically. To this end, we introduce the change rate of the models' weight divergence (WDR) to judge the aggregation timing. Set WDR in the t-th round as:

$$WDR_t = \left| \frac{WD_t - WD_{t-1}}{WD_t} \right| \tag{4}$$

WD_t is the models' weight divergence (WD) in the t-th round:

$$WD_t = \frac{1}{K} \sum_{0 \leq i < j < K} \| w_t^i - w_t^j \| \tag{5}$$

K is the number of selected devices per round. The model weight W_t^i is updated for the selected i-th device after local training in the t-th round. In this way, the server recycles updated weights, it judges the timing of aggregation according to (4) calculated WDR. When the WDR is less than the threshold, the server aggregates and forms a new global model, otherwise, the server would randomly distribute collected and updated weights to the next round of devices. When the model weights are not aggregated, the models' WD becomes larger and larger, but their WDR becomes smaller and smaller. Because, with the increase in continuous sampling, the updated directions of the models' weights are gradually consistent with the global optimization direction.

Zhao et al. [19] have theoretically proved that the WD between FedAvg and vanilla SGD is related to the probabilistic distance of the local and global distribution. Since the global data distribution is unknown as usual, the WD focuses on the divergence of the models' weights, which are to be aggregated in this paper.

$$\begin{aligned}
\| w_t^i - w_t^j \| \leq & (1 + \eta\beta) \| w_{t-1}^i - w_{t-1}^j \| \\
& + \eta g_{max}(w_{t-1}^j) \sum_{k=1}^{C} \| p^i(y = k) - p^j(y = k) \|
\end{aligned} \tag{6}$$

where $0 < \beta = \sum_k^C p^i(y=k)\lambda_{x|y=k}$ and $g_{max}(w_{t-1}^j) = \max\limits_{0 \leq k < C} \nabla_w E_{x|y=k}[log f_k(x, w_{t-1}^j)]$ and η is the fixed learning rate (LR) of local SGD in our research. We can conclude that the WD mainly comes from the difference in the initial weights and sampling distributions. The former can be reduced by synchronous aggregation, which is the focus of previous works, but this paper has shifted its attention to the latter. For the aggregation, the result of multi-epochs training with a small LR isn't different from an epoch training with a lager LR on the device side, and multiple devices with few samples train sequentially which is equivalent to a device training with a large number of samples. Therefore, we can generalize the conclusion of the inequality (6). The sampling range is extended from batch to device.

With sampling, the WD growth caused by distribution difference disappears gradually. Therefore, monitoring WDR is a great way to judge whether sampling is necessary. If WDR is greater than the threshold, the current round of weights won't be aggregated but forwarded to the next round of selected devices. On the contrary, we consider that the sampling is sufficient. Only when their whole distributions are consistent with the global distribution will the WD growth slow down. On this basis, this paper proposes a federated lazy aggregate (FedLA) as shown in Algorithm 1:

Algorithm 1: Federated Lazy Aggregate (FedLA)

Input: T, K, E, B, w_0, d_0, ε

Initialize $n_0^0, n_0^1 \ldots n_0^{k-1} \leftarrow 0$; $w_0^0, w_0^1 \ldots w_0^{K-1} \leftarrow w_0$

for t = 0, 1, 2, \ldots , $T - 1$ **do**

    Sample devices $S_t \in [N]$, $|S_t| = K \leq N$

    Transmit $w_t^0 \, w_t^1, \ldots, w_t^{K-1}$ to selected K devices, respectively

    for each device k $\in S_t$ in parallel **do**

        $n_k, w_{t+1}^k = ClientUpdate\left(w_t^k, E, B\right)$

        Transmit device n_k, w_{t+1}^k to server

    end for

    for k = 0, 1, 2, \ldots , $T - 1$ **do**

        $n_{t+1}^k \leftarrow n_t^k + n_k$

    end for

    $d_{t+1} = WD\left(w_{t+1}^0, w_{t+1}^1, \ldots w_{t+1}^{K-1}\right)$

    if $\left|\frac{d_{t+1} - d_t}{d_{t+1}}\right| \leq \varepsilon$ **do**

        $w_{t+1} \leftarrow \frac{1}{\sum_{k=0}^{K-1} n_{t+1}^k} \sum_{k=0}^{K-1} n_{t+1}^k w_{t+1}^k$; $w_{t+1}^0, w_{t+1}^1 \ldots w_{t+1}^{K-1} \leftarrow w_{t+1}$

        $n_{t+1}^0, n_{t+1}^1 \ldots n_{t+1}^{K-1} \leftarrow 0$; $d_{t+1} = 0$

    else

        $w_{t+1} \leftarrow w_t$

    end if

end for

Output: w_T

Where E represents the number of local training epochs, B represents the batch size for loading data, and w^k represents the weight of k-th device which is initialized to the received weight in the t-th round. T represents the number of total rounds; w_0 represents initial global weight; d_0 indicates the initial WD value, generally assign 0; ε is the threshold of WDR; and n_t^k indicates the number of cumulative samples in the t-th round. FedLA adopts the same local solver as FedAvg, which optimizes the received weight by mini-batch SGD, and then the updated weights are sent back to server. Unlike FedAvg, aggregation doesn't occur every round, but is determined according to WDR. The impact of heterogeneous data could be effectively mitigated through aggregation with intervals. In addition to adding WDR (computational complexity $O(K^2 d)$, where d is the number of parameters in the model) and postponing aggregation, no other optimizations are introduced. Thus, FedLA has the same convergence rate $O(1/T)$ as FedAvg [4] for strongly convex and smooth problems in non-IID. In practice, only classifier weights are needed for judgment, and the weights of feature layers are transmitted when aggregation is possible which ensures efficiency and reduces redundant aggregation. Moreover, it is compatible with other optimization and privacy strategies.

3.3. FedLAM

FedLA eliminates redundant aggregations and allows more devices to train sequentially, enabling more data to be sampled and alleviating data sparsity (e.g., few samples, class imbalance, and poor quality). Continuous training sequentially without aggregation occurs more jitters than vanilla FL on vertical optimization. It is more urgent for momentum mechanism, especially for cross-devices. It can correct current gradients by previous gradients and raise the utility of historical gradients. Firstly, we assign K groups of gradient momentums on the server side, the initial value is zero gradient $\hat{0}$ and its transfer equation is:

$$m_t^k \leftarrow \mu m_{t-1}^k + \Delta w_t^k \tag{7}$$

where μ is the momentum coefficient, the bigger coefficient means that historical gradients are dominant and have a longer survival period; m_t^k is the momentum of the k-th group in the t-th round; and Δw_t^k indicates the k-th updated gradient in the t-th round. Therefore, we propose the FedLA with momentum (FedLAM) that adds the CDM mechanism. Algorithm 2 is as follows:

Algorithm 2: Federated Lazy Aggregate with Momentum (FedLAM)

Input: T, K, E, B, w_0, d_0, ε, μ
Initialize $m_0^0, m_0^1 \ldots m_0^{K-1} \leftarrow \hat{0}$; $n_0^0, n_0^1 \ldots n_0^{K-1} \leftarrow 0$; $w_0^0, w_0^1 \ldots w_0^{K-1} \leftarrow w_0$
for t = 0, 1, 2, … , $T-1$ **do**
    Sample devices $S_t \in [N], |S_t| = K \leq N$
    Transmit $w_t^0\, w_t^1, \ldots, w_t^{K-1}$ to selected K devices, respectively
    for each device $k \in S_t$ in parallel **do**
        $n_k,\ \Delta w_{t+1}^k = ClientUpdate\left(w_t^k,\ E,\ B\right)$
        Transmit device $n_k,\ \Delta w_{t+1}^k$ to server
    end for
    for k = 0, 1, 2, … , $T-1$ **do**
        $n_{t+1}^k \leftarrow n_t^k + n_k$
        $m_{t+1}^k \leftarrow \mu m_t^k + \Delta w_{t+1}^k$
        $w_{t+1}^k \leftarrow w_t^k + m_{t+1}^k$
    end for
    $d_{t+1} = WD\left(w_{t+1}^0,\ w_{t+1}^1,\ \ldots\ w_{t+1}^{K-1}\right)$
    if $\left|\frac{d_{t+1} - d_t}{d_{t+1}}\right| \leq \varepsilon$ **do**
        $w_{t+1} \leftarrow \frac{1}{\sum_{k=0}^{K-1} n_{t+1}^k} \sum_{k=0}^{K-1} n_{t+1}^k w_{t+1}^k;\ w_{t+1}^0, w_{t+1}^1 \ldots w_{t+1}^{K-1} \leftarrow w_{t+1}$
        $m_{t+1} \leftarrow \frac{1}{\sum_{k=0}^{K-1} n_{t+1}^k} \sum_{k=0}^{K-1} n_{t+1}^k m_{t+1}^k;\ m_0^0, m_0^1 \ldots m_0^{K-1} \leftarrow m_{t+1}\ (*)$
        $n_{t+1}^0, n_{t+1}^1 \ldots n_{t+1}^{K-1} \leftarrow 0;\ d_{t+1} = 0$
    else
        $w_{t+1} \leftarrow w_t$
    end if
end for
Output: w_T

It is noteworthy that the selected devices send their gradients instead of weights, and the weights of each group are updated by corrected gradients which enhance the horizontal optimization and offset the vertical optimization jitters. In non-IID settings, the local models fall easily into the local optimum [7], which is further passed to the global model through their gradients [27]. The step of momentum aggregation is optional in Algorithm 2(*). It is suggested to train early and close it later to expand the search optimization space, which contributes to the walking out of the saddle point. CDM could help FL alleviate gradient disappearance and take the global performance to a higher level. Compared with [28], our scheme doesn't pass momentum on the device, which undoubtedly reduces the traffic between the device side and server side.

4. Evaluation

In this section, we will evaluate the actual performance of FedLA and FedLAM. First, we implement the above and several benchmark algorithms based on PyTorch and Ray frameworks. Three public datasets are selected for testing. In addition, we set up IID and two kinds of non-IID scenarios (label distribution skew, [5] Dirichlet [29]) to simulate actual environments in FL.

4.1. Experiment Setup

To compare different benchmark algorithms simply and fairly, this section selects three kinds of image classification datasets, Mnist [30], Emnist [31], and Cifar10 [32], and simulates IID and non-IID scenarios by partitioning the datasets. All experimental datasets are divided into 100 copies; each device holds one, and the number of selected devices $K = 10$. It should be noted that the local training doesn't use momentum, weight decay options of SGD optimizer and gradient clipping technologies, and the constant learning rate is used by local solver. The datasets, partitions and models used in our experiments are shown in Table 1.

4.1.1. Datasets and Models

- Mnist, a widely used image classification task, contains a training dataset of 60,000 handwritten digit pictures and a test dataset of 10,000. Its elements are 28×28 pixel black and white pictures with 10 classification labels. Since the dataset is relatively simple and pure, and complex models (e.g., CNN, RNN) are little discriminated, for comparison, only single-layer MLP is used in this section.
- Emnist (Extend Mnist) expands 402,000 numbers and 411,000 26 letters samples on the basis of Mnist. Due to the huge amount of dataset, six types of splits (e.g., by class, by merge, letters etc.) are introduced. To distinguish from Mnist, the letters split is used, and a two-layer simple CNN is used for the test model.
- Cifar10 contains 60,000 32×32 pixel color images in 10 categories, and its classes are completely mutually exclusive (e.g., car and airplane), while the samples of the same category are quite different (e.g., car and truck). The difficulty of classification tasks has undoubtedly increased significantly, so we choose a three-layer convolutional layer and a two-layer fully connected layer CNN.

4.1.2. Data Partitions

As shown in Figure 2, data distributions of the clients are under three federated data partitions.

- IID: The dataset is randomly shuffled and divided into several subsets of the same size.
- Label distribution skew: Zhao et al. [5] proposed a more demanding non-IID setting, the outstanding feature of which is that each subset only holds a few classes of samples.
- Dirichlet: Wang et al. [29] proposed a partition scheme based on the Dirichlet distribution, where each subset is also class-imbalanced with different amounts of samples. It is closer to a partition scheme than the real FL environments.

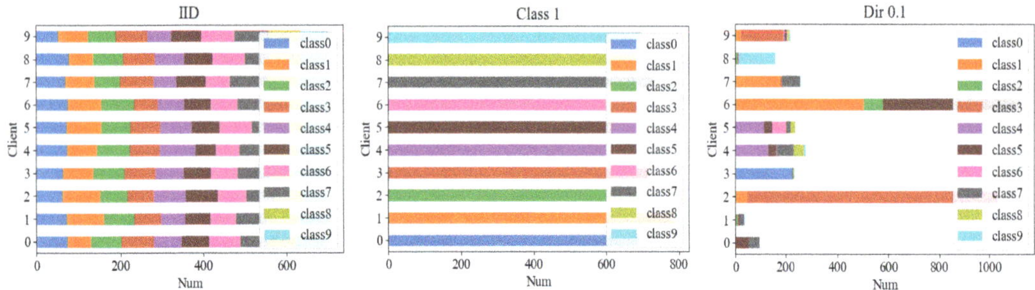

Figure 2. Schematic diagram of class distribution of data partitions.

Table 1. Summary of federated datasets and models.

Dataset	Samples	Classes	Devices	Partitions	Models	
MNIST	69,035			IID	MLP	101,770
CIFAR10	60,000	10	100	Class 1/3	CNN	553,514
				Dir-0.1/0.3		
EMNIST (letters)	145,600	26		Class 3/9	SimpleCNN	28,938

4.1.3. Baselines

1. FedAvg: The vanilla FL Algorithm [3].
2. FedProx: A popular FL adaptive optimization scheme that limits excessive offset of the device-side model by adding a proximal term [7].
3. FedDyn: A dynamic regularization method for FL, and local objectives are dynamically updated to ensure that local optimums are asymptotically consistent with the global optimum [10].
4. CFL: A clustering federated learning framework based on recursive bi-partition that uses cosine similarity to separate the devices with gradient conflicts and form their respective cluster center models [14].

4.1.4. Evaluation Metric

In order to comprehensively evaluate the FL algorithms' performances, we take top-1 classification accuracy of the devices' test dataset as the main indicator. Note that all original training and test samples are merged into a new dataset which is divided into 100 copies, including 80% training dataset and 20% test dataset, which is to ensure that the training and test dataset is the same distribution on the devices. As it involves multi-party evaluation and local data are inaccessible, we use weighted accuracy on the test dataset of selected devices as verification accuracy and we take the weighted accuracy of all devices as test accuracy. There are several cluster center models in CFL, and it cannot be directly compared with other algorithms. To facilitate comparison, except for the weighted accuracies of cluster models, the pseudo-global model aggregated by all cluster models is introduced [17]. Note that this paper selects the benchmark algorithms without introducing extra synchronized parameters, so their communication traffics are at the same level.

4.2. Effects of the Proposed Algorithm
4.2.1. Performance

There are several points to explain: (1) The highest score of testing is taken after training with 300 rounds on Mnist and Emnist or 500 rounds on Cifar10. (2) The learning rates of the experiments are unified 0.002. (3) Other hyperparameters are as follows: FedProx: offset penalty coefficient u = 0.01; FedDyn: regularity coefficient α = 0.01; CFL: mean-norm threshold eps1 = 0.035, max-norm threshold eps2 = 0.5; FedLA: WDR threshold ε = 0.02; and FedLAM: historical gradient momentum coefficient μ = 0.5. Differences from the above will be noted. We found that the FedLA and FedLAM algorithms perform well in most scenarios, as shown in Table 2, compared with the baseline algorithms.

Table 2. Comparisons of FedAvg, FedProx, FedDyn, CFL, FedLA, and FedLAM on Mnist, Emnist, and Cifar10 (lr = 0.01) in IID and four kinds of non-IID settings. Local epoch E = 5, batch size B = 32.

		FedAvg	FedProx	FedDyn	FedLA	FedLAM	CFL
Mnist	IID	91.5	91.5	93.7	91.4	**95.3**	$91.6_1 91.6$
	Class 1	83.3	83.2	85.8	87.6	90.6	$\mathbf{93.4_2} 70.8$
	Class 3	90.0	90.0	88.9	89.8	**92.4**	$90.0_1\ 90.0$
	Dir 0.1	90.0	90.0	88.7	88.8	**91.5**	$91.2_2 88.8$
	Dir 0.3	89.8	89.8	89.1	90.3	**90.4**	$89.8_1\ 89.8$
Emnist	IID	89.4	89.7	**91.4**	89.9	89.4	$89.9_1\ 89.9$
	Class 3	79.1	79.6	84.9	79.3	**85.3**	$79.2_1\ 79.2$
	Class 9	87.6	87.4	**90.0**	85.7	89.1	$87.6_1\ 87.6$
	Dir 0.1	84.3	84.9	**88.5**	84.2	87.8	$85.1_1\ 85.1$
	Dir 0.3	89.1	89.3	90.7	89.1	89.7	$89.1_1\ 89.1$
Cifar10	IID	70.8	69.6	71.5	70.4	**76.3**	$70.9_1\ 70.9$
	Class 1	27.0	27.0	**41.8**	$28.0_{0.03}$	$30.0_{0.07,0.2}$	$\mathbf{100.0_{10}}\ 17.1$
	Class 3	65.3	66.2	64.1	67.3	**71.1**	$66.0_1\ 66.0$
	Dir 0.1	62.5	62.1	59.3	60.4	**64.6**	$62.7_1\ 62.7$
	Dir 0.3	66.0	66.1	65.6	68.1	**71.5**	$66.5_1\ 66.5$

On Mnist, FedLAM is ahead of other algorithms. Compared with FedAvg, it leads by 3.8% in IID, and leads comprehensive by 2.9% in non-IID, especially by 7.3% in class 1. For the more difficult task, Cifar10, it increases by 5.5% in IID and there is an overall increase of 3.2% in non-IID. In most cases, FedProx is very close to FedAvg. The method of adding a proximal term doesn't overcome gradient conflicts caused by heterogeneous data but only reduces excessive gradient offset. As for the straggler devices mentioned in Section 5.2 [7], we will not explore this case in this article. FedDyn performs well on Emnist by aligning the local targets with the global target in real-time through an attached dynamic regularizer. As shown in Figure 3, FedDyn has a fast convergence speed in the early stage, but the accuracy of the global model is stagnant in the later stage, and its loss fluctuates significantly. The main reason is that it doesn't adapt to the constant learning rate which needs to be adjusted flexibly. CFL separates the devices with gradient conflicts to build a flesh model and utilizes multiple models to adapt to devices with different distributions to improve comprehensive accuracy. For Cifar10 in class 1, the performances of other algorithms are not satisfactory, and CFL uses recursive dual partition to separate the different classes of devices and bring test accuracies to 100%. We also observed that the pseudo-global model has a poor effect, which verifies the great differences between classes. Although FedLA mitigates the gradient conflict caused by insufficient sampling in a way, it is limited by the aggregation method and it's inability to fuse gradients well in Cifar10 (class 1). Gradient projection [33] technology solves the impact of opposing gradients. In non-IID scenarios, FedLA has a better performance than FedAvg. It reduces redundant aggregation and finds parallel groups of devices which are consistent with global distribution by computing WDR and overcomes the impact of data sparsity on a single device. Furthermore, the collected gradients could be used to correct vertical oscillation and walk out of the steep fall. FedLAM supplemented by CDM has the best performance in most cases, which is an exciting optimization.

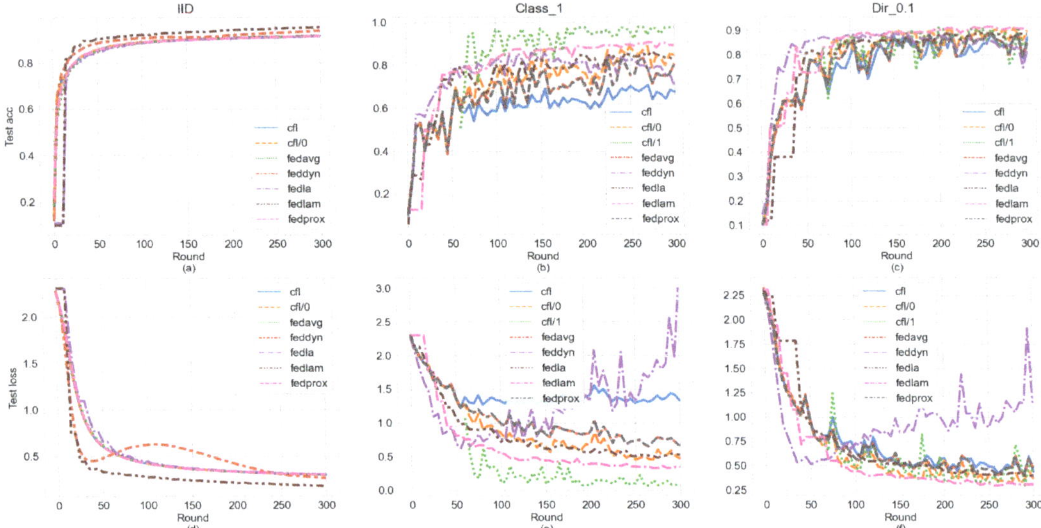

Figure 3. Mnist-MLP: Test top 1 accuracy and test loss under IID, class 1, dir 0.1. (**a–c**) test accuracy curves of FedLA(M) and baselines, (**d–f**) corresponding loss curves under three scenarios. To evaluate the convergence efficiency and stability.

4.2.2. WDR and Cross-Device Momentum

In order to explore the effectiveness of WDR and cross-device momentum, we added fixed-interval aggregations and different proportions of momentum based on Mnist (class 1).

As shown in Figure 4a,c, we sampled 2, 5, 10, and 20 devices for sequential training before aggregation. We found that sampling five devices is enough to obtain unbiased gradients in this experiment. Oversampling will not further improve the performance of the global model and reduces learning efficiency, or even has negative effects (knowledge forgetting, resource waste). According to inequality 3, WD mainly comes from the difference in the initial weights and sampling distribution. We prefer to sacrifice part of the former for the latter in non-IID. However, the number of sampling rounds is unknown. Fortunately, it could be observed that the growth rate of WD slowed down significantly after five rounds. Apart from the difference caused by initial weight, sampling distribution gradually tends to be consistent, and the second term of the inequality (6) approaches 0. This is why this paper proposes that WDR is used to find sufficient sampling in a timely manner. It can be used to adjust aggregated intervals dynamically, which makes the performance curve of the global model smoother and accelerates convergence.

In Figure 4b,d, we explored the influence of proportions of historical gradients in FL. In early training, a higher momentum will hinder the convergence of the global model, but it can bring FL performance to a higher level in the later stage. The gradients of devices will gradually disappear with FL training; nonetheless, the main reason is that the device model falls into the local optimum. Whether increasing sampled devices or the proportions of momentum, their gradient qualities are enhanced remarkably, which is undoubtedly helpful to the promotion of the global model. Inspired by the experiment, the momentum mechanism with dynamic adjustment deserves further exploration in the future. To sum up, FedLA determines aggregation timing by monitoring WDR, flexibly handles complex FL environments, and CDM mechanism improves FL performance in both IID and non-IID.

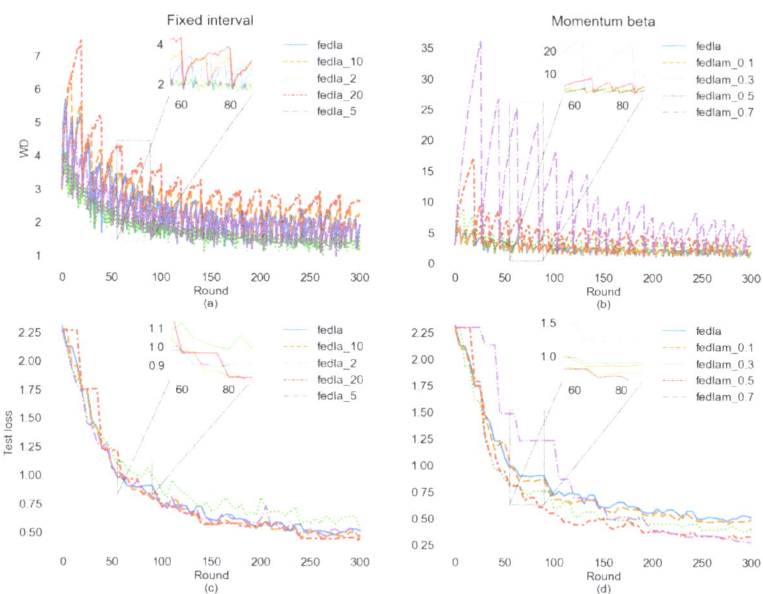

Figure 4. WD and loss under class 1 on Mnist. (**a**,**b**) model weight divergence during training, (**c**,**d**) model convergence performance.

5. Conclusions

This paper studies the training problem with heterogeneous data in FL. Due to sparse data and different distributions of devices in FL, sampling only one round of devices is not enough and creates a poor model. Therefore, this paper proposes a novel FL framework, FedLA, which allows us to sample more devices by putting off aggregation, making aggregation more robust. We also note that local optimization is easier to fall into the saddle point. Thus, the cross-device momentum mechanism is added to FedLA to further release the performance of the training model in FL. Compared with benchmark algorithms (e.g., FedAvg, FedProx, FedDyn, and CFL), FedLAM has the best performance in most scenarios. In the future, we plan to conduct the theoretical analysis of FedLA in detail, and study more advanced aggregation and sampling judgment strategies. Moreover, the dynamic scheduling strategies of learning rate and momentum are introduced to further accelerate the convergence speed of FedLA.

Author Contributions: Conceptualization, D.-L.K. and G.X.; methodology, D.-L.K.; software, D.-L.K.; validation, D.-L.K.; formal analysis, D.-L.K.; investigation, G.X., X.-B.C. and D.-L.K.; re-sources, D.-L.K.; data curation, D.-L.K.; writing original draft preparation, D.-L.K.; writing review and editing, X.L.; visualization, X.-B.C.; supervision, G.X.; project administration, X.L.; funding acquisition, G.X., X.-B.C. and X.L. All authors have read and agreed to the published version of the manuscript.

Funding: This research was funded by NSFC (Grant Nos. 92046001), the Fundamental Research Funds for Beijing Municipal Commission of Education, Beijing Urban Governance Research Base of North China University of Technology, the Natural Science Foundation of Inner Mongolia (2021MS06006), Baotou Kundulun District Science and technology plan project (YF2020013), and Inner Mongolia discipline inspection and supervision big data laboratory open project fund (IM-DBD2020020).

Institutional Review Board Statement: Not applicable.

Informed Consent Statement: Not applicable.

Data Availability Statement: Not applicable.

Conflicts of Interest: The authors declare no conflict of interest.

References

1. Li, T.; Sahu, A.K.; Talwalkar, A.; Smith, V. Federated learning: Challenges, methods, and future directions. *IEEE Signal Processing Mag.* **2020**, *37*, 50–60. [CrossRef]
2. Zhu, H.; Xu, J.; Liu, S.; Jin, Y. Federated learning on non-IID data: A survey. *Neurocomputing* **2021**, *465*, 371–390. [CrossRef]
3. McMahan, H.B.; Moore, E.; Ramage, D.; Hampson, S.; Arcas, B.A. Communication-Efficient Learning of Deep Networks from Decentralized Data. In Proceedings of the 20th International Conference on Artificial Intelligence and Statistics, Fort Lauderdale, FL, USA, 20–22 April 2017.
4. Li, X.; Huang, K.; Yang, W.; Wang, S.; Zhang, Z. On the Convergence of FedAvg on Non-IID Data. In Proceedings of the 8th International Conference on Learning Representations, Virtual, 26–30 April 2020.
5. Zhao, Y.; Li, M.; Lai, L.; Suda, N.; Civin, D.; Chandra, V. Federated learning with non-iid data. *arXiv* **2018**, arXiv:1806.00582. [CrossRef]
6. Zeng, S.; Li, Z.; Yu, H.; He, Y.; Xu, Z.; Niyato, D.; Yu, H. Heterogeneous federated learning via grouped sequential-to-parallel training. In *International Conference on Database Systems for Advanced Applications*; Springer: Berlin/Heidelberg, Germany, 2022; pp. 455–471.
7. Li, T.; Sahu, A.K.; Zaheer, M.; Sanjabi, M.; Talwalkar, A.; Smith, V. Federated optimization in heterogeneous networks. *Proc. Mach. Learn. Syst.* **2020**, *2*, 429–450.
8. Shamir, O.; Srebro, N.; Zhang, T. Communication-efficient distributed optimization using an approximate newton-type method. In Proceedings of the 31th International Conference on Machine Learning, Beijing, China, 21–26 June 2014.
9. Karimireddy, S.P.; Kale, S.; Mohri, M.; Reddi, S.J.; Stich, S.U.; Suresh, A.T. SCAFFOLD: Stochastic Controlled Averaging for Federated Learning. In Proceedings of the 37th on International Conference on Machine Learning, Virtual, 12–18 July 2020.
10. Acar, D.A.E.; Zhao, Y.; Navarro, R.M.; Mattina, M.; Whatmough, P.N.; Saligrama, V. Federated learning based on dynamic regularization. In Proceedings of the 9th International Conference on Learning Representations, Virtual, 3–7 May 2021.
11. Finn, C.; Abbeel, P.; Levine, S. Model-agnostic meta-learning for fast adaptationof deep networks. In Proceedings of the 34th on International Conference on Machine Learning, Sydney, NSW, Australia, 6–11 August 2017.
12. Fallah, A.; Mokhtari, A.; Ozdaglar, A. Personalized federated learning with theoretical guarantees: A model-agnostic meta-learning approach. *Adv. Neural Inf. Processing Syst.* **2020**, *33*, 3557–3568.
13. Arivazhagan, M.G.; Aggarwal, V.; Singh, A.K.; Choudhary, S. Federated learning with personalization layers. *arXiv* **2019**, arXiv:1912.00818.
14. Sattler, F.; Müller, K.-R.; Samek, W. Clustered federated learning: Model-agnostic distributed multitask optimization under privacy constraint. *IEEE Trans. Neural Netw. Learn. Syst.* **2020**, *32*, 3710–3722. [CrossRef] [PubMed]
15. Ghosh, A.; Chung, J.; Yin, D.; Ramchandran, K. An efficient framework for clustered federated learning. *Adv. Neural Inf. Processing Syst.* **2020**, *33*, 19586–19597. [CrossRef]
16. Long, G.; Xie, M.; Shen, T.; Zhou, T.; Wang, X.; Jiang, J. Multi-center federated learning: Clients clustering for better personalization. *World Wide Web* **2022**, *6*, 1–20.
17. Duan, M.; Liu, D.; Ji, X.; Liu, R.; Liang, L.; Chen, X.; Tan, Y. FedGroup: Efficient clustered federated learning via decomposed data-driven measure. *arXiv* **2020**, arXiv:2010.06870.
18. Li, C.; Li, G.; Varshney, P.K. Federated Learning with Soft Clustering. *IEEE Internet Things.* **2021**, *9*, 7773–7782.
19. Yao, X.; Huang, T.; Zhang, R.-X.; Li, R.; Sun, L. Federated learning with unbiased gradient aggregation and controllable meta updating. *arXiv* **2019**, arXiv:1910.08234.
20. Tuor, T.; Wang, S.; Ko, B.J.; Liu, C.; Leung, K.K. Overcoming noisy and irrelevant data in federated learning. In Proceedings of the 25th International Conference on Pattern Recognition, Virtual, 10–15 January 2021.
21. Tanner, M.A.; Wong, W.H. The calculation of posterior distributions by data augmentation. *J. Am. Stat. Assoc.* **1987**, *82*, 528–540. [CrossRef]
22. Duan, M.; Liu, D.; Chen, X.; Tan, Y.; Ren, J.; Qiao, L.; Liang, L. Astraea: Self-balancing federated learning for improving classification accuracy of mobile deep learning applications. In Proceedings of the 2019 IEEE 37th International Conference on Computer Design, Abu Dhabi, United Arab Emirates, 17–20 November 2019.
23. Buciluă, C.; Caruana, R.; Niculescu-Mizil, A. Model compression. In Proceedings of the 12th ACM SIGKDD International Conference on Knowledge Discovery and Data Mining, Philadelphia, PA, USA, 20–23 August 2006.
24. Lin, T.; Kong, L.; Stich, S.U.; Jaggi, M. Ensemble distillation for robust model fusion in federated learning. *Adv. Neural Inf. Processing Syst.* **2020**, *33*, 2351–2363.
25. Konečný, J.; McMahan, H.B.; Ramage, D.; Richtárik, P. Federated optimization: Distributed machine learning for on-device intelligence. *arXiv* **2016**, arXiv:1610.02527.
26. Khaled, A.; Mishchenko, K.; Richtárik, P. Tighter theory for local SGD on identical and heterogeneous data. In Proceedings of the 23rd International Conference on Artificial Intelligence and Statistic, Parerlmo, Sicily, Italy, 26–28 August 2020.
27. Kingma, D.P.; Ba, J. Adam: A method for stochastic optimization. *arXiv* **2014**, arXiv:1412.6980.
28. Liu, W.; Chen, L.; Chen, Y.; Zhang, W. Accelerating federated learning via momentum gradient descent. *IEEE Trans. Parallel Distrib. Syst.* **2020**, *31*, 1754–1766. [CrossRef]
29. Wang, H.; Yurochkin, M.; Sun, Y.; Papailiopoulos, D.; Khazaeni, Y. Federated learning with matched averaging. *arXiv* **2020**, arXiv:2002.06440.

30. Yann, L.; Corest, C.; Burges, C.J.C. The MNIST Database of Handwritten Digits. Available online: http://yann.lecun.com/exdb/mnist (accessed on 10 October 2021).
31. Cohen, G.; Afshar, S.; Tapson, J.; Van Schaik, A. EMNIST: Extending MNIST to handwritten letters. In Proceedings of the 2017 International Joint Conference on Neural Networks, Anchorage, AK, USA, 14–19 May 2017.
32. Doon, R.; Rawat, T.K.; Gautam, S. Cifar-10 classification using deep convolutional neural network. In Proceedings of the 2018 IEEE Punecon, Pune, India, 30 November–2 December 2018.
33. Wang, Z.; Fan, X.; Qi, J.; Wen, C.; Wang, C.; Yu, R. Federated learning with fair averaging. *arXiv* **2021**, arXiv:2104.14937.

Article

Privacy and Security in Federated Learning: A Survey

Rémi Gosselin [1,†], **Loïc Vieu** [1,†], **Faiza Loukil** [2,*] and **Alexandre Benoit** [2,*]

1   Polytech Annecy-Chambéry, Savoie Mont Blanc University, F-74944 Annecy, France
2   LISTIC, Savoie Mont Blanc University, F-74944 Annecy, France
*   Correspondence: faiza.loukil@univ-smb.fr (F.L.); alexandre.benoit@univ-smb.fr (A.B.)
†   These authors contributed equally to this work.

Abstract: In recent years, privacy concerns have become a serious issue for companies wishing to protect economic models and comply with end-user expectations. In the same vein, some countries now impose, by law, constraints on data use and protection. Such context thus encourages machine learning to evolve from a centralized data and computation approach to decentralized approaches. Specifically, Federated Learning (FL) has been recently developed as a solution to improve privacy, relying on local data to train local models, which collaborate to update a global model that improves generalization behaviors. However, by definition, no computer system is entirely safe. Security issues, such as data poisoning and adversarial attack, can introduce bias in the model predictions. In addition, it has recently been shown that the reconstruction of private raw data is still possible. This paper presents a comprehensive study concerning various privacy and security issues related to federated learning. Then, we identify the state-of-the-art approaches that aim to counteract these problems. Findings from our study confirm that the current major security threats are poisoning, backdoor, and Generative Adversarial Network (GAN)-based attacks, while inference-based attacks are the most critical to the privacy of FL. Finally, we identify ongoing research directions on the topic. This paper could be used as a reference to promote cybersecurity-related research on designing FL-based solutions for alleviating future challenges.

Keywords: survey; federated learning; deep learning; machine learning; distributed learning; privacy; security; blockchain; deep learning security and privacy threats

Citation: Gosselin, R.; Vieu, L.; Loukil, F.; Benoit, A. Privacy and Security in Federated Learning: A Survey. *Appl. Sci.* **2022**, *12*, 9901. https://doi.org/10.3390/app12199901

Academic Editors: George Drosatos and Gianluca Lax

Received: 30 August 2022
Accepted: 27 September 2022
Published: 1 October 2022

Publisher's Note: MDPI stays neutral with regard to jurisdictional claims in published maps and institutional affiliations.

Copyright: © 2022 by the authors. Licensee MDPI, Basel, Switzerland. This article is an open access article distributed under the terms and conditions of the Creative Commons Attribution (CC BY) license (https://creativecommons.org/licenses/by/4.0/).

1. Introduction

Machine Learning (ML) approaches can be considered as classical methods to address complex problems when the underlying physical model is not perfectly known. They are also a hot topic when considering a high semantic level analysis of complex data, such as object recognition on images and anomaly detection on time series. It is now possible to learn complex non-linear models directly from large quantities of data and deploy them for a variety of domains, including sensitive ones such as autonomous driving and medical data analysis. Numerous domains indeed generate ever-increasing quantities that thus allow for the application of ML methods. As an illustration, connected edge devices being integrated into most domains are expected to increase their number of collected data by more than 75% by 2025 [1], encouraged by new wireless technologies, namely 5G [2].

As a counterpart, such a successful model optimization requires an extended processing power to process large quantities of training data to improve robustness and generalization behaviors. As illustrated in Figure 1a, traditional machine learning approaches generally rely on centralized systems that gather both computing resources and the entirety of the data. However, this strategy raises confidentiality issues when transferring, storing, and processing data. In addition, it implies high communication costs that may forbid its use in a variety of sensitive application domains.

Decentralized learning is an alternative approach that aims to optimize models locally to reduce communication costs and preserve privacy. This strategy is challenging, since the end-user expects at least similar performances and generalization behaviors as those that originate from a centralized approach while dealing with smaller and potentially biased local data collections, i.e., non-IID (Identically and Independently Distributed) data. Then, collaborative approaches are developed to introduce communication between local learner agents to look for a robust, general model. More specifically, the recently introduced Federated Learning (FL) [3] has been subject to a growing interest to address complex problems while never sharing raw data between collaborative agents, only exchanging model parameters. As illustrated in Figure 1b,c, a variety of system infrastructures are possible. Agents can communicate in a peer-to-peer (P2P) fashion or with a centralized parameter server. Further, hierarchical structures are possible in order to consider different model aggregation scales [4]. With such an approach, sensitive data are strictly kept on the client's device, at the edge, thus initiating privacy. In addition, communication costs can be significantly reduced, since model parameters are significantly smaller than the raw data. This, however, expects edge devices to be powerful enough to conduct local learning. Nevertheless, the quantity of data is smaller than with the centralized approach, and thus allows for cheaper hardware. In conclusion, federated learning is an attractive solution for multiple application domains and technologies, from medical applications to the Internet of Things, and is subject to intensive research.

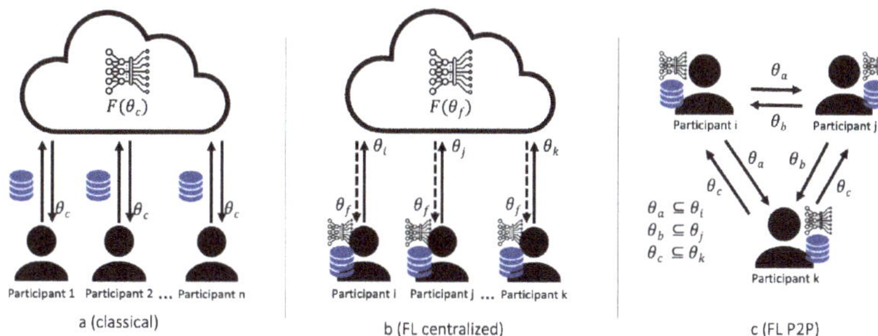

Figure 1. Comparison between classical centralized ML approach and centralized or P2P FL.

Nevertheless, despite being more privacy friendly than the centralized approach, FL is still subject to attacks that may impact the learned model's relevance, as well as the data integrity. A typical example is poisoning attacks [5,6], which aim to distort the model by sending corrupted training data. This attack type is facilitated by the centralized federated learning topology, since the parameter server cannot directly verify the data. User data integrity can also be compromised by relying on generative adversarial networks (GANs)-based attacks [7] to reconstruct data from local node models. Thus, despite the improvements compared to the centralized approaches, data security and privacy with federated learning are still burning issues that must be resolved.

The current state-of-the-art research surveys and reviews in the field provide remarkable work from various perspectives of FL, while focusing only on categorizing existing FL models [8], summarizing implementation details in the vertical FL approach [9] and open issues in FL [10]. While research already exists on the FL topic, the examination of FL security and privacy has not been sufficiently addressed in the literature. This work discusses the privacy and security risks of FL in terms of FL formal definitions, attacks, defense techniques, and future challenges, in an attempt to help the community and newcomers by providing in-depth information and knowledge of FL security and privacy.

Thus, the aim of this paper is to conduct a comprehensive study that highlights open security and privacy issues related to federated learning by answering several research questions. Figure 2 depicts the proposed taxonomy of security and privacy studies in FL. Following this structure, the remainder of this paper is then organized as follows. Section 2 presents the FL approach, the main infrastructure topology, and the related most common aggregation techniques. Section 3 provides an overview of the security flaws inherent to FL and the defensive techniques proposed in the state of the art. Section 4 provides an outlook on the various privacy breaches and how advisers can minimize them. Section 5 discusses current issues and future trends. Finally, Section 6 concludes this survey paper.

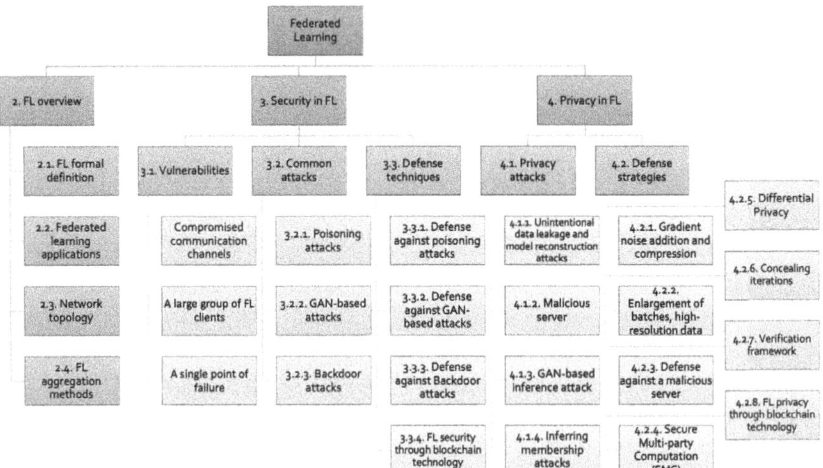

Figure 2. Taxonomy of security and privacy studies in FL. We report the sections of the paper that discuss each of the mentioned items.

2. An Overview of Federated Learning

In order to present the main security and privacy threats related to federated learning, this section first presents the main principles of FL compared to the traditional centralized learning approaches. Figure 1 illustrates the discussed approaches. Then, some real-world scenarios illustrate the FL relevance. Finally, the main features, topology, and model aggregation methods are presented.

2.1. From Centralized to Federated Learning

In a traditional centralized approach, the aim is to learn a single model $f(\theta_C)$ from a large data collection built from the aggregation of data coming from a variety of sources. Despite its impressive success, several issues must be highlighted. Indeed, when dealing with very large data collections and complex models, server-level data and communication costs must be reduced to efficiently distribute the data across computing devices [11]. Moreover, advanced machine learning techniques require large computing resources, which induce costs that may not be sustainable depending on the application economic model and may create dependencies on such computation power in the long term.

To overcome traditional machine learning weaknesses, federated learning was introduced in 2016 by McMahan et al. [3] as a learning technique that allows users to collectively reap the benefits of shared models trained from their data, without the need to centrally store them. Federated learning aims at identifying a general model $f(\theta_{FL})$ by aggregating local ones $f(\theta_i)$ trained by a set of participating clients that keep their data, but occasionally share their parameters. Indeed, FL introduces a new paradigm that reduces communication costs and pushes forward data privacy approaches in several application domains [12].

2.2. Federated Learning Applications

The first large-scale demonstration of FL was Google predictive keyboard (Gboard) on Android smartphones. The smartphone's processing power is used to train a local model from the typing data of the device owner. By occasionally communicating local models to a Google central server that performs aggregation, the general word prediction model improves and is distributed to all the users. This approach allows integration of both global user behaviors, and the specialization of a local model [13]. Compared to the centralized approach, personal typing data should have been sent to the cloud and would have led to privacy issues. Further, this use case is interesting in terms of understanding the limits of the classic ML approach. Indeed, a smartphone keyboard has to fit the user's language. This results in a very personal ML model; however, to be efficient and benefit from the richness of the language, this model should be trained at a larger scale, taking advantage of other people making use of similar languages. However, in a privacy concern, it is impossible to collect the user's typing data. FL then enables taking advantage of both the global knowledge and the specialization of a local model [13]. Similarly, healthcare is another typical case study. The COVID-19 crisis illustrated the use of FL in predicting the future oxygen needs of patients for the EXAM model. In [14], data collected from over 20 institutions around the world were used to train a global FL model. Although no raw data were communicated to a central server, FL allowed for the optimization of a general model relying on a large amount of data.

2.3. Network Topology

The studied state-of-the-art papers mainly report two typical FL communication schemes that impact the network topology, which we summarize as follows.

2.3.1. Centralized FL

This approach is the standard one. The central server is the cornerstone of the architecture. It manages clients, centralizes their local models, and updates the global model. FL optimization is an iterative process wherein each iteration improves the global ML model. It consists of three main steps:

- Model initialization: FL is based on an initial model generally prepared on the server side. Initial weights can rely on a pretraining step. The model parameters are then distributed to the participating clients and will be updated along the next steps in accordance with clients' feedback.
- Local model training: a selection of participating clients is defined. Each of them receives the global model and fine-tuning parameters, which rely on their local data for a set of training epochs. Then, the locally updated model weights are sent to the central server to update the global model.
- Aggregation: the central server collects the participating clients' updated models. Then, an aggregation of their parameters yields an updated general model. This step is critical and should integrate several factors, including client confidence and participation frequency, in order to reduce bias.

Steps 2 and 3 constitute a single round that is repeated until a stop condition is reached.

2.3.2. Peer-to-Peer FL

In a fully decentralized approach, there is no longer a central server that acts as an initiator, coordinator, and model aggregator. Communication with the central server is replaced by peer-to-peer communication as shown in Figure 1c. In this type of topology, network agents learn personalized models. Communication with other members with a common goal is essential in order to increase the quality of their model. The gossip communication protocol is one of the most widely used and efficient protocols today [15,16].

2.4. FL Aggregation Methods

The aggregation of the local models should result in an improved and more general model. Some of the main state-of-the-art methods are presented below:

- The first proposal was the Federated Averaging Algorithm (FedAvg), introduced in [3]. Considered the default approach for centralized FL, the central server will generate the global model by averaging all the participating client models. This approach can be considered as gradient descent on the server side. Extensions have been proposed to adapt efficient optimization strategies, such as Adam and Adagrad, to this context [17].

- FedProx [18] was proposed as a generalization of the FedAvg method. It takes into account the variable amount of work to be performed by each client. This also depends on global data distributions across clients, local computational resources, energy, and communication limits. It has been shown that FedProx results in better averaging accuracy than FedAvg in heterogeneous settings.
- The Federated Matched Averaged (FedMa) method [19] aims to update the global model via layer-wise matching and aggregation of inner model components, namely neurons. This approach only works on neural networks due to its specificities. Moreover, it only works on simple neural networks such as Convolutional Neural Networks (CNNs) and Long Short-Term Memory (LSTM)-based models. This method presents good results with heterogeneous learners and better performances than FedAvg and FedProx within a few training iterations.

3. Security in Federated Learning

Clients participating in a federated learning model can be numerous. This opens doors to a variety of attacks on the client, server, and communication sides. Thus, the development of models via this technology must follow and take into account the main concepts of information security confidentiality, integrity, and availability.

Current studies that explore vulnerabilities and provide existing defensive techniques for security attacks of FL are very limited. Thus, we define the following research questions on the security aspect of the FL:

- RQ1: What are the major vulnerabilities in the FL domain?
- RQ2: What security threats and attacks do FL models face?
- RQ3: What are the defensive techniques against security attacks in the FL ecosystem?

In the rest of this section, we answer each research question based on the studied publications in the FL domain.

3.1. Vulnerabilities in the Ecosystem

A vulnerability can be defined as a weakness in a system that provides an opportunity for curious/malicious attackers to gain unauthorized access [20]. Scanning all sources of vulnerabilities and tightening defenses are thus mandatory steps needed to build up a federated learning-based system while ensuring the security and privacy of the data. Based on the studied publications, we answer the RQ1 question below. Vulnerability sources are actually similar to those of distributed applications. We categorize them as follows:

- Compromised communication channels: an insecure communication channel is an open vulnerability in the FL process that could be addressed using a cryptography public key, which keeps message content secure and safe throughout the communication.
- A large group of FL clients: The number of clients participating in the model is large, so the general model is likely to receive models from Byzantine nodes. This type of safety threat is called Data Poisoning [21].
- A single point of failure: The central parameter server, which is at the heart of the network, must be robust and secure to prevent intrusions. Then, its vulnerabilities must be checked to ensure that they are not exploited. Furthermore, the security updates and all the security recommendations must be followed to limit the risk of intrusions [22].

3.2. Common Attacks in FL

A cyber-attack consists of any action aiming at undermining a computer element: network, software, or data with a financial, political, or security purpose [23]. The number of attacks against a deep learning model is high, such that we focus on the three main ones in terms of impact on the system, frequency, and relevance concerning federated learning. Interested readers can find more details in [24]. These attacks are derivations of the so-called Poisoning Attacks we mentioned previously. Next, we provide the answer to the RQ2 question by introducing some specific and challenging approaches.

3.2.1. Poisoning Attacks

A Poisoning Attack is one of the most common techniques used in the FL. This type of attack can be of a different nature, but the principle remains essentially the same. A malicious client will send rigged data to affect the global model [22,25]. Such an attack can be conducted by two means:

- Data poisoning: The aim is to incorporate malicious data points to create bias in the global model. To be undetectable, attackers slightly modify the model parameters they send but repeat this operation over several training iterations or rounds. This makes it difficult to determine whether a local parameter set is poisoned or not. With several model updates of such a corrupted client, the global model becomes biased enough to degrade its task performance [26].
- Model poisoning: This type of poisoning is more direct; it seeks to manipulate the global model without going through the insertion of malicious data points. It is generally permitted thanks to an intrusion at the server level.

3.2.2. Generative Adversarial Network (GAN) Based Attacks

Poisoning attacks have evolved and new methodologies for creating poisoned models have been proposed. One of them is called PoisonGAN [27]. This attack uses a generative adversarial network (GAN) to generate realistic datasets that are controlled at will by the attacker. A GAN is optimized on a given client side, relying on model updates with the aim of manipulating the parameters of the global model, as shown in Figure 3. Attacks using GANs have several advantages. Indeed the attacker does not need to have a dataset before making the attack [6]. Such kind of attack is actually facilitated by the federated learning infrastructure since the attacker has access to the local model, which would be more difficult with centralized learning.

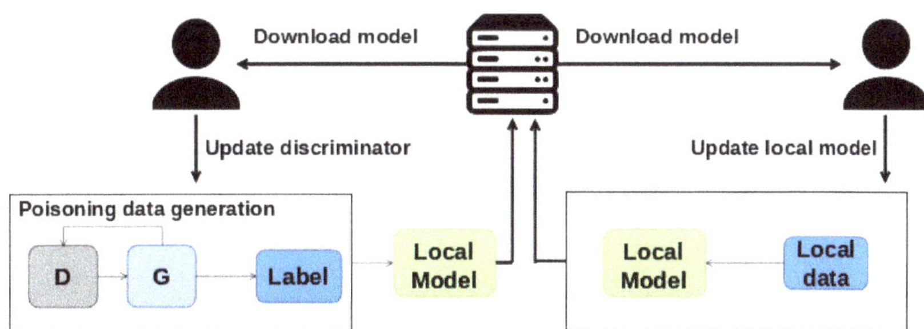

Figure 3. Overview of GAN-based poisoning attack in FL.

Finally, such GAN-based attacks are the most difficult to detect since the generated data are realistic. Its application can yield disastrous consequences for the model's accuracy by introducing strong and controlled bias.

3.2.3. Backdoor Attacks

Bagdasaryan et al. [22] show that it is possible to create a backdoor by poisoning the models. This type of attack does not aim to reduce the global model accuracy, but introduces a very specific bias focused on certain labels. A classic example of a backdoor in FL is the image detection models that we find in [27], where the image classifier is distorted to assign a specific label chosen by the attacker.

3.3. Techniques for Defending against Attacks in FL

As a general rule, good security practices for information systems and networks such as encryption of communications must be put in place. No computer system is impenetrable; however, protecting against the various known methods can greatly reduce the number of attacks and their impact. There are two main defense approaches: (i) the proactive one, which is upstream of the attack, and which looks for ways to detect and blocking the latter, and (ii) the reactive one, which is set up when the attack has already taken place and aims to mitigate the consequences and make the patch. With the emergence of federated learning and its specific attacks, dedicated protections are proposed. To answer the RQ3 question, strategies against the most common vulnerabilities and attacks presented in Sections 3.1 and 3.2) are explained below.

3.3.1. Defense against Poisoning Attacks

The proposed methods are mainly reactive ones which continuously monitor client behaviors. Rodríguez-Barroso et al. [28] have proposed a method to screen out malicious clients. This method works by using artificial intelligence that detects model changes or nonconforming data distributions. Another method following the same principle but applied to an attack by several malicious users is proposed in [29]. This is called sniper. Another defense against poisoning attacks is proposed in [21]. This technique consists of checking the performance of the global model at each new model update.

3.3.2. Defense against GAN-Based Attacks

Those specific poisoning attacks require dedicated approaches, such as advanced Byzantine actor detection algorithms [30]. Additionally, defensive techniques are enabled via heterogeneous federated learning via model distillation, which is detailed in [31]. However, defense techniques against this type of attack are still poorly developed and documented.

3.3.3. Defense against Backdoor Attacks

The main approach consists of minimizing the size of the model to reduce its complexity and capacity while potentially improving its accuracy. This technique is called pruning [32]. Since the resulting model is less expressive, backdoor attacks are more complex to carry out. Such a method also introduces some beneficial side effects. The reduced number of parameters indeed reduces communication costs and reduces message interception probability.

One can highlight the fact that the absence of a central server avoids the risk of an attack at the heart of the system. Thus, peer-to-peer federated learning infrastructure could be an interesting solution. However, this would reduce global monitoring capacities and delegate a fraction of this task to each client. Peer-to-peer approaches then introduce additional constraints on the potentially limited capacity edge nodes, thus limiting its application.

3.3.4. FL Security through the Blockchain Technology

As defined by [33], blockchain technology can be represented as a distributed database—a ledger that can be consulted by everyone and everything. Each new element in the database is verified and forgery-proofed [34].

Technologies such as blockchain can be used to secure federated learning and introduce device and model trust. This has been proven in [35,36]. Blockchain technology would act at two levels. The first would be to encourage users to participate in the development of the global model by rewarding contributors for their involvement. The second would be to save the evolution of the parameters in the blockchain. This can then increase user confidence and reduce the risk of poisoning attacks. One of the most cited models following this approach is called blockFL [36]. It proposes a way to trust devices in which a blockchain network replaces the central server of a classical centralized FL infrastructure and conducts local model validation, but model aggregation is performed on the client side. In further detail, each client sends its updates to an associated miner in the network. Miners are in charge of exchanging and verifying all the model updates. For a given operation, a miner runs a Proof of Work (PoW), which aims to generate a new block where the model updates are stored. Finally, the generated block stores the aggregated local model updates that can be downloaded by the other members of the network. Then, each client can compute the global model locally. This strategy makes poisoning attacks more difficult by certifying model updates. As a counterpart, communication and client computation costs are increased. Additionally, in order to keep attackers from tampering with local models, Kalapaaking et al. [37] proposed a blockchain-based Federated Learning framework with an Intel Software Guard Extension (SGX)-based Trusted Execution Environment to securely aggregate local models in Industrial Internet of Things. Indeed, each blockchain node could (i) host an SGX-enabled processor to securely perform the FL-based aggregation tasks, (ii) verify the authenticity of the aggregated model, and (iii) add it to the distributed ledger for tamper-proof storage and integrity verification.

4. Privacy in Federated Learning

Some of the most frequent concerns about traditional ML include privacy weaknesses. In the industrial area, companies invest to protect their intellectual property. However, traditional ML and, more specifically, deep learning model optimization often go against those privacy requirements. It is indeed necessary to store potentially large quantities of data close to the large processing power to train, validate, and test models. Therefore, such data collections, when sensitive, must be communicated and centralized at a high-security level to prevent any data leakage or attacks. However, threats remain. The traditional ML approach thus has topology threats, which require alternatives to be found. As an answer to these issues, federated learning promotes a new topology to limit the data transfers and, consequently, the data footprint. However, some privacy issues are already identified and must be combated in this new setting.

Current studies exploring privacy attacks in FL and providing existing privacy-preserving defense techniques are very limited. Thus, we define the following research questions on the privacy aspect of the FL:

- RQ4: What are the privacy threats and attacks in the FL ecosystem?
- RQ5: What are the privacy-preserving techniques that tackle each type of the identified attacks in RQ1'?
- RQ6: What new technology could enhance the general privacy-preserving feature of FL?

In the rest of this section, we answer each research question based on the studied publications in the FL domain.

4.1. Privacy Attacks in FL

Federated learning [3] introduces the assumption that it is safer for user data privacy thanks to its topology. Each client shares its model updates instead of its dataset, preventing user data interception while communicating with the server or peer [38]. However, several studies have shone light on many attacks that still compromise data privacy. This section provides the answer to the RQ4 question by presenting several popular attacks, allowing user data to be inferred or the local model to be reconstructed.

4.1.1. Unintentional Data Leakage and Model Reconstruction Attacks

Despite the non-communication of data, Nasr et al. [39] have shown that data could be inferred by considering only model weights. Indeed, an honest-but-curious adversary eavesdrops communications of clients and the server, allowing for the reconstruction of the client model. Ref. [40] defines an honest-but-curious client as a client, which aims to store and process data on its own. Unlike other attacks such as data poisoning, the client may not interfere with the collaborative learning algorithm.

Several studies demonstrated the possibility of reconstructing local data collections by inverting the model gradients sent by the clients to the server [40,41]. Nonetheless, this attack has some limitations. It performs well on models that have been trained on small batches of data or data points with poor diversity. It is not suitable when multiple stochastic gradient updates have been performed locally before communicating with the server. Other recent studies [42] go further by reconstructing the local client model without interfering in the training process. This attack, called the model reconstruction attack, copes with the previous limits, allowing for high-quality local model reconstruction.

4.1.2. Malicious Server

In a centralized federated learning setting, the central server is one of the most critical parts of the architecture. If the server is malicious or is accessed by unauthorized persons, then all local models sent by clients can be intercepted and studied in order to reconstruct a part of the original client data. The malicious server can then analyze the model shared by the clients in a discontinuous way (passive mode) or follow the chronological evolution of the model shared by the victim (active mode) [43].

4.1.3. GAN-Based Inference Attack

This kind of attack has already been experienced in [40,44–46] and may be passive or active. The passive attack aims to analyze the user inputs at the server level, and the active one works on sending the global update to only an isolated client.

Experiments in [47,48] demonstrated that with only some of the computed gradients of a neural network, training data could be inferred. With only 3.89% of the gradients, Aono et al. [47] have been able to reconstruct an image close enough to the original one to infer the information. Further, Zhao et al. [48] were able to reconstruct nearly 100% of the original data by inference. Their Deep Leaked Gradients algorithm was the first algorithm capable of rebuilding a pixel-wise image and token-wise matching texts. These works prove that is it possible to infer data from only leaked gradients using an honest-but-curious client with only a few iterations or a small number of gradients.

4.1.4. Inferring Membership Attacks

The purpose of the Inferring Membership is to determine if data have already been seen by the model. This attack can be carried out in two ways: actively or passively. During a passive attack, the user will only observe the updates of the global model [49]. For the active attack, the adversary participates in the creation of the model, which allows him to recover more information. In this type of attack, the goal is to follow the evolution and the behavior of the global model. This type of attack exploits the stochastic gradient descent (SGD) algorithm to extract information about training sets. According to [39], during training, the SGD aims to make the gradient of the loss leaning zero for information extraction.

4.2. Privacy-Preserving Defense Strategies

To mitigate the aforementioned attacks and answer the RQ5 question, the state of the art already reports some defensive strategies, which are summarized in the following subsections.

4.2.1. Gradient Noise Addition and Compression

Despite the performance of the Deep Leakage from Gradients (DLG) algorithm to infer data from gradients, Zhu et al. [7] demonstrated that in FL, and more generally in ML, the addition of noise to the gradients makes the inference more difficult. The authors show that from a variance larger than 10^{-2}, the accuracy drops significantly and leads to an inability to infer the original data. Another solution proposed in [50] is the compression of the model gradients. This approach sparsifies gradients, which impacts the sensitivity of algorithms such as DLG. The authors of [50] show that the gradients can be compressed by $300\times$ before affecting the accuracy of the model. Conversely, the tolerance of the DLG algorithm is around 20% of sparsity, and the compression makes it ineffective.

4.2.2. Enlargement of Batches, High-Resolution Data

Despite the good performances of the DLG algorithm, the technique is not yet generalizable enough to present a significant threat to private data. One of the limits of the DLG algorithm is related to large batch size and image resolution. In their study, Zhu et al. [7] used a batch size of up to eight images and an image resolution of up to 64×64, which illustrates that some FL configurations rely on small datasets. In another larger-scale setting still addressable by FL, more data and processing power would allow for large batch training and/or high-resolution data, such as the data used for classical centralized machine learning. In such a context, DLG would be out of its operating conditions and would not allow for private data inference.

4.2.3. Defense against a Malicious Server

Due to privacy concerns and critical communication bottlenecks, it can become impractical to send the FL updated models to a centralized server. Thus, a recent study [51] proposed an optimized solution for user assignment and resource allocation over hierarchical FL architecture for IoT heterogeneous systems. According to the study's results, the proposed approach could significantly accelerate FL training and reduce communication overhead by providing 75–85% reduction in the communication rounds between edge nodes and the centralized server, for the same model accuracy.

4.2.4. Secure Multi-Party Computation (SMC)

Initially, the secure multi-party computation aims to jointly compute a function on different parties over their personal data. One of the benefits of this approach is the possibility of keeping the inputs private thanks to cryptography. According to [47], SMC is currently used in FL but in a different version, which only needs to encrypt the parameters instead of the large volume of data inputs. Although this approach prevents leaks from a malicious central server, the encryption is expensive to use and may have an impact on a larger scale. Thus, the main cost of this solution is efficiency loss due to encryption.

4.2.5. Differential Privacy

Differential Privacy (DP) is a technique widely used to preserve privacy in industry and academic domains. The main concept of DP is to preserve privacy by adding noise to sensitive personal attributes. In FL, DP is introduced to add noise to clients' uploaded parameters in order to avoid inverse data retrievals. For instance, the DPFedAvgGAN framework [52] uses DP to make GAN-based attacks inefficient in training data inference of other users for FL-specific environments. Additionally, Ghazi et al. [53] improve privacy guarantees of the FL model by combining the shuffling technique with DP and masking user data with an invisibility cloak algorithm. However, such solutions bring uncertainty into the uploaded parameters and may harm the training performance.

4.2.6. Concealing Iterations

Xu et al. [54] have suggested iterations concealing as a method of avoiding privacy threats. In typical FL, the client model iterations are visible to multiple actors of the system, such as the central server and the clients chosen to participate in each round. Thus, a malicious server may be able to infer data through iterations thanks to GAN-based attacks. By concealing iterations, each client should make the learning phase inside a Trusted Execution Environment (TEE). A TEE provides the ability to run code on a remote trusty machine, regardless of the trust placed in the administrator. Indeed, a TEE limits the abilities of any party, including the administrator, to interact with the machine. This specificity results in three properties that make TEE trusty:

- Confidentiality: The running code and its execution state are not shared outside of the machine and cannot be accessed.
- Integrity: Because of access limitations, the running code cannot be poisoned.

- Attestation: The running virtual machine can prove to an external entity what code is executing and the inputs that were provided.

Thus, thanks to TEE, the model iterations can be concealed, and the model parameters may be encrypted inside the TEE before being shared. Nonetheless, this approach might not be used across clients, especially with devices such as smartphones. In fact, TEE may not be powerful enough for training and may involve too many computational and memory costs. These weaknesses may have a significant impact on devices' autonomy, performance, and user experience. For these reasons, in its current state, TEE cannot yet be used on limited-capacity edge devices.

4.2.7. Verification Framework

VerifyNet [54] is a FL framework that provides the possibility for clients to verify the central server results to ensure its reliability. To preserve privacy, it provides a double masking protocol, which makes it difficult for attackers to infer training data. Moreover, VerifyNet is robust enough to handle clients' dropouts due to the user's battery or hardware issues. However, this framework imposes more communication with the clients, which involves a decrease in performance and additional costs.

4.2.8. FL Privacy through the Blockchain Technology

As discussed in Section 3.3.4, and to provide an answer to the RQ6 question, we argue that blockchain technology brings a new FL network topology. Above the security benefits, it can also provide privacy benefits. Indeed, the decentralized network allows each participant to access all transactions made on the network. This transparency makes the network more reliable and increases trust in the learning process. Thus, instead of sharing gradients to a centralized server, the model parameters and clients' updates are stored on the blockchain. In order to introduce privacy in blockchain, the authors of [55] presented a specific design adapted to IoT that relies on differential privacy. Experiments showed that robustness against GAN and curious server-based attacks is increased while maintaining significant task performance levels. Moreover, the blockchain allows for client updates audits in order to maintain model reliability and cope with malicious participants. Thus, blockchain-based models such as BlockFL [36] provide a layer of security that guarantees the model parameters' integrity. Nonetheless, such an approach does not prevent privacy attacks, and it must be enhanced with privacy-oriented methods such as differential privacy.

5. Open Issues and Future Trends

Federated learning is a recent and continuously evolving approach in machine learning that can already benefit many economic, medical, and sociological domains. Industrial sectors and, more specifically, the so-called Industry 4.0 that introduces the data processing paradigm to improve the manufacturing process can push the implications of federated learning [56] further by integrating contextual factors, such as decentralized production models. FL thus appears as an interesting general solution that effectively associates global and local machine learning with privacy by design. However, despite overcoming some classical machine learning issues, some remaining ones and new challenges must be faced in this context. This paper therefore provides an overview of recent trends in the specific problems related to privacy and security. Nevertheless, as reported in recent publications such as [10], other critical challenges are related to system architecture and learning bias. As an opening discussion, we thus elaborate on the following subsections, connections between security, privacy, and bias, and identify research directions.

5.1. Security, Privacy, and Bias Mitigation Dependency Management for Trustfully and Fair Federated Learning

Machine learning model bias is a general problem encountered by machine learning methods. It is critical, since it actually impacts model decisions and can yield discrimination of data populations that have a direct impact on our life, including sexism and racism. More specifically, learning bias can in fact be aggravated by Federated Learning. As reported in recent studies, such as [10], bias is indeed mostly related to the non-IID nature of the data distributed across federated participants. Among other bias sources, participants' involvement frequency in the learning process is a significant factor. Bias mitigation techniques have long been proposed in classical machine learning [57], but they must be adapted to the specific context of Federated Learning [58]. Indeed, we would like to point out in this discussion that bias mitigation, privacy, and security, in the context of federated learning, are highly associated and thus should not be considered separately. For instance, the search for bias in the data, as well as in the model predictions, generally relies on global knowledge that conflicts with client data privacy. Additionally, Poisoning Attacks typically introduce bias, such that their detection allows for security flaws to be counterfeit. Those connections allow us to identify some promising research directions that should, in our opinion, increase federated learning trust and fairness.

5.2. Research Directions

Recent contributions such as [59] propose innovative approaches that consist of detecting communities of clients that share similar behaviors. The aim is next to build up a general model that integrates all those populations in an equal way and thus reduces bias. Such an approach could also help detect specific outliers such as poisoning sources, thus increasing security. Such a direction is indeed promising; however, the identification of those communities may lead to new issues related to user privacy. In addition, the semantics and ethics behind the identified clusters should also be clearly reported. Addressing those challenges is, therefore, of real interest.

A complementary research direction relates to partial federated learning recently introduced in [60]. Such an approach considers the fact that clients may need to share only a subset of their knowledge and still improve each other. As for real applications related to healthcare, autonomous driving, and home automation, clients may have different sensors but try to solve the same problem or try to solve different tasks with the same sensors. A given model may thus be considered as a set of cascading functions, some being local and private while others are shared. Such an approach has many potential advantages. It better corresponds to real-life applications and could also reduce global costs while improving generalization behaviors for a variety of complementary problems. In addition, it provides means to increase security and privacy by sharing only a fraction of the knowledge. We believe that this research direction is promising, and suggest associating this with bias detection and mitigation research.

Finally, new models of artificial intelligence approaches that specialize in blockchain technology seem relevant, sharing similar ideas. For instance, Swarm Learning (SL) [61] takes the main principle of FL, where the data are not shared with a central server. However, the computation of the global model is deferred to the customers. In each training cycle, the operation falls to a node that the others have elected. SL implementations have been applied to the medical field [61,62]. In such a sensitive context, it has been shown that SL can push the BlockFL approach further by using smart contracts to protect the model from misleading participants and keep data and learning safe [63]. Nevertheless, some security issues related to computation and leader node selection still remain, but show interest in blockchain-based methods. This direction, which maintains connections with federated learning while providing an innovative communication paradigm, is therefore of interest for further research.

6. Conclusions

Despite the recent improvements and the growing interest in the federated learning field, it is still in its infancy. Recent studies demonstrated that federated learning is a serious alternative to traditional machine learning methods. While existing surveys have already studied FL from various perspectives, sufficient progress has not been made concerning understanding FL for its security and privacy risks. This paper intends to fill this gap by presenting a comprehensive study of privacy and security issues related to federated learning. Thus, we introduced an overview of the FL applications, topology, and aggregation methods. Then, we discussed the existing FL-based studies in terms of security and privacy defensive techniques that aim to counteract FL vulnerabilities. Finally, FL open issues and future trends are identified to be addressed in further studies. Based on the findings from the study, we conclude that FL fixed several privacy and security issues of classical ML. Thus, it opens up new possibilities, especially in areas in which data are sensitive, such as the medical area or personal user data. Already used by some companies, such as Google, its uses are bound to develop and become more democratic. However, FL is confronted with privacy and security issues inherited from the distributed architecture and the traditional ML. As a consequence, several research works try to go further and integrate the FL principles into other architectures, such as blockchain architectures, to skirt threats.

Author Contributions: Conceptualization, R.G. and L.V.; methodology, F.L. and A.B.; formal analysis, F.L. and A.B.; investigation, R.G. and L.V.; resources, R.G. and L.V.; writing—original draft preparation, R.G. and L.V.; writing—review and editing, F.L. and A.B.; visualization, R.G. and L.V.; supervision, F.L. and A.B. All authors have read and agreed to the published version of the manuscript.

Funding: This research received no external funding.

Institutional Review Board Statement: Not applicable.

Informed Consent Statement: Not applicable.

Conflicts of Interest: The authors declare no conflict of interest.

References

1. Al Hayajneh, A.; Bhuiyan, M.Z.A.; McAndrew, I. Improving internet of things (IoT) security with software-defined networking (SDN). *Computers* **2020**, *9*, 8. [CrossRef]
2. Wang, J.; Lim, M.K.; Wang, C.; Tseng, M.L. The evolution of the Internet of Things (IoT) over the past 20 years. *Comput. Ind. Eng.* **2021**, *155*, 107174. [CrossRef]
3. McMahan, B.; Moore, E.; Ramage, D.; Hampson, S.; y Arcas, B.A. Communication-efficient learning of deep networks from decentralized data. In Proceedings of the Artificial Intelligence and Statistics, Fort Lauderdale, FL, USA, 20–22 April 2017; pp. 1273–1282.
4. Shayan, M.; Fung, C.; Yoon, C.J.; Beschastnikh, I. Biscotti: A blockchain system for private and secure federated learning. *IEEE Trans. Parallel Distrib. Syst.* **2020**, *32*, 1513–1525. [CrossRef]
5. Biggio, B.; Nelson, B.; Laskov, P. Poisoning Attacks against Support Vector Machines. *arXiv* **2012**, arXiv:1206.6389.
6. Zhang, J.; Chen, J.; Wu, D.; Chen, B.; Yu, S. Poisoning attack in federated learning using generative adversarial nets. In Proceedings of the 2019 18th IEEE International Conference On Trust, Security And Privacy In Computing And Communications/13th IEEE International Conference On Big Data Science And Engineering (TrustCom/BigDataSE), Rotorua, New Zealand, 5–8 August 2019; pp. 374–380.
7. Zhu, L.; Liu, Z.; Han, S. Deep leakage from gradients. *Adv. Neural Inf. Process. Syst.* **2019**, *32*, 14774–14784.
8. Li, Q.; Wen, Z.; He, B. Federated Learning Systems: Vision, Hype and Reality for Data Privacy and Protection. *arXiv* **2019**, arXiv:1907.09693.
9. Yang, Q.; Liu, Y.; Cheng, Y.; Kang, Y.; Chen, T.; Yu, H. Federated learning. *Synth. Lect. Artif. Intell. Mach. Learn.* **2019**, *13*, 1–207.
10. Kairouz, P.; McMahan, H.B.; Avent, B.; Bellet, A.; Bennis, M.; Bhagoji, A.N.; Bonawitz, K.; Charles, Z.; Cormode, G.; Cummings, R.; et al. Advances and open problems in federated learning. *Found. Trends® Mach. Learn.* **2021**, *14*, 1–210. [CrossRef]
11. Tang, Z.; Shi, S.; Chu, X.; Wang, W.; Li, B. Communication-Efficient Distributed Deep Learning: A Comprehensive Survey. *arXiv* **2020**, arXiv:2003.06307.
12. Sun, Z.; Strang, K.D.; Pambel, F. Privacy and security in the big data paradigm. *J. Comput. Inf. Syst.* **2020**, *60*, 146–155. [CrossRef]
13. McMahan, B.; Ramag, D. Federated Learning: Collaborative Machine Learning without Centralized Training Data. 2017. Available online: https://ai.googleblog.com/2017/04/federated-learning-collaborative.html (accessed on 29 September 2022).
14. Dayan, I.; Roth, H.R.; Zhong, A.; Harouni, A.; Gentili, A.; Abidin, A.Z.; Liu, A.; Costa, A.B.; Wood, B.J.; Tsai, C.S.; et al. Federated learning for predicting clinical outcomes in patients with COVID-19. *Nat. Med.* **2021**, *27*, 1735–1743. [CrossRef] [PubMed]
15. Hu, C.; Jiang, J.; Wang, Z. Decentralized Federated Learning: A Segmented Gossip Approach. *arXiv* **2019**, arXiv:1908.07782.
16. Vanhaesebrouck, P.; Bellet, A.; Tommasi, M. Decentralized collaborative learning of personalized models over networks. In Proceedings of the Artificial Intelligence and Statistics, Fort Lauderdale, FL, USA, 20–22 April 2017; pp. 509–517.

17. Reddi, S.J.; Charles, Z.; Zaheer, M.; Garrett, Z.; Rush, K.; Konečný, J.; Kumar, S.; McMahan, H.B. Adaptive Federated Optimization. In Proceedings of the International Conference on Learning Representations (ICLR), Virtual Conference, 2020.
18. Li, T.; Sahu, A.K.; Zaheer, M.; Sanjabi, M.; Talwalkar, A.; Smith, V. Federated optimization in heterogeneous networks. *Proc. Mach. Learn. Syst.* **2020**, *2*, 429–450.
19. Wang, H.; Yurochkin, M.; Sun, Y.; Papailiopoulos, D.; Khazaeni, Y. Federated learning with matched averaging. In Proceedings of the International Conference on Learning Representations (ICLR), Virtual Conference, 2020.
20. OWSAP. OWSAP Defination for Vulnerability. 2018. Available online: https://owasp.org/www-community/vulnerabilities/ (accessed on 29 September 2022).
21. Bhagoji, A.N.; Chakraborty, S.; Mittal, P.; Calo, S. Analyzing federated learning through an adversarial lens. In Proceedings of the International Conference on Machine Learning, Long Beach, CA, USA, 9–15 June 2019; pp. 634–643.
22. Bagdasaryan, E.; Veit, A.; Hua, Y.; Estrin, D.; Shmatikov, V. How to backdoor federated learning. In Proceedings of the International Conference on Artificial Intelligence and Statistics, Palermo, Sicily, Italy, 26–28 August 2020; pp. 2938–2948.
23. Hathaway, O.A.; Crootof, R.; Levitz, P.; Nix, H. The law of cyber-attack. *Calif. Law Rev.* **2012**, *100*, 817.
24. Mothukuri, V.; Parizi, R.M.; Pouriyeh, S.; Huang, Y.; Dehghantanha, A.; Srivastava, G. A survey on security and privacy of federated learning. *Future Gener. Comput. Syst.* **2021**, *115*, 619–640. [CrossRef]
25. Lyu, L.; Yu, H.; Yang, Q. Threats to federated learning: A survey. *arXiv* **2020**, arXiv:2003.02133.
26. Tolpegin, V.; Truex, S.; Gursoy, M.E.; Liu, L. Data poisoning attacks against federated learning systems. In *Proceedings of the European Symposium on Research in Computer Security, Guildford, UK, 14–18 September 2020*; Springer: Berlin/Heidelberg, Germany, 2020; pp. 480–501.
27. Zhang, J.; Chen, B.; Cheng, X.; Binh, H.T.T.; Yu, S. Poisongan: Generative poisoning attacks against federated learning in edge computing systems. *IEEE Internet Things J.* **2020**, *8*, 3310–3322. [CrossRef]
28. Rodríguez-Barroso, N.; Martínez-Cámara, E.; Luzón, M.; González Seco, G.; Ángel Veganzones, M.; Herrera, F. Dynamic Federated Learning Model for Identifying Adversarial Clients. *arXiv* **2020**, arXiv:2007.15030.
29. Cao, D.; Chang, S.; Lin, Z.; Liu, G.; Sun, D. Understanding distributed poisoning attack in federated learning. In Proceedings of the 2019 IEEE 25th International Conference on Parallel and Distributed Systems (ICPADS), Tianjin, China, 4–6 December 2019; pp. 233–239.
30. Hayes, J.; Ohrimenko, O. Contamination attacks and mitigation in multi-party machine learning. *Adv. Neural Inf. Process. Syst.* **2018**, *31*, 6604–6616.
31. Li, D.; Wang, J. FedMD: Heterogenous Federated Learning via Model Distillation. *arXiv* **2019**, arXiv:1910.03581.
32. Liu, K.; Dolan-Gavitt, B.; Garg, S. Fine-pruning: Defending against backdooring attacks on deep neural networks. In *Proceedings of the International Symposium on Research in Attacks, Intrusions, and Defenses, Crete, Greece, 10–12 September 2018*; Springer: Berlin/Heidelberg, Germany, 2018; pp. 273–294.
33. Nakamoto, S. Bitcoin: A peer-to-peer electronic cash system. *Decentralized Bus. Rev.* **2008**, 21260. [CrossRef]
34. Crosby, M.; Pattanayak, P.; Verma, S.; Kalyanaraman, V. Blockchain technology: Beyond bitcoin. *Appl. Innov.* **2016**, *2*, 71.
35. Majeed, U.; Hong, C.S. FLchain: Federated learning via MEC-enabled blockchain network. In Proceedings of the 2019 20th Asia-Pacific Network Operations and Management Symposium (APNOMS), Matsue, Japan, 18–20 September 2019; pp. 1–4.
36. Kim, H.; Park, J.; Bennis, M.; Kim, S.L. Blockchained on-device federated learning. *IEEE Commun. Lett.* **2019**, *24*, 1279–1283. [CrossRef]
37. Kalapaaking, A.P.; Khalil, I.; Rahman, M.S.; Atiquzzaman, M.; Yi, X.; Almashor, M. Blockchain-based Federated Learning with Secure Aggregation in Trusted Execution Environment for Internet-of-Things. *IEEE Trans. Ind. Inform.* **2022**. [CrossRef]
38. Truong, N.; Sun, K.; Wang, S.; Guitton, F.; Guo, Y. Privacy preservation in federated learning: An insightful survey from the GDPR perspective. *Comput. Secur.* **2021**, *110*, 102402. [CrossRef]
39. Nasr, M.; Shokri, R.; Houmansadr, A. Comprehensive privacy analysis of deep learning: Passive and active white-box inference attacks against centralized and federated learning. In Proceedings of the 2019 IEEE Symposium on Security and Privacy (SP), San Francisco, CA, USA, 19–23 May 2019; pp. 739–753.
40. Melis, L.; Song, C.; De Cristofaro, E.; Shmatikov, V. Exploiting unintended feature leakage in collaborative learning. In Proceedings of the 2019 IEEE Symposium on Security and Privacy (SP), San Francisco, CA, USA, 19–23 May 2019; pp. 691–706.
41. Geiping, J.; Bauermeister, H.; Dröge, H.; Moeller, M. Inverting gradients-how easy is it to break privacy in federated learning? *Adv. Neural Inf. Process. Syst.* **2020**, *33*, 16937–16947.
42. Xu, C.; Neglia, G. What else is leaked when eavesdropping Federated Learning? In Proceedings of the CCS workshop Privacy Preserving Machine Learning, Virtual Event, USA, 2021.
43. Song, M.; Wang, Z.; Zhang, Z.; Song, Y.; Wang, Q.; Ren, J.; Qi, H. Analyzing user-level privacy attack against federated learning. *IEEE J. Sel. Areas Commun.* **2020**, *38*, 2430–2444. [CrossRef]
44. Wang, Z.; Song, M.; Zhang, Z.; Song, Y.; Wang, Q.; Qi, H. Beyond inferring class representatives: User-level privacy leakage from federated learning. In Proceedings of the IEEE INFOCOM 2019-IEEE Conference on Computer Communications, Paris, France, 29 April–2 May 2019; pp. 2512–2520.
45. Truong, N.; Sun, K.; Wang, S.; Guitton, F.; Guo, Y. Privacy preservation in federated learning: Insights from the gdpr perspective. *arXiv* **2020**, arXiv:2011.05411.

46. Hitaj, B.; Ateniese, G.; Perez-Cruz, F. Deep models under the GAN: Information leakage from collaborative deep learning. In Proceedings of the 2017 ACM SIGSAC Conference on Computer and Communications Security, Dallas, TX, USA, 30 October–3 November 2017; pp. 603–618.

47. Aono, Y.; Hayashi, T.; Wang, L.; Moriai, S. Privacy-preserving deep learning via additively homomorphic encryption. *IEEE Trans. Inf. Forensics Secur.* **2017**, *13*, 1333–1345.

48. Zhao, B.; Reddy Mopuri, K.; Bilen, H. iDLG: Improved Deep Leakage from Gradients. *arXiv* **2020**, arXiv:2001.02610.

49. Zari, O.; Xu, C.; Neglia, G. Efficient passive membership inference attack in federated learning. *arXiv* **2021**, arXiv:2111.00430.

50. Lin, Y.; Han, S.; Mao, H.; Wang, Y.; Dally, W.J. Deep Gradient Compression: Reducing the Communication Bandwidth for Distributed Training. *arXiv* **2017**, arXiv:1712.01887.

51. Abdellatif, A.A.; Mhaisen, N.; Mohamed, A.; Erbad, A.; Guizani, M.; Dawy, Z.; Nasreddine, W. Communication-efficient hierarchical federated learning for IoT heterogeneous systems with imbalanced data. *Future Gener. Comput. Syst.* **2022**, *128*, 406–419. [CrossRef]

52. Augenstein, S.; McMahan, H.B.; Ramage, D.; Ramaswamy, S.; Kairouz, P.; Chen, M.; Mathews, R.; y Arcas, B.A. Generative Models for Effective ML on Private, Decentralized Datasets. In Proceedings of the International Conference on Learning Representations, New Orleans, LA, USA, 6–9 May 2019.

53. Ghazi, B.; Pagh, R.; Velingker, A. Scalable and Differentially Private Distributed Aggregation in the Shuffled Model. *arXiv* **2019**, arXiv:1906.08320.

54. Xu, G.; Li, H.; Liu, S.; Yang, K.; Lin, X. Verifynet: Secure and verifiable federated learning. *IEEE Trans. Inf. Forensics Secur.* **2019**, *15*, 911–926. [CrossRef]

55. Zhao, Y.; Zhao, J.; Jiang, L.; Tan, R.; Niyato, D.; Li, Z.; Lyu, L.; Liu, Y. Privacy-preserving blockchain-based federated learning for IoT devices. *IEEE Internet Things J.* **2020**, *8*, 1817–1829. [CrossRef]

56. Hao, M.; Li, H.; Luo, X.; Xu, G.; Yang, H.; Liu, S. Efficient and privacy-enhanced federated learning for industrial artificial intelligence. *IEEE Trans. Ind. Inform.* **2019**, *16*, 6532–6542. [CrossRef]

57. Mehrabi, N.; Morstatter, F.; Saxena, N.; Lerman, K.; Galstyan, A. A survey on bias and fairness in machine learning. *ACM Comput. Surv. (CSUR)* **2021**, *54*, 1–35. [CrossRef]

58. Ferraguig, L.; Djebrouni, Y.; Bouchenak, S.; Marangozova, V. Survey of Bias Mitigation in Federated Learning. In Proceedings of the Conférence Francophone D'informatique en Parallélisme, Architecture et Système, Lyon, France, July 2021.

59. Briggs, C.; Fan, Z.; Andras, P. Federated learning with hierarchical clustering of local updates to improve training on non-IID data. In Proceedings of the 2020 International Joint Conference on Neural Networks (IJCNN), Glasgow, UK, 19–24 July 2020; pp. 1–9.

60. Singhal, K.; Sidahmed, H.; Garrett, Z.; Wu, S.; Rush, J.; Prakash, S. Federated Reconstruction: Partially Local Federated Learning. In *Proceedings of the Advances in Neural Information Processing Systems, Virtual Conference*; Ranzato, M., Beygelzimer, A., Dauphin, Y., Liang, P., Vaughan, J.W., Eds.; Curran Associates, Inc.: Red Hook, NY, USA, 2021; Volume 34, pp. 11220–11232.

61. Saldanha, O.L.; Quirke, P.; West, N.P.; James, J.A.; Loughrey, M.B.; Grabsch, H.I.; Salto-Tellez, M.; Alwers, E.; Cifci, D.; Ghaffari Laleh, N.; et al. Swarm learning for decentralized artificial intelligence in cancer histopathology. *Nat. Med.* **2022**, *28*, 1232–1239. [CrossRef]

62. Becker, M. Swarm learning for decentralized healthcare. *Der Hautarzt* **2022**, *73*, 323–325. [CrossRef]

63. Warnat-Herresthal, S.; Schultze, H.; Shastry, K.L.; Manamohan, S.; Mukherjee, S.; Garg, V.; Sarveswara, R.; Händler, K.; Pickkers, P.; Aziz, N.A.; et al. Swarm learning for decentralized and confidential clinical machine learning. *Nature* **2021**, *594*, 265–270. [CrossRef] [PubMed]

Article

Comparisons Where It Matters: Using Layer-Wise Regularization to Improve Federated Learning on Heterogeneous Data

Ha Min Son [1,2], **Moon Hyun Kim** [1,2] and **Tai-Myoung Chung** [1,2,*]

[1]   Department of Computer Science and Engineering, Sungkyunkwan University, Seoul 03063, Korea
[2]   Hippo T&C, Suwon 16419, Korea
*   Correspondence: tmchung@skku.edu

Abstract: Federated Learning is a widely adopted method for training neural networks over distributed data. One main limitation is the performance degradation that occurs when data are heterogeneously distributed. While many studies have attempted to address this problem, a more recent understanding of neural networks provides insight to an alternative approach. In this study, we show that only certain important layers in a neural network require regularization for effective training. We additionally verify that Centered Kernel Alignment (CKA) most accurately calculates similarities between layers of neural networks trained on different data. By applying CKA-based regularization to important layers during training, we significantly improved performances in heterogeneous settings. We present FedCKA, a simple framework that outperforms previous state-of-the-art methods on various deep learning tasks while also improving efficiency and scalability.

Keywords: federated learning; heterogeneity; non-IID; regularization; layer-wise similarity

1. Introduction

The success of deep learning in a plethora of fields has led to a countless number of research studies conducted to leverage its strengths [1]. One main outcome resulting from this success is the mass collection of data [2]. As the collection of data increases at a rate much faster than that of the computing performance and storage capacity of consumer products, it is becoming progressively difficult to deploy trained state-of-the-art models within a reasonable budget. In light of this, Federated Learning [3] has been introduced as a method to train a neural network with massively distributed data. The most widely used and accepted approach for the training and aggregation process is FedAvg [3]. FedAvg typically progresses with the repetition of four steps, as shown in Figure 1. (1) A centralized or de-centralized server broadcasts a model (the global model) to each of its clients. (2) Each client trains its copy of the model (the local model) with its local data. (3) Clients upload their trained model to the server. (4) The server aggregates the trained models into a single model and prepares it to be broadcast in the next round. These steps are repeated until convergence or other criteria are met.

Federated learning is appealing for many reasons, such as negating the cost of collecting data into a centralized location and effective parallelization across computing units [4]. Thus, it has been applied to a wide range of research studies, including a distributed learning framework on vehicular networks [5], IoT devices [6], and even as a privacy-preserving method for medical records [7]. However, one major issue with the application of Federated Learning is the performance degradation that occurs with heterogeneous data. This refers to settings in which data are not independent and identically distributed (non-IID) across clients. The drop in performance is observed to be caused by a disagreement in local optima. That is, because different clients train its copy of the neural network according to its individual local data, the resulting average can stray from the true optimum. Unfortunately, it is realistic to expect non-IID data in many real-world applications [8,9]. While many

Citation: Son, H.M.; Kim, M.H.; Chung, T.-M. Comparisons Where It Matters: Using Layer-Wise Regularization to Improve Federated Learning on Heterogeneous Data. *Appl. Sci.* **2022**, *12*, 9943. https://doi.org/10.3390/app12199943

Academic Editor: George Drosatos

Received: 23 August 2022
Accepted: 28 September 2022
Published: 3 October 2022

Publisher's Note: MDPI stays neutral with regard to jurisdictional claims in published maps and institutional affiliations.

Copyright: © 2022 by the authors. Licensee MDPI, Basel, Switzerland. This article is an open access article distributed under the terms and conditions of the Creative Commons Attribution (CC BY) license (https://creativecommons.org/licenses/by/4.0/).

studies have attempted to address this problem by regularizing the entire model during the training process, we argue that a more recent understanding of neural networks suggests regularizing every layer may limit performance.

In this study, we present FedCKA to address these limitations. First, we show that regularizing the first two naturally similar layers is most important for improving performance in non-IID settings. Previous studies had regularized each individual layers. Not only is this ineffective for training, it also limits scalability as the number of layers in a model increases. By regularizing only these important layers, performance improves beyond previous studies. Efficiency and scalability also improved, as we do not need to calculate regularization terms for every layer. Second, we show that Centered Kernel Alignment (CKA) is most suitable when comparing the representational similarity between layers of neural networks. Previous studies added a regularization term by comparing the representation of neural networks with the l2-distance (FedProx) or cosine similarity (MOON). By using CKA, we improve performances, as representations between important layers can accurately be regularized regardless of dimension or rank [10]; hence, the name FedCKA. Our contributions are summarized as follows:

- We improve performances in heterogeneous settings. By building on the most up-to-date understanding of neural networks, we apply layer-wise regularization to only important layers.
- We improve the efficiency and scalability of regularization. By regularizing only important layers, we exclusively show training times that are comparable to FedAvg.

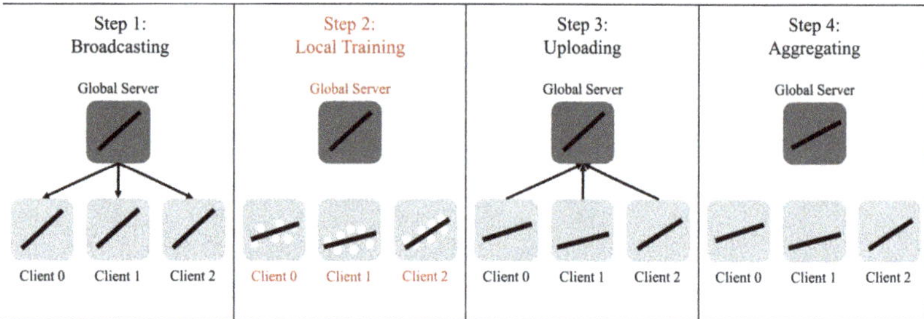

Figure 1. The typical steps of Federated Learning (FedAvg).

2. Related Works

Layers in Neural Networks

Understanding the function of layers in a neural network is an under-researched field of deep learning. It is, however, an important prerequisite for the application of layer-wise regularization. We build our study based on findings of two relevant papers.

The first study [11] showed that there are certain 'critical' layers that define a model's performance. In particular, when trained layers were re-initialized back to their original weights, 'critical' layers heavily decreased performance, while 'robust' layers had minimal impact. This study drew several relevant conclusions. First, the very first layer of neural networks is the most sensitive to re-initialization. Second, robustness is not correlated with the l2-norm or l∞-norm between initial weights and trained weights. Third, while 'robust' layers did not affect performance when changed to their initial weights, most layers heavily decreased performance when changed to non-initial random weights. Considering these conclusions, we understand that certain layers are not important in defining performance. Regularizing these non-important layers would be ineffective and may even hurt performance.

The second study [10] introduced Centered Kernel Alignment (CKA) as a metric for measuring the similarity between layers of neural networks. In particular, the study showed that metrics that calculate the similarity between representations of neural networks should be invariant to orthogonal transformations and isotropic scaling while being invertible to linear transformations. This study drew one very relevant conclusion. For neural networks trained on different datasets, early layers, but not late layers, learn similar representations. Considering this conclusion, if we were to properly regularize neural networks trained on different datasets, we should focus on layers that are naturally similar and not on those that are naturally different.

Studies that improve performances on non-IID data generally fall into two categories. The first focuses on regularizing or modifying the client training process (step 2). The second focuses on modifying the aggregation process (step 4). Some studies employ knowledge distillation techniques [12,13], while others use data sharing [14]. Other works such as FedMA [15] aggregates each layer separately, starting with the layer closest to the input. After a single layer is aggregated, it is frozen, and each client trains all subsequent layers. This process is repeated until all layers are individually aggregated. FedBN [16] aggregates all layers except for batch normalization layers, which are locally stored to reduce local data bias. While these approaches are significant and relevant to our work, they focus more on the aggregation process. Thus, we focus on works that regularize local training as it is more closely related. Namely, we focus on FedProx [17], SCAFFOLD [18], and MOON [19], all of which add a regularization term to the default FedAvg [3] training process.

FedAvg was the first study to introduce Federated Learning. Each client trains a model using a gradient descent loss function, and the server averages the trained model based on the number of data samples that each client holds. However, due to the performance degradation in non-IID settings, many studies added a regularization term to the default FedAvg training process. The objective of these methods is to decrease the disagreement in local optima by limiting local updates that stray too far from the global model. FedProx adds a proximal regularization term that calculates the l2-distance between the local and global model. SCAFFOLD adds a control variate regularization term that induces variance reductions on local updates based on the updates of other clients. Most recent and most similar to our work is MOON. MOON adds a contrastive regularization term that calculates the cosine similarity between the MLP projections of the local and global model. The study takes inspiration from contrastive learning, particularly SimCLR [20]. The intuition is that the global model is less biased than local models; thus, local updates should be more similar to the global model than past local models. One difference to note is that while contrastive learning trains a model using the projections of one model on many different images (i.e., one model, different data), MOON regularizes a model using the projections of different models on the same images (i.e., three models, same data).

Overall, these studies add a regularization term by comparing all layers of the neural network. However, we argue that only important layers should be regularized. Late layers are naturally dissimilar when trained on different datasets. Regularizing a model based on these naturally dissimilar late layers would be ineffective. Rather, it may be beneficial to focus only on the naturally similar early layers of the model.

3. FedCKA

3.1. Regularizing Naturally Similar Layers

FedCKA is designed on the principle that naturally similar, but not naturally dissimilar, layers should be regularized. This is based on the premise that early layers, but not late layers, develop similar representations when trained on different datasets [10]. We verify this in a Federated Learning environment. Using a small seven-layer convolutional neural network and the ResNet-50 model [21], we trained 10 clients for 20 communications rounds on independently and identically distributed (IID) subsets of the CIFAR-10 [22] dataset. After training, we viewed the similarity between the layers of local models, calculated by the Centered Kernel Alignment [10] on the CIFAR-10 test set.

3.2. Federated Learning with Non-IID Data

Figure 2 shows the similarity of layers between local models after training. We report all layers for the simple CNN. For the ResNet-50 model, we report the accuracy of the initial convolution layer, each of the four blocks of the ResNet, the two layers in the projector, and the output layer. We verify that early layers, but not late layers, develop similar representations in the ideal Federated Learning setting, where the distribution of data across clients are IID. For convolutional neural networks without residual blocks, the first two naturally similar layers are the two layers closest to the input. For ResNets, it is the initial convolutional layer and the first post-residual block. As also mentioned in Kornblith et al. [10], post-residual layers, but not layers within residuals, develop similar representations.

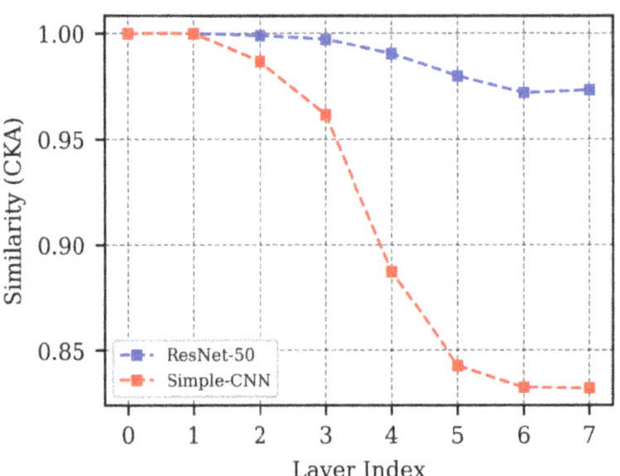

Figure 2. CKA similarity between clients at the end of training (refer to *Experimental Setup* for more information on training).

The objective of regularization in Federated Learning is to penalize local updates that stray from the global model. Since late layers are naturally dissimilar in an IID Federated Learning setting, all settings, including sub-ideal non-IID settings, should emulate the representational similarity of the ideal IID setting—achieved by regularizing only naturally similar layers. Furthermore, regularizing the first two naturally similar layers is unique from previous studies, which had regularized local updates based on all layers. This allows FedCKA to be much more scalable than other methods. The computational overhead for previous studies increases rapidly in proportion to the number of parameters, because all layers are regularized. FedCKA keeps the overhead nearly constant, as only two layers close to the input are regularized.

3.3. Measuring Layer-Wise Similarity

FedCKA is designed to regularize dissimilar updates in layers that should naturally be similar. However, there is currently no standard for measuring the similarity of layers between neural networks. While there are classical methods of applying univariate or multivariate analysis for comparing matrices, these methods are not suitable for comparing the similarity of layers and representations of different neural networks [10]. As for norms, Zhang et al. [11] concluded that a layer's robustness to re-initialization is not correlated with the l2-norm or l∞-norm. This suggests that using these norms to regularize dissimilar updates, as in previous works, may be inaccurate.

Kornblith et al. [10] concluded that similarity metrics for comparing the representation of different neural networks should be invariant to orthogonal transformations and isotropic scaling, while they are invertible to linear transformation. The study introduced centered kernel alignments (CKAa) and showed that the metric is most consistent in measuring the similarity between representation of neural networks. Thus, FedCKA regularizes local updates using the CKA metric as a similarity measure.

3.4. Modifications to FedAvg

FedCKA adds a regularization term to the local training process of the default FedAvg algorithm, keeping the entire framework simple. Algorithm 1 and Figure 3 shows the FedCKA framework in algorithm and figure form, respectively. More formally, we add ℓ_{cka} as a regularization term to the FedAvg training algorithm. The local loss function is as shown in Equation (1).

$$\ell = \ell_{sup}(w_{l_i}^t; D^i) + \mu \ell_{cka}(w_{l_i}^t; w_g^t; w_{l_i}^{t-1}; D^i) \tag{1}$$

Here, ℓ_{sup} is the cross-entropy loss, μ is a hyper-parameter to control the strength of the regularization term, ℓ_{cka}, in proportion to ℓ_{sup}. ℓ_{cka} is shown in more detail in Equation (2).

$$\ell_{cka} = \frac{1}{M} \sum_{n=1}^{M} -\log\left(\frac{e^{(CKA(a_{l_{in}}^t, a_{g_n}^t))}}{e^{(CKA(a_{l_{in}}^t, a_{g_n}^t))} + e^{(CKA(a_{l_{in}}^t, a_{l_{in}}^{t-1}))}} \right) \tag{2}$$

Algorithm 1 FedCKA

Input: number of communication rounds R, number of clients C, number of local epochs E, loss weighting variable μ, learning rate η
Output: The trained model w
1: Initialize w_g^0
2: **for** each round $r \in [0, R-1]$ **do**
3:     **for** each client $i \in [0, C]$ **do**
4:         $w_{l_i}^t \leftarrow$ LocalUpdate(w_g^t)
5:     **end for**
6:     $W_l^t \leftarrow [w_{l_0}^t, w_{l_1}^t, \dots, w_{l_{C-1}}^t]$
7:     $w_g^t \leftarrow$ WeightedAvg(W_l^t)
8: **end for**
9: **return** w_g^t

10: **LocalUpdate**(w_g^t):
11: $w_{l_i}^t \leftarrow w_g^t$
12: **for** each epoch $e \in [0, E-1]$ **do**
13:     **for** each batch $b \in D^i$ **do**
14:         $\ell_{sup} \leftarrow CrossEntropyLoss(w_{l_i}^t; b)$
15:         $\ell_{cka} \leftarrow CKALoss(w_{l_i}^t; w_{l_i}^{t-1}; w_{l_i}^t; b)$
16:         $\ell \leftarrow \ell_{sup} + \mu \ell_{cka}$
17:         $w_{l_i}^t \leftarrow w_{l_i}^t - \eta \nabla \ell$
18:     **end for**
19: **end for**
20: **return** solution

21: **WeightedAvg**(W_l^t):
22: Initialize w_g^t
23: **for** each client $i \in [0, C]$ **do**
24:     $w_g^t \mathrel{+}= \frac{|D^i|}{|D|} W_{l_i}^t$
25: **end for**
26: **return** w_g^t

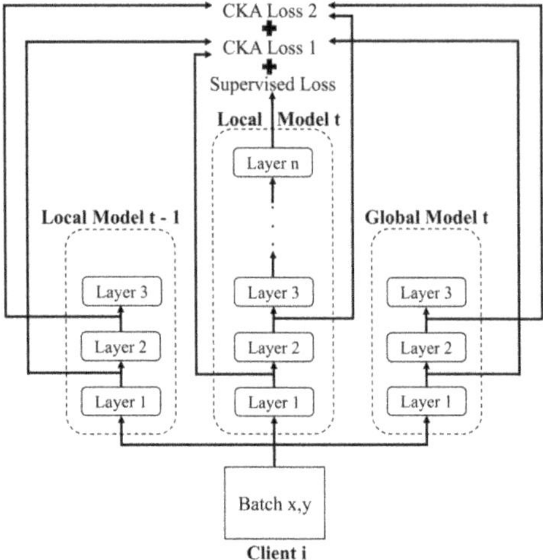

Figure 3. Training process of FedCKA

The formula of ℓ_{cka} is a slight modification to the contrastive loss that is used in SimCLR [20]. There are four main differences. First, SimCLR uses the representations of one model on different samples in a batch to calculate contrastive loss. FedCKA uses the representation of three models on the same samples in a batch to calculate ℓ_{cka}. $a^t_{l_i}$, $a^{t-1}_{l_i}$, and a^t_g are the representations of client i's current local model, client i's previous round local model, and the current global model, respectively. Second, SimCLR uses the temperature parameter τ to increase performance on difficult samples. FedCKA excludes τ, as it was not seen to help performance. Third, SimCLR uses cosine similarity to measure the similarity between the representations of difference datasets. FedCKA uses CKA as its measure of similarity. Fourth, SimCLR calculates contrastive loss once per batch, using the representations of the projection head. Here, M represents the number of layers being regularized . FedCKA use calculates ℓ_{cka} M times per batch, using the representations of the first M naturally similar layers, indexed by n, and averages the loss based on the number of layers to regularize. M is set to two by default unless otherwise stated.

As per Kornblith et al. [10], CKA is shown in Equation (3). Here, the ith eigenvalue of XX^T is λ^i_X. While Kornblith et al. [10] also presented a method to use kernels with CKA, we use the linear variant, as it is more computationally efficient, while having minimal impact on accuracy.

$$
\begin{aligned}
CKA(XX^T, YY^T) &= \frac{\left\|Y^T X\right\|^2_F}{\left\|X^T X\right\|_F \left\|Y^T Y\right\|_F} \\
&= \frac{\sum_{i=1}^{p1}\sum_{j=1}^{p2} \lambda^i_X \lambda^i_Y \left\langle \mathbf{u}^i_X, \mathbf{u}^j_Y \right\rangle^2}{\sqrt{\sum_{i=1}^{p1}(\lambda^i_X)^2}\sqrt{\sum_{j=1}^{p2}(\lambda^j_Y)^2}}
\end{aligned}
\tag{3}
$$

4. Experimental Results and Discussion

4.1. Experiment Setup

We compare FedCKA with the current state-of-the-art method, MOON [19], as well as FedAvg [3], FedProx [17], and SCAFFOLD [18]. We purposefully use a similar experimental setup to MOON, both because it is the most recent study and also reports the highest performance. In particular, CIFAR-10, CIFAR-100 [22], and Tiny ImageNet [23] datasets are used to test the performance of all methods.

For CIFAR-10, we use a small convolutional neural network. Two 5×5 convolutional layers comprise the base encoder, with 16 and 32 channels and two 2×2 max-pooling layers following each convolutional layer. A projection head of four fully connected layers follow the encoder, with 120, 84, 84, and 256 neurons. The final layer is the output layer with the number of classes. Although FedCKA and other studies can perform without this projection head, we include it because MOON shows a high discrepancy in performance without it. For CIFAR-100 and Tiny ImageNet, we use ResNet-50 [21]. We also add the projection head before the output layer, as per MOON.

We use the cross-entropy loss and SGD as our optimizer with a learning rate of 0.1, momentum of 0.9, and weight decay of 0.00001. Local epochs are set to 10. These are also the parameters used in MOON. Some small changes we made were with the batch size and communication rounds. We use a constant 128 for the batch size and trained for 100 communication rounds on CIFAR-10, 40 communication rounds on CIFAR-100, and 20 communication rounds on Tiny ImageNet. We used a lower number of communication rounds for the latter two datasets, because the ResNet-50 model overfit quite quickly.

As with many previous studies, we use the Dirichlet distribution to simulate heterogeneous settings [9,19,24]. The α parameter controls the strength of heterogeneity, with $\alpha = 0$ being the most heterogeneous, and $\alpha = \infty$ being non-heterogeneous. We report results for $\alpha \in [5.0, 1.0, 0.1]$, similarly to MOON. Figure 4 shows the distribution of data across clients on the CIFAR-10 dataset with different α. Figure 4A shows $\alpha := 5.0$, Figure 4B shows $\alpha := 1.0$, and Figure 4C shows $\alpha := 0.1$. All experiments were conducted using the PyTorch [25] library on a single GTX Titan V and four Intel Xeon Gold 5115 processors.

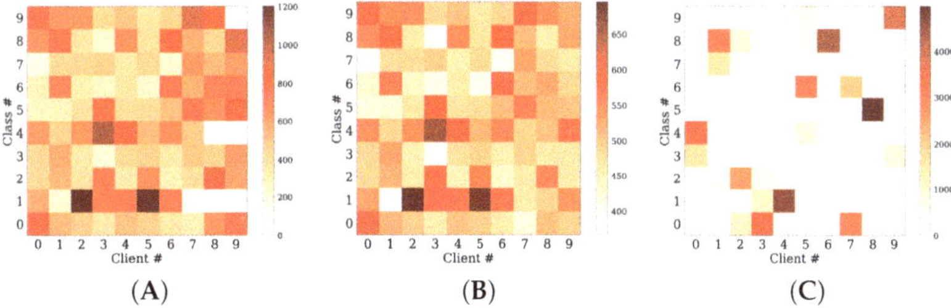

(A)   (B)   (C)

Figure 4. Distribution of the CIFAR-10 dataset across 10 clients according to the Dirichlet distribution. The x-axis shows the index of the client, and the y-axis shows the index of the class (label). (**A**–**C**) shows the data distribution of $\alpha := 5.0$, $\alpha := 1.0$, and $\alpha := 0.1$, respectively. As parameter α approaches 0, the heterogeneity of class distribution increases.

4.2. Accuracy

FedCKA adds a hyperparameter μ to control the strength of ℓ_{cka}. We tune μ from [3, 5, 10] and report the best results. MOON and FedProx also have a μ term. We also tune the hyperparameter μ with these methods. For MOON, we tune μ from [0.1, 1, 5, 10], and for FedProx, we tune μ from [0.001, 0.01, 0.1, 1], as used in each work. In addition, for MOON, we use $\tau = 0.5$ as reported in their work.

Table 1 shows the performance across CIFAR-10, CIFAR-100, and Tiny ImageNet with $\alpha = 5.0$. For FedProx, MOON, and FedCKA, we report performances with the best μ. For FedCKA, the best μ is 3, 10, and 3 for CIFAR-10, CIFAR-100, and Tiny ImageNet, respectively. For MOON, the best μ is 10, 5, and 0.1. For FedProx, the best μ is 0.001, 0.1, and 0.1. Table 2 shows the performance across increasing heterogeneity on the CIFAR-10 dataset with $\alpha \in [5.0, 1.0, 0.1]$. For FedCKA, the best μ is 5, 3, and 3 for each $\alpha \in [5.0, 1.0, 0.1]$, respectively. For MOON, the best μ is 0.1, 10, and 10. For FedProx, the best μ is 0.001, 0.1, and 0.001.

Table 1. Accuracy across datasets ($\alpha = 5.0$).

Method	CIFAR-10	CIFAR-100	Tiny ImageNet
FedAvg	64.37%	37.41%	19.49%
FedProx	64.58%	37.81%	20.93%
SCAFFOLD	64.33%	39.16%	21.18%
MOON	65.25%	38.37%	21.29%
FedCKA	**67.86%**	**40.07%**	**21.46%**

Table 2. Accuracy across $\alpha \in [5.0, 1.0, 0.1]$ (CIFAR-10).

Method	$\alpha = 5.0$	$\alpha = 1.0$	$\alpha = 0.1$
FedAvg	64.37%	62.49%	50.43%
FedProx	64.58%	62.51%	51.07%
SCAFFOLD	64.33%	63.31%	40.53%
MOON	65.25%	62.60%	51.63%
FedCKA	**67.86%**	**66.19%**	**52.35%**

We observe that FedCKA consistently outperforms previous methods across different datasets and across different α. FedCKA improves performances in heterogeneous settings due to regularizing layers that are naturally similar and not layers that are naturally dissimilar. It is also interesting to see that FedCKA performs better by a larger margin when α is substantial. This is likely because the global model can more effectively regularize updates as it is less biased when data distribution approaches IID settings. However, we also observe that other studies consistently improve performance, albeit by a smaller margin than FedCKA. FedProx and SCAFFOLD improve performances, likely due to their inclusion of naturally similar layers in regularization. The performance gain is lower, as they also include naturally dissimilar layers in regularization. MOON generally improves performance compared to FedProx and SCAFFOLD likely due to their use of a contrastive loss. That is, MOON shows that neural networks should be trained to be more similar to the global model *than past local model* , rather than only being blindly similar to the global model. By only regularizing naturally similar layers using a contrastive loss based on CKA, FedCKA outperforms all methods.

Note that across most methods and settings, there are discrepancies to the accuracy reported by MOON [19]. In particular, MOON reports higher accuracy across all methods, although the model's architecture are similar if not equivalent. We verify that this discrepancy is caused by the data augmentation used in experiments with MOON. We disclude augmentation as it would be unfair to generalize results across different non-IID settings if augmentations were to be used. The reported non-IIDness would decrease, as clients could create a more balanced distribution.

4.3. Communication Rounds and Local Epochs

We study the effects of regularization on the performance improvement per communication rounds. Results are shown in Figure 5. As expected with any regularization methods, we find that the accuracy for FedCKA is lower for the 40 communication rounds. However, we also find that after 40 communication epochs, FedCKA improves performances due to effective regularization. FedCKA decreases the bias that would have otherwise limited performance by penalizing weight updates that are not in agreement with the global model.

We also explore the effects of the number of local epochs on overall performance. We find that the performance of both FedAvg and FedCKA increases slightly when the number of local epochs increased. However, when further increasing the number of local epochs, we find that accuracy decreases, suggesting overfitting. The small increase in performance does not warrant additional local epochs. Clients are limited in their computational budget. Thus, computation cannot be used sparingly.

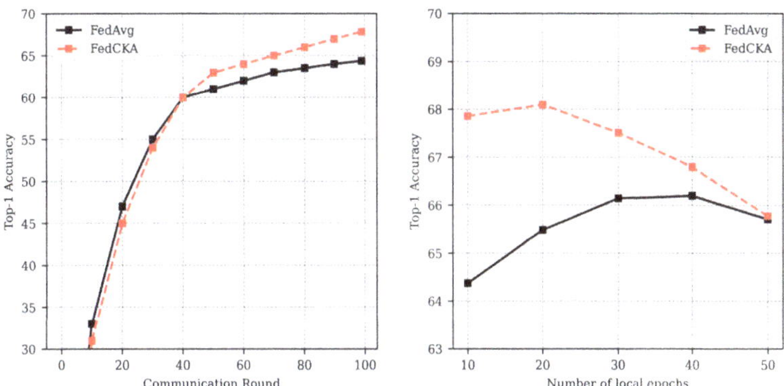

Figure 5. Effects of communication round and local epochs.

4.4. Regularizing Only Important Layers

We study the effects of regularizing different number of layers. Using the CIFAR-10 dataset with $\alpha = 5.0$, we change the number of layers to regularize through ℓ_{cka}. Formally, we change M in Equation (2) by scaling $M \in [1, 2, 3, 4, 5, 6, 7]$, and report the accuracy in Figure 6. Accuracy is the highest when only the first two layers are regularized. Note the dotted line representing the upper bound for Federated Learning. When the same model is trained on a centralized server with the entire CIFAR-10 dataset, accuracy is 70%. FedCKA with regularization on the first two naturally similar layers nearly reaches this upper bound.

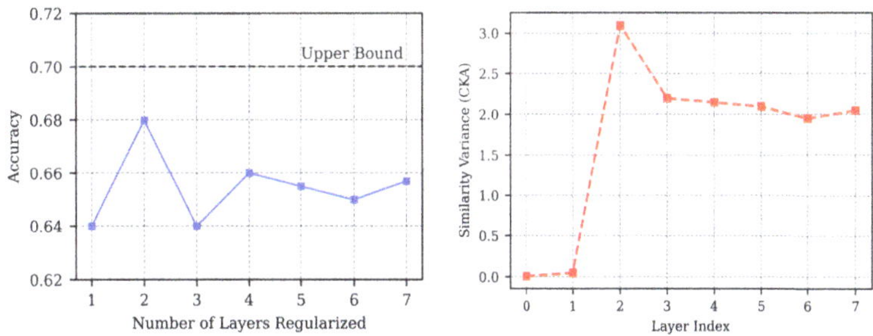

Figure 6. Accuracy with respect to the number of layers regularized and variance in similarity measures between clients on CIFAR-10 and $\alpha = 5.0$.

This verifies our hypothesis that only naturally similar, but not naturally dissimilar layers, should be regularized. By regularizing only one layer, a naturally similar layer (the second layer) would be excluded, thus decreasing performance. By regularizing three layers, a non-similar layer (the third layer) with the highest variance of non-similarity (see Figure 6) would be regularized, thus decreasing performance. The performance increase when 4–7 layers are regularized may seem anomalous, yet is valid when considering that both the first two naturally similar layers are included, and the weight of regularization of the third layer decreases with a higher M.

4.5. Using the Best Similarity Metric

We study the effects of regularizing the first two naturally similar layers with different similarity metrics. Using the CIFAR-10 dataset with $\alpha = 5.0$, we change the similarity metric to regularize through ℓ_{cka}. Formally, we change $CKA(a_1, a_2)$ in Equation (2) to three other similarity metrics: first, the kernel CKA, introduced in Kornblith et al. [10] ($CKA_k(a_1, a_2)$); second, the squared Frobenius norm ($\|a_1 - a_2\|_F^2$); third, the vectorized cosine similarity ($\|vec(a_1)\| \|vec(a_2)\| \cos\theta$). We compare the results with these different metrics as well as the baseline, FedAvg. The results are shown in Table 3.

Table 3. Accuracy and training duration with FedCKA with different similarity metrics (CIFAR-10).

Similarity Metric	Accuracy	Training Duration (s)
None (FedAvg)	64.37%	54.82
Frobenius Norm	65.54%	64.73
Vectorized Cosine	66.67%	65.75
Kernel CKA	67.93%	122.41
Linear CKA	67.86%	104.17

We observe that performance is the highest with CKA due to the increased accuracy of measuring similarity. Only truly dissimilar updates are penalized, thus improving performance. Note Equation (3) includes an inner product in the numerator. While the l2-distance or cosine similarity are also inner products, CKA is more suitable for measuring similarities between matrices of higher dimension than the number of data points, as is the case for neural network representations [10].

Furthermore, while kernel CKA slightly outperforms linear CKA, we opt to use linear CKA considering the computational overhead. We also observe that the squared Frobenius norm and vectorized cosine similarity decreased performance only slightly. These methods outperformed most previous works. This verifies that while it is important to use an accurate similarity measure, it is more important to focus on regularizing naturally similar layers.

4.6. Efficiency and Scalability

Efficient and scalable local training procedures are important engineering principles in Federated Learning. That is, for Federated Learning to be applied to real-world applications, we must assume that clients have limited computing resources. Thus, we analyze the local training time of all methods, as shown in Table 4. Note that FedAvg is the lower bound for training time, since all other methods add a regularization term.

Table 4. Average Training Duration Per Communication Round (in seconds).

Method	7 Layers	Time Extended	50 Layers	Time Extended
FedAvg	54.82	-	638.79	-
SCAFFOLD	57.19	2.37	967.04	328.25
FedProx	57.20	2.38	862.12	223.33
MOON	97.58	42.76	1689.28	1050.49
FedCKA	104.17	**49.35**	750.97	**112.18**

For a seven-layer CNN trained on CIFAR-10, the training time for all methods are fairly similar. FedCKA extends training by the largest amount, as the matrix multiplication operation to calculate the CKA similarity is proportionally expensive to the forward and back propagation of the small model. However, for ResNet-50 trained on Tiny ImageNet, we see that the training time of FedProx, SCAFFOLD, and MOON increased substantially. Only FedCKA has comparable training times to FedAvg. This is because FedProx and SCAFFOLD perform expensive operations on the weights of each layer, and MOON performs forward propagation on three models until the penultimate layer. All these operations scale substantially as the number of layers increase. While FedCKA also performs forward propagation on three models, the number of layers remains static, thus being most efficient with medium-sized models.

We emphasize that regularization must remain scalable for Federated Learning to be applied to state-of-the-art models. Even on ResNet-50, which is no longer considered a large model, other Federated Learning regularization methods lack scalabililty. This causes difficulty in testing these methods with the current state-of-the-art models, such as ViT [26] posessing 1.843 billion parameters, or slightly older models, such as EfficientNet-B7 [27] posessing 813 layers.

5. Conclusions and Future Work

Improving the performance of Federated Learning on heterogeneous data is a widely researched topic. However, many previous studies suggested that regularizing every layer of neural networks during local training is the best method to increase performance. We propose FedCKA, an alternative approach built on the most up-to-date understanding of neural networks. By regularizing naturally similar, but not naturally dissimilar layers, during local training, performance improves beyond previous studies. We also show that FedCKA is currently the one of the best regularization methods with adequate scalability when trained with a moderatly sized model.

FedCKA shows that the proper regularization of important layers improves the performance of Federated Learning on heterogeneous data. However, standardizing the comparison of neural networks is an important step in a deeper understanding of neural networks. Moreover, there are questions as to the accuracy of CKA in measuring similarity in models such as Transformers or Graph Neural Networks. These are some topics we leave for future studies.

Author Contributions: Conceptualization, H.M.S.; formal analysis, H.M.S., M.H.K. and T.-M.C.; funding acquisition, M.H.K. and T.-M.C.; Investigation, H.M.S. and M.H.K.; methodology, H.M.S.; project administration, M.H.K. and T.-M.C.; software, H.M.S.; supervision, T.-M.C.; validation, H.M.S. and M.H.K.; visualization, H.M.S.; writing—original draft, H.M.S.; writing—review and editing, H.M.S., M.H.K. and T.-M.C. All authors have read and agreed to the published version of the manuscript.

Funding: This work was supported by the Institute of Information and communications Technology Planning and Evaluation (IITP) grant funded by the Korea government (MSIT) (No. 2020-0-00990, Platform Development and Proof of High Trust and Low Latency Processing for Heterogeneous·Atypical·Large Scaled Data in 5G-IoT Environment). The APC was funded by the Sungkyunkwan University and the BK21 FOUR (Graduate School Innovation) funded by the Ministry of Education (MOE, Korea) and National Research Foundation of Korea (NRF).

Institutional Review Board Statement: Not applicable.

Data Availability Statement: The CIFAR-10/CIFAR-100 dataset is available at https://www.kaggle.com/c/tiny-imagenet and the Tiny ImageNet dataset is available at https://www.kaggle.com/c/tiny-imagenet.

Conflicts of Interest: The authors declare no conflict of interest. The funders had no role in the design of the study; in the collection, analyses, or interpretation of data; in the writing of the manuscript; or in the decision to publish the results.

Abbreviations

The following abbreviations are used in this manuscript:

FL   Federated Learning
IID   Independent and Identically Distributed
CKA   Centered Kernel Alignment

References

1. Lecun, Y.; Bengio, Y.; Hinton, G. Deep Learning. *Nature* **2015**, *521*, 436–444. [CrossRef] [PubMed]
2. Sejnowski, T.J. (Ed.) *The Deep Learning Revolution*; MIT Press: Cambridge, MA, USA, 2018.
3. McMahan, H.B.; Moore, E.; Ramage, D.; Hampson, S.; Arcas, B.A.Y. Communication-Efficient Learning of Deep Networks from Decentralized Data. In Proceedings of the 20th International Conference on Artificial Intelligence and Statistics (AISTATS), Fort Lauderdale, FL, USA, 20–22 April 2017.
4. Verbraeken, J.; Wolting, M.; Katzy, J.; Kloppenburg, J.; Verbelen, T.; Rellermeyer, J.S. A Survey on Distributed Machine Learning. *arXiv* **2019**, arXiv:1912.09789.

5. Samarakoon, S.; Bennis, M.; Saad, W.; Debbah, M. Distributed Federated Learning for Ultra-Reliable Low-Latency Vehicular Communications. *IEEE Trans. Commun.* **2020**, *68*, 1146–1159. [CrossRef]

6. Yang, K.; Jiang, T.; Shi, Y.; Ding, Z. Federated Learning via Over-the-Air Computation. *IEEE Trans. Wirel. Commun.* **2020**, *19*, 2022–2035. [CrossRef]

7. Brisimi, T.S.; Chen, R.; Mela, T.; Olshevsky, A.; Paschalidis, I.C.; Shi, W. Federated learning of predictive models from federated Electronic Health Records. *Int. J. Med. Inform.* **2018**, *112*, 59–67. [CrossRef] [PubMed]

8. Kairouz, P.; McMahan, H.B.; Avent, B.; Bellet, A.; Bennis, M.; Bhagoji, A.N.; Bonawitz, K.; Charles, Z.; Cormode, G.; Cummings, R.; et al. Advances and Open Problems in Federated Learning. *arXiv* **2021**, arXiv:1912.04977.

9. Hsu, T.M.H.; Qi, H.; Brown, M. Measuring the Effects of Non-Identical Data Distribution for Federated Visual Classification. *arXiv* **2019**, arXiv:1909.06335.

10. Kornblith, S.; Norouzi, M.; Lee, H.; Hinton, G. Similarity of Neural Network Representations Revisited. In Proceedings of the 36th International Conference on Machine Learning, Beach, CA, USA, 9–15 June 2019; Volume 97, pp. 3519–3529.

11. Zhang, C.; Bengio, S.; Singer, Y. Are All Layers Created Equal? In Proceedings of the ICML 2019 Workshop on Identifying and Understanding Deep Learning Phenomena, Long Beach, CA, USA, 15 June 2019.

12. Zhang, L.; Shen, L.; Ding, L.; Tao, D.; Duan, L.Y. Fine-tuning global model via data-free knowledge distillation for non-iid federated learning. In Proceedings of the IEEE/CVF Conference on Computer Vision and Pattern Recognition, New Orleans, LA, USA, 19–20 June 2022; pp. 10174–10183.

13. Shen, Y.; Zhou, Y.; Yu, L. CD2-pFed: Cyclic Distillation-guided Channel Decoupling for Model Personalization in Federated Learning. In Proceedings of the IEEE/CVF Conference on Computer Vision and Pattern Recognition, New Orleans, LA, USA, 19–20 June 2022; pp. 10041–10050.

14. Yoon, T.; Shin, S.; Hwang, S.J.; Yang, E. Fedmix: Approximation of mixup under mean augmented federated learning. *arXiv* **2021**, arXiv:2107.00233.

15. Wang, H.; Yurochkin, M.; Sun, Y.; Papailiopoulos, D.; Khazaeni, Y. Federated Learning with Matched Averaging. In Proceedings of the International Conference on Learning Representations, Addis Ababa, Ethiopia, 26–30 April 2020.

16. Li, X.; JIANG, M.; Zhang, X.; Kamp, M.; Dou, Q. FedBN: Federated Learning on Non-IID Features via Local Batch Normalization. In Proceedings of the International Conference on Learning Representations, Virtual Event, Austria, 3–7 May 2021.

17. Li, T.; Sahu, A.K.; Zaheer, M.; Sanjabi, M.; Talwalkar, A.; Smith, V. Federated Optimization in Heterogeneous Networks. *arXiv* **2020**, arXiv:1812.06127.

18. Karimireddy, S.P.; Kale, S.; Mohri, M.; Reddi, S.; Stich, S.; Suresh, A.T. SCAFFOLD: Stochastic Controlled Averaging for Federated Learning. In Proceedings of the 37th International Conference on Machine Learning, Virtual, 13–18 July 2020; Volume 119, pp. 5132–5143.

19. Li, Q.; He, B.; Song, D. Model-Contrastive Federated Learning. In Proceedings of the IEEE/CVF Conference on Computer Vision and Pattern Recognition (CVPR 2021), Nashville, TN, USA, 20–25 June 2021; pp. 10713–10722.

20. Chen, T.; Kornblith, S.; Norouzi, M.; Hinton, G. A Simple Framework for Contrastive Learning of Visual Representations. *arXiv* **2020**, arXiv:2002.05709.

21. He, K.; Zhang, X.; Ren, S.; Sun, J. Deep Residual Learning for Image Recognition. In Proceedings of the IEEE/CVF Conference on Computer Vision and Pattern Recognition (CVPR 2016), Las Vegas, NV, USA, 27–30 June 2016; pp. 770–778.

22. Krizhevsky, A. Learning Multiple Layers of Features from Tiny Images. 2009. Available online: https://www.bibsonomy.org/bibtex/cc2d42f2b7ef6a4e76e47d1a50c8cd86 (accessed on 27 September 2022).

23. Li, F.F.; Karpathy, A.; Johnson, J. Tiny ImageNet. 2014. Available online: https://www.kaggle.com/c/tiny-imagenet (accessed on 27 September 2022).

24. Lin, T.; Kong, L.; Stich, S.U.; Jaggi, M. Ensemble Distillation for Robust Model Fusion in Federated Learning. *arXiv* **2021**, arXiv:2006.07242.

25. Paszke, A.; Gross, S.; Massa, F.; Lerer, A.; Bradbury, J.; Chanan, G.; Killeen, T.; Lin, Z.; Gimelshein, N.; Antiga, L.; et al. PyTorch: An Imperative Style, High-Performance Deep Learning Library. In *Advances in Neural Information Processing Systems 32*; Wallach, H., Larochelle, H., Beygelzimer, A., d'Alché-Buc, F., Fox, E., Garnett, R., Eds.; Curran Associates, Inc.: Red Hook, NY, USA, 2019; pp. 8024–8035.

26. Dosovitskiy, A.; Beyer, L.; Kolesnikov, A.; Weissenborn, D.; Zhai, X.; Unterthiner, T.; Dehghani, M.; Minderer, M.; Heigold, G.; Gelly, S.; et al. An Image is Worth 16x16 Words: Transformers for Image Recognition at Scale. In Proceedings of the International Conference on Learning Representations (ICLR 2021), Vienna, Austria, 4 May 2021.

27. Tan, M.; Le, Q. EfficientNet: Rethinking Model Scaling for Convolutional Neural Networks. In Proceedings of the 36th International Conference on Machine Learning, Beach, CA, USA, 9–15 June 2019; Volume 97, pp. 6105–6114.

Article

Intelligent Arabic Handwriting Recognition Using Different Standalone and Hybrid CNN Architectures

Waleed Albattah * and Saleh Albahli

Department of Information Technology, College of Computer, Qassim University, Buraydah 52571, Saudi Arabia
* Correspondence: w.albattah@qu.edu.sa

Abstract: Handwritten character recognition is a computer-vision-system problem that is still critical and challenging in many computer-vision tasks. With the increased interest in handwriting recognition as well as the developments in machine-learning and deep-learning algorithms, researchers have made significant improvements and advances in developing English-handwriting-recognition methodologies; however, Arabic handwriting recognition has not yet received enough interest. In this work, several deep-learning and hybrid models were created. The methodology of the current study took advantage of machine learning in classification and deep learning in feature extraction to create hybrid models. Among the standalone deep-learning models trained on the two datasets used in the experiments performed, the best results were obtained with the transfer-learning model on the MNIST dataset, with 0.9967 accuracy achieved. The results for the hybrid models using the MNIST dataset were good, with accuracy measures exceeding 0.9 for all the hybrid models; however, the results for the hybrid models using the Arabic character dataset were inferior.

Keywords: classification; convolutional neural network; recurrent neural networks; MNIST; model selection

Citation: Albattah, W.; Albahli, S. Intelligent Arabic Handwriting Recognition Using Different Standalone and Hybrid CNN Architectures. *Appl. Sci.* **2022**, *12*, 10155. https://doi.org/10.3390/app121910155

Academic Editors: George Drosatos, Pavlos S. Efraimidis and Avi Arampatzis

Received: 16 August 2022
Accepted: 4 October 2022
Published: 10 October 2022

Publisher's Note: MDPI stays neutral with regard to jurisdictional claims in published maps and institutional affiliations.

Copyright: © 2022 by the authors. Licensee MDPI, Basel, Switzerland. This article is an open access article distributed under the terms and conditions of the Creative Commons Attribution (CC BY) license (https://creativecommons.org/licenses/by/4.0/).

1. Introduction

Despite massive technological advances, many people's textual compositions are still handwritten. Using pen and paper for writing is essential to people's work. Handwriting has different sizes and styles, making the creation of automatic techniques for recognizing texts a challenging task in computer vision [1]. Text-recognition systems utilize automated techniques for text recognition by converting text included in images into matching digital formats. Such systems can discover typed or handwritten characters and are used in different application domains.

A handwritten-character-recognition system is a computer-vision system that is intended to classify and recognize handwritten characters [2]. Character recognition is still critical and challenging in many computer-vision tasks [3]. With the increased interest in handwriting recognition and the developments in machine-learning and deep-learning algorithms, researchers have made significant improvements and advances in this field. English-handwriting-recognition methodologies have received significant interest from researchers [4]; however, Arabic has not yet received enough interest. With more than 315 million native Arabic speakers [5], the need for Arabic-handwriting-recognition systems is critical. Arabic is one of the most popular spoken languages in the world, with twenty-eight alphabets and different letter styles, based on geography.

In general, handwriting is a pattern-recognition research area with different applications. For each application domain, specific constraints should be considered [6], otherwise the recognition process will be complicated due to the wide range of handwriting styles and sizes. For example, recognizing characters on car license plates is more straightforward than recognizing Arabic handwriting due to the different styles of handwriting. Therefore, researchers are making great efforts to improve recognition systems using various techniques, deep-learning algorithms being at the top of the list.

For a while, Arabic handwritten character recognition (AHCR) has been an area of research in pattern recognition and computer vision. Several machine-learning (ML) algorithms, such as support vector machines (SVMs), have improved AHCR. Such models are still limited and cannot outperform convolutional neural networks (CNNs) on different Arabic handwriting datasets.

Several Arabic handwriting datasets have been used in the literature to create convolutional neural network (CNN) models [1,7,8] that automatically extract features from images and outperform classical machine-learning techniques (such as [9,10]), especially when large datasets with a large number of classes are used. As Niu and Suen claim [11], better classification results can be achieved when an SVM is replaced with an MLP in deep learning. This is because MLPs are based on empirical risk minimization, which tries to minimize errors in the training set. As a result, the training procedure is terminated when the back-propagation algorithm finds the first separating hyperplane.

Several deep-learning architectures have been used in the literature in different applications (deep neural networks (DNNs), convolutional neural networks (CNNs), deep belief network (DBNs), recurrent neural networks (RNNs), and generative adversarial networks (GANs)) [12]. In this work, a CNN, a commonly used deep-learning algorithm, was used to recognize Arabic handwritten characters.

Despite all the progress that has been made, there are still some challenges and a demand for new methods to overcome these limitations [13]. Compared to the English language, the quantity of available datasets of Arabic handwritten characters is relatively small. Additionally, some of the published datasets include few records. There is a requirement for a vast dataset containing a variety of font sizes, styles, illuminations, users, and texts [14]. Some of the collected records (pictures) may have unnecessary data noise that must be eliminated, or misclassification might occur. Especially for massive datasets, it is necessary to identify a simple and rapid method for removing noise automatically rather than manually [15,16].

The rest of the paper is organized as follows. Section 2 explains the background of the previous research work on handwriting recognition. Section 3 presents the materials and methods of the study. Section 4 describes the experimental results and analysis, while Section 5 presents the discussion and comparison. Finally, the conclusions are drawn and future works are considered.

2. Related Works and Motivation

Handwriting recognition using CNNs has received attention in terms of research work in different languages, such as English [17–21], Arabic [1,3,7–10,22], Bangla [2], and Chinese [23–28]. Recently, the focus on Arabic handwriting recognition has increased [29]. Researchers have developed different techniques to enhance recognition outcomes.

According to El-Sawy et al. [3], CNN techniques outperform other feature-extraction and classification methods, especially with big datasets. This is not applicable for most of the studies on Arabic language recognition. Therefore, the authors created the Arabic Handwritten Characters Dataset (AHCD) and proposed a CNN model that achieved an accuracy of 94.9. Similarly, Altwaijry et al. [1] released Arabic handwritten alphabets in what is called "the Hijja" dataset. The dataset consists of samples written by children aged 7 to 12 years old. The researchers conducted handwriting-recognition experiments using a CNN trained on two datasets, the Hijja dataset and the Arabic Handwritten Character Dataset (AHCD). The proposed model achieved accuracies of 97% on the AHCD dataset and 88% on the Hijja dataset. These results indicate that they have outperformed the achieved results of El-Sawy et al. [3].

Using a different approach, Alrobah and Albahli [13] merged an ML model with a deep-learning model to create a hybrid, taking advantage of CNN models in feature extraction and ML models in classification. The study achieved an accuracy of 96.3, proving the hybrid model's effectiveness. The result was better than those obtained in the original experiment performed on the same dataset (the Hijja) by Altwaijry and Al-Turaiki [1].

A CNN architecture known as Alexnet was utilized by Boufenar et al. [7], which includes three layers of max pooling followed by three layers of fully connected convolutions. They investigated the impact of preprocessing on model improvement. The Alexnet model was trained and evaluated using two datasets, OIHACDB-40 and AHCD, and three learning strategies: training the CNN model from scratch, utilizing a transfer-learning technique, and fine-tuning the weights of the CNN architecture. The experimental outcomes demonstrated that the first technique outperformed the others, with 100 percent and 99.98 percent accuracy for the OIHACDB-40 and AHCD datasets, respectively.

Balaha et al. [30] established a vast, complicated dataset of Arabic handwritten characters (HMBD). They implemented a deep-learning (DL) system with two convolutional-neural-network (CNN) architectures (called HMB1 and HMB2), using optimization, regularization, and dropout techniques. They employed Elsawy et al. [3] as a controlled study throughout their 16 experiments with the HMBD, CMATER, and AIA9k datasets. The study suggested that data augmentation helped increase testing accuracy and reduce overfitting. Data augmentation increased the volume of input data; as a result, the architectures were learned and trained using more data. The top results for HMBD, CMATER, and AIA9k were 90.7%, 97.3%, and 98.4%, respectively. Younis [31] and Najadat et al. [32] built CNN models that were trained and tested on AHCD to improve AHCD performance. Three convolutional layers and one fully connected layer comprised the model of [30]. In addition, two regularization methods, dropout and batch normalization, with distinct data-augmentation techniques, were employed to enhance the model's performance. Using the AIA9k and AHCD datasets to train and test the model, the accuracies were 94.8 and 97.6 percent, respectively. Similarly, the CNN design suggested in [31] comprised four convolutional layers, two max-pooling layers, and three fully connected layers. The authors examined various epochs and batch sizes and found that 40 epochs with a batch size of 16 produced the best results for training of the model. Based on empirical findings, the model accuracy achieved was 97.2%.

Considering various model architectures, subsequent investigations attempted to identify more successful instances. Alyahya et al. [33] examined the performance of the ResNet-18 architecture when an FCL and dropout were added to the original architecture for recognizing handwritten Arabic characters. Two models utilized a fully connected layer with/without a dropout layer following all convolutional layers. The other two models used two fully connected layers with/without a dropout layer. They used the AHCD dataset to train and evaluate the CNN-based ResNet-18 model, with the original ResNet-18 achieving the best test result of 98.30 percent. Almansari et al. [34] examined the performance of a CNN and a multilayer perceptron (MLP) in detecting Arabic characters from the AHCD dataset. To reduce model overfitting, they examined various dropout levels, batch sizes, neuron counts in the MLP model, and filter sizes in the CNN model. According to the experimental results, the CNN model and the MLP model achieved the highest test accuracies of 95.3% and 72.08%, respectively, demonstrating that CNN models are more suitable for Arabic handwritten character recognition.

Similarly, to improve the detection of Arabic digits, Das et al. [35] provided a collection of 88 features of handwritten Arabic number samples. An MLP classifier was constructed with three layers (input, single hidden layer, and output). Back-propagation was performed to train the multi-layer perceptron, which was subsequently used to classify Arabic numerals from the CMATERDB 3.3.1 dataset. According to testing results, the model achieved an average accuracy of 94.93 percent on a database of 3000 samples.

Musa [36] introduced datasets that include Arabic numerals, isolated Arabic letters, and Arabic names. The vast majority of these datasets are offline. In addition, the report described published results and an upcoming study. Noubigh et al. [37] examined the issue of Arabic handwriting recognition. They offered a new architecture based on a character-model approach and a CTC decoder that combined a CNN and a BLSTM. For experimentation, the handwriting Arabic database KHATT was used. The results demonstrated a net performance benefit for the CNN–BLSTM combined method compared to the methods employed in the literature.

De Sousa [8] suggested two deep CNN-based models for Arabic handwritten character and digit recognition: VGG-12 and REGU. The VGG-16 model was derived from the VGG-12 model by removing the fifth convolutional block and adding a dropout layer prior to the SoftMax FCL classifier. In contrast, the REGU model was created from scratch by adding dropout and batch-normalization layers to both the CNN and fully connected layers. The two models were trained, one with and the other without data augmentation. The predictions of each of the four models were then averaged to construct an ensemble of the four models. The ensemble model's best test accuracy was 99.47 percent. Mudhsh et al. [10] proposed a model for recognizing Arabic handwritten numerals and characters. The model consisted of thirteen convoluted layers, followed by two max-pooling layers and three completely connected layers. To reduce model complexity and training time, the suggested model employed only one-eighth of the filters in each layer of the original VGG-16. The model was trained and evaluated using two distinct datasets: ADBase for the digit-recognition task and HACDB for the character-recognition task. To prevent overfitting, they utilized dropout and data augmentation. The model's attained accuracy was 99.66% when using the ADBase dataset and 97.32% when using the HACDB dataset.

Al-Taani et al. [38] built a ResNet architecture to recognize handwritten Arabic characters. The suggested method included pre-processing, training ResNets on the training set, and testing trained ResNets on the datasets. Using MADBase, AIA9K, and AHCD, this method achieved 99.8 percent, 99.05 percent, and 99.55 percent accuracies, respectively.

AlJarrah et al. [39] established a CNN model to detect printed Arabic letters and numbers. Using an AHCD dataset, the model was trained. This study used data-augmentation techniques to improve model performance and detection outcomes. The experiment demonstrated that the proposed strategy might achieve a success rate of 97.2 percent. The model's accuracy increased to 97.7 percent once data augmentation was implemented. Elkhayati et al. [40] created a method for segmenting Arabic words for recognition purposes using a convolutional neural network (CNN) and mathematical morphology operations (MMOs). In their study, the authors offered a directed CNN and achieved better performance than a standard CNN. Elleuch et al. [41] presented a deep-belief neural network (DBNN) for identifying handwritten Arabic characters/words. The proposed model began with the row data before proceeding to the unsupervised learning technique. This model's performance on the HACDB dataset was 97.9 percent accurate. Kef et al. [42] developed a fuzzy classifier with structural properties for Arabic-handwritten-word-recognition offline systems based on segmentation procedures. Before extracting features using invariant pseudo-Zernike moments, the model splits characters into five distinct categories. According to the study's findings, the proposed model achieved a high level of accuracy for the IFN/ENIT database of 93.8%.

Based on the generic-feature–independent-pyramid multilevel model (GFIPML), Korichi et al. [43] developed a method for recognizing Arabic handwriting. To evaluate their system's performance, the authors utilized the AHDB dataset and obtained better outcomes. They combined local phase quantization (LPQ) with multiple binarized statistical image features (BSIfs) to enhance the recognition. The proposed system achieved an accuracy of 98.39%. However, in a previous study, Korichi et al. [44] compared the performance of multiple CNN networks to statistical descriptors derived from the PML model. The highest recognition rate attained by the LPQ descriptor was 91.52%. Another common application of handwriting recognition is Arabic handwritten literal amount recognition. Korichi et al. [45] performed numerous experiments with convolutional neural networks (CNNs), such as basic CNN, VGG-16, and ResNet, which were developed using regularization approaches, such as dropout and data augmentation. The results demonstrated that CNN architectures are more effective than previous approaches based on handmade characteristics.

Since pre-trained models are trained on a general dataset and may not be suitable for Arabic handwriting classification, it is evident from the articles mentioned above that models created from scratch tend to produce better results. Models that have been pre-trained include machine-learning and deep-learning models that have been created and trained on a wide range of data. For example, the ImageNet dataset is used to train models and comprises thousands of images; however, models constructed utilizing the image dataset from the beginning of their training make them far more fit for the classification task due to their comprehensive understanding of the dataset. Later, adjusting the hidden layers of the deep-learning models makes them more suitable for the classification challenge. In this work, we tried to overcome this limitation by building the model specifically for Arabic handwritten character classification. Table 1 summarizes the CNN models developed in the literature for Arabic handwriting recognition. Table 2 presents a summary of the Arabic handwriting datasets.

Table 1. Summary of the CNN models for Arabic handwriting recognition.

Paper	Year	Model	Dataset	Accuracy (%)
El-Sawy et al. [3]	2017	CNN	AHCD	94.9
Altwaijry et al. [1]	2017	CNN	AHCD	97
			Hijja	88
Alrobah and Albahli [29]	2021	Hybrid CNN models: CNN + FCL,	AHCD	94.5
		CNN + SVM, CNN + XGBoost	Hijja	96.3
		CNN from scratch		100
		CNN as features ex-tractor	OIHACDB-40	82.38
Boufenar et al. [7]	2018	Fine-tuned CNN		98.86
		AHCD CNN from scratch		99.98
		CNN as feature extractor	AHCD	82.53
		Fine-tuned CNN		96.78
			HMBD	90.7
Balaha et al. [30]	2021	CNN	AHCD	97.3
			AIA9k	98.4
Younis et al. [31]	2017	CNN	AHCD	97.6
			AIA9k	94.8
Najadat et al. [32]	2019	CNN	AHCD	97.2
Alyahya et al. [33]	2020	CNN	AHCD	98.3
Almansari et al. [34]	2019	CNN	AHCD	95.27
		MPL		72.08
Das et al. [35]	2010	MPL	CMATERDB	94.93
Musa [36]	2011	CNN	SUST-ALT	
Noubigh et al. [37]	2021	CNN and BLSTM	KHATT	
De Sousa [8]	2017	VG16-based CNN	AHCD	97.32
Mudhsh et al. [10]	2018			98.42
Al-Taani et al. [38]	2021	ResNet	AHCD	99.5
AlJarrah et al. [39]	2021	CNN	AHCD	97.7
Elleuch et al. [43]	2015	DBNN	HACDB	97.9
Kef et al. [42]	2016	5 Fuzzy ARTMAP	IFN/ENIT	93.8

Table 2. Summary of the Arabic handwriting datasets.

Dataset	Type	Amount
AHCD	Characters	16,800
Hijja	Characters	47,434
OIHACDB-40	Characters	30,000
HMBD	Characters	54,115
AIA9k	Alphabet	8737
CMATERDB	Digits	3000

Table 2. *Cont.*

Dataset	Type	Amount
SUST-ALT	Digits Letters Words	47,988
KHATT	Characters	4000
HACDB	Characters	6600
IFN/ENIT	Characters	212.211

3. Materials and Methods

The convolutional neural network (CNN) is the leading technique applied in automatic character and digit detection using computer systems. Various deep-learning models are being tested for multiple languages. As one of the most spoken languages in the world, Arabic is no exception. This section discusses the methods and techniques used to create the system for detecting handwritten Arabic characters.

There have been many approaches used for handwritten character recognition in different languages, but proper techniques have not yet been developed for the Arabic language. Arabic handwritten character recognition is now needed. The CNN approach was best suited for other language datasets, so the proposed system tried this technique with the complete setup.

3.1. Brief Overview of Convolutional Neural Networks

A typical convolutional neural network (CNN) is an artificial neural network that tries to mimic the way the human brain detects, recognizes, and interprets images. It does so by processing pixel data to find features that stand out and serve as identification points. It works by assigning importance (learnable biases and weights) to certain parts of inputted images to differentiate them from one another, which ultimately leads to recognition of what the images contain. The various parts of a typical CNN are further elaborated below.

A CNN automatically detects the available features in a dataset. These features may be statistical, texton, curvature, along with many others. The features used depends on the problem that needs to be addressed, but this process was performed automatically in this model. According to the proposed system, the data were required to know these character types. Here, the CNN model also detected the curvature features or image contours.

3.1.1. Convolutional Layer

A convolutional layer constitutes the foundation and main building block of a CNN. It works by converting an input image into a feature map, also known as an activation map. The convolutional layer has a kernel, or filter, a two-dimensional array of weights that is responsible for carrying out the task of feature extraction, which leads to the creation of a feature map. The filter works by moving from one image stride to another while performing a dot product and feeding the result to an output array. This output is the feature map. The filter needs to be configured before the operation begins and maintained throughout. The parameters to be configured are:

The number of filters: This affects the number of feature maps to be obtained, as the number of feature maps increases with the number of filters.

Stride: This is the number of pixels the filter travels for each operation. Usually, a stride of one or two is used because a larger number of strides leads to a smaller output and missing key features. The stride and feature map are shown in Figure 1.

Zero padding: This is crucial, since most input images have elements that fall outside the input matrix, which might be ignored. Zero padding covers boundary pixels, thereby producing larger outputs of high quality.

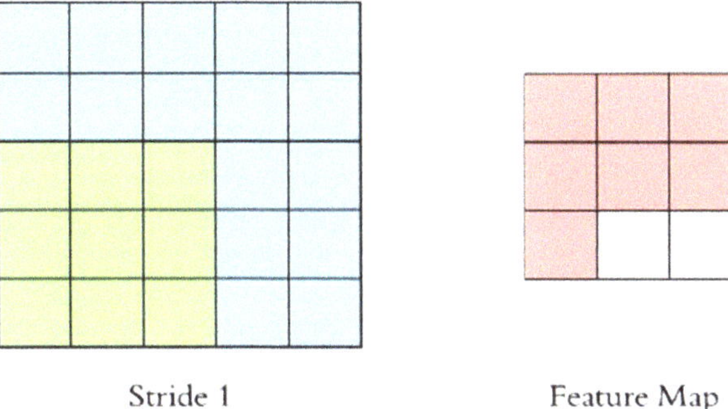

Figure 1. Stride and feature map.

In Figure 1, the stride image shows the input image being turned into a matrix. The filter is then applied to the green section, and a dot product is computed between the input pixels and the filter. After this, the filter is then moved by one stride, repeating the initial process until the kernel has covered the entire matrix formed by the image. The final result is a new, smaller matrix called the feature map or activation map.

3.1.2. Pooling Layer

The pooling layer, or down sampling, performs the task of reducing the parameters in the input image, thereby resulting in dimensionality reduction. It also sweeps through the entire input just like the filter; however, unlike the convolutional layer, it does not carry any weights. Instead, it applies an aggregation function to the image to populate the output array. Two types of pooling are usually used:

Max pooling selects the pixel with the maximum value as the filter moves across the input image and sends it to the output array.

Average pooling: Here, the average value within the receptive field is calculated as the filter moves through the image and sends it to the output array.

The pooling used for this project is illustrated in Figure 2.

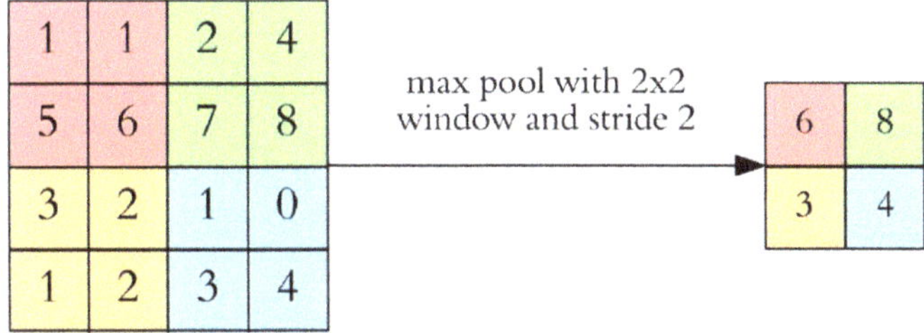

Figure 2. The application of max pooling on the input.

As shown in Figure 2, we applied max pooling, which returns the maximum value in the filter, then moves by one stride and repeats the process. This is repeated until the entire image has been covered. In the above figure, we can see how 6 is chosen from the first stride. The same applies to 8, 3, and 4.

3.1.3. Fully Connected Layer

The convolutional and pooling layers perform the task of feature extraction, using the filter to create the feature map [46]. The fully connected layer performs the detection and recognition tasks by matching patterns in the feature maps of the images being operated on. The fully connected layer is arranged so that each node in the output layer connects directly to a node in the previous layer. Figure 3 shows a typical convolutional neural network and how all the parts are interconnected.

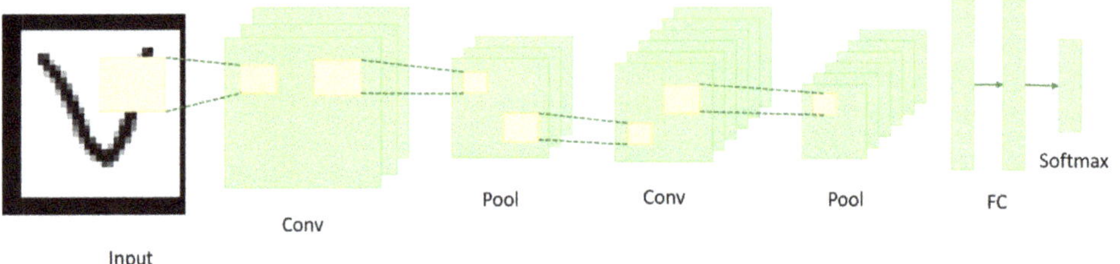

Figure 3. A typical convolutional neural network.

3.2. Architecture of the Applied CNN Models

This section outlines the machine-learning and deep-learning algorithms used in this experiment. All artificial intelligence experiments follow the same procedure. First, relevant data are collected and preprocessed to ensure that the raw data are suitable for training and testing. Second, certain features are extracted from the data and used to train and test the models. Finally, the extracted features are used for prediction, depending on the purpose of the research. Our research is a classification problem, so the extracted data will be used to classify the handwritten Arabic characters when we get to the final stage.

From the literature review, it can be seen that machine-learning models have not been used as extensively as deep-learning models. Most of the papers examined were generally focused on feature ANN and CNN approaches. This research explores advanced ensemble methods of classification which have not previously been used in related experiments.

The two datasets used for all the models were split into training and test sets at a ratio of 70% to 30%. The first dataset used was the Arabic MNIST dataset, which contains 10 classes for 0–9 numerical digits; the other dataset contains handwritten Arabic characters and has 28 classes resembling the 20 Arabic characters.

Two different strategies were used to conduct a series of experiments. The first strategy involved standalone models, while the second strategy utilized hybrid models, which were designed by combining two models. The two strategies are discussed below:

Standalone Models

- XGBoost stands for extreme gradient boosting. It is an ensemble machine that uses trees for boosting. It makes use of gradient boosting for the decision tree, thereby increasing speed and performance. These trees are built sequentially to reduce errors from the preceding tree, and each new tree learns from the previous one. Hence, as new trees grow, more knowledge is passed on. To enhance accuracy, the model, after the iterations, tries to minimize the following objective function, which comprises a loss function and regularization. There are three main forms of boosting:

  1. A gradient-boosting algorithm uses a gradient algorithm and the learning rate;
  2. A stochastic gradient-boosting algorithm uses sampling at the row and column per split levels;
  3. A regularized gradient-boosting algorithm uses L1 and L2 regularization.

- Random forest is a meta-estimator that fits several different decision trees on various subsamples, and the output is averaged so as to control overfitting. Random forest was used for this experiment because the error generated is always lesser than the decision tree due to out-of-bag error. Decision trees are a popular method for machine-learning tasks. Tree learning derives from the shell method for data mining because of its invariant behavior when it comes to scaling and other transformation methods for feature values, which are robust given the inclusion of feature values. The error generated by a random forest is always lesser than that generated by a decision tree because of the out-of-bag error, which is also called the out-of-bag estimate. This error-estimation technique is a method of estimating the prediction errors of random forests, which involves boosted decision trees that utilize bootstrap aggregation to subsample data samples required for training. The parameters used for training were max_depth = 12, random_state = 0, and n_estimators = 100.

- CatBoost stands for category boosting, which was developed by Yandex [47]. It uses the gradient-boosting technique and does not require conversion of the dataset to a specific format, unlike other machine-learning algorithms, making it more reliable and easier to implement. The parameters used for this experiment were iterations = 100, depth = 4, and learning_rate = 0.1.
- Logistic regression uses L1, L2, and ElasticNet as regularization techniques and then calculates the probability of a particular set of data points belonging to either of those classes. For this experiment, the log was used as the cost function.
- A support vector machine (SVM) is a supervised machine-learning algorithm for classification. It takes data points as inputs and outputs a hyperplane that separates the classes with the aim of achieving a hyperplane that maximizes the margin between the classes. This is the best hyperplane. For this experiment, two kernels were tested, the RBF and the linear kernel.
- A feed-forward neural network is an artificial neural network wherein connections between the nodes do not form a cycle. This means that information moves only in the forward direction, from the input nodes, through the hidden nodes, to the output nodes [48]. Four optimization methods were experimented on, including Adam Optimizer, RMSprop, Adagrad, and stochastic gradient descent. Table 3 shows the feed-forward network architecture and all the parameters.

Table 3. The feed-forward architecture design.

Feed-Forward Network		
Architecture Design		
Layer (type)	Output Shape	Param #
Dense_30 (Dense)	(None, 512)	401,920
Dense_31 (Dense)	(None, 512)	262,656
Dense_32 (Dense)	(None, 10)	5130
Total params: 669,706 **Trainable params: 669,706** **Non-trainable params: 0**		

- All these proposed algorithms were used for the recognition of handwritten Arabic characters, where statistical type features, shape-based features (curvatures), and categorical features with indexes were used.
- A convolutional neural network is a deep-learning algorithm that can take in an input image, assign importance (learnable weights and biases) to various aspects/objects in the image, and differentiate one from another. The convolutional layer is first applied to create a feature map with the right stride and padding. Next, pooling is performed to lessen the dimensionality and properly adjust the quality of parameters used for the training, thereby reducing preparation time and battling overfitting. This experiment used max pooling, which takes the maximum incentive in the pooling window. Transfer learning was also carried out. This focuses on storing knowledge gained while solving one problem and applying it to a different, related problem; it is particularly useful in our case, since there is a limited number of data. A sequential model was used while using different optimization algorithms, including RMSProp, Adagrad, stochastic gradient descent, and Adam Optimizer. Table 4 shows the architecture of the convolutional neural network, with all the layers, the shapes expected, and the number of parameters.

Table 4. The CNN architecture design.

CNN		
Architecture Design		
Layer (type)	Output Shape	Param #
Conv2d_7	(None, 26, 26, 32)	320
Max_pooling2d_7	(None, 13, 13, 32)	0
Batch_normalization_7	(None, 13, 13, 32)	128
Conv2d_8	(None, 11, 11, 64)	18,496
Max_pooling2d_8	(None, 5, 5, 64)	0
Batch_normalization_8	(None, 5, 5, 64)	256
Conv2d_9	(None, 3, 3, 128)	73,856
Max_pooling2d_9	(None, 1, 1, 128)	0
Batch_normalization_9	(None, 1, 1, 128)	512
Flatten_3	(None, 128)	0
Dense_5	(None, 256)	33,024
Dense_6	(None, 10)	2570
Total params: 129,162 **Trainable params: 128,714** **Non-trainable params: 448**		

CNN parameters were selected, and 129,162 parameters were chosen during the model training. Trainable parameters were set at 128,714, whereas the non-trainable parameters numbered 448 during the model training. The number of epochs was 20, the batch size was 32, and there were 3600 training samples and 2400 validation samples.

3.3. Hybrid Models

Hybrid models are models that combine two or more models. Those models integrate machine-learning models and other soft-computing, deep-learning, or optimization techniques.

In this research, we used CNN as a base-feature extractor, and these extracted features were fed into machine-learning models to see how the models performed, as shown in Figure 4. The architecture of the feature extractor was the same as that of the CNN model and the various machine-learning models mentioned above. The following are the hybrid models that were experimented on:

1. CNN + SVM;
2. CNN + Random Forest;
3. CNN + AdaBoost;
4. CNN + XGBoost;
5. CNN + CatBoost;
6. CNN + Logistic Regression.

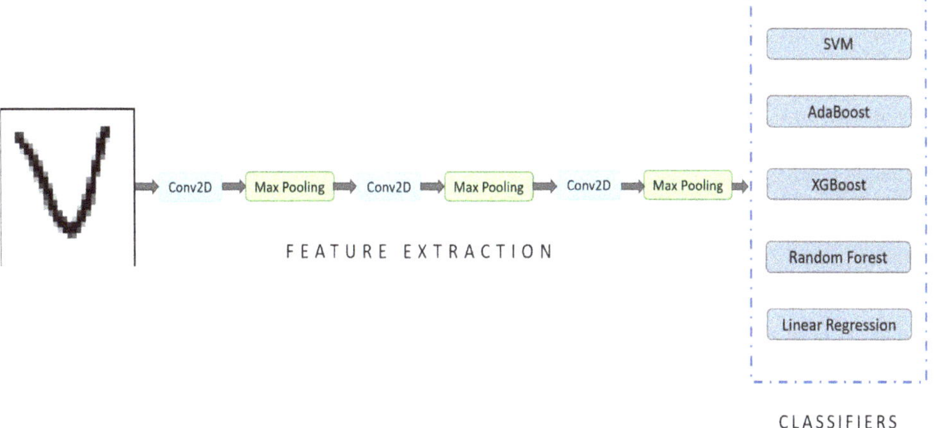

Figure 4. The architecture of the hybrid models.

The models were trained using both the Arabic MNIST dataset and the Arabic character dataset. The CNN's only task is to extract relevant features from the handwritten images, which are its outputs. It then passes the features to the machine-learning models, which use these features to find patterns, thereby classifying the images.

The proposed approaches for the system have their benefits and drawbacks. A CNN was proposed because this approach has been used for handwritten character recognition with many other languages, so adopting this method could achieve the best performance. The CNN approach involved two steps. The first was the extraction of the required features; the second was classification. The standalone approach was proposed to check the individual impact on the dataset and see whether it could perform the same procedure for handwriting recognition. The hybrid approach was designed because it helped to achieve more reliable results than the standalone approach. It works like the ensemble approach in solving the Arabic handwriting recognition problem. This was decided after reviewing the multiple approaches to handwriting recognition reported in the literature.

4. Experimental Results and Analysis

This section discusses the experimental setup for this work, including all the hardware and software requirements. Then, the results obtained from all the models are reported to show how they all performed. We compared the models to see which ones did well and which did not. Finally, we took the best performance we achieved and compared it with state-of-the-art models for Arabic handwriting recognition to see how well our model performed compared to the others.

The proposed system used two approaches: deep learning and conventional machine learning. Both methods involve some signs; as with the conventional machine learning approach, where the model can be trained with an available dataset, this approach worked based on the extraction of various features, such as statistical, curvature, and multiple different features. The machine-learning approach must extract features manually and pass them to the model for training. The deep-learning approach has some other aspects compared with this approach. In deep-learning models, there is no need to extract features manually because the models can extract the required or available features automatically. These features are passed for the model training and classification. The CNN approach extracts features and gives them to the model for classification. The proposed system used both to check where the model can deliver the best results for the problem.

The deep-learning approach requires more time for training due to the large volumes of datasets, whereas the conventional machine-learning approach takes less time with small datasets. Deep-learning models are more complex than conventional learning approaches and need devices with high computational power to execute them.

4.1. Datasets

The first dataset used for this research was an Arabic MNIST dataset obtained from the MNIST database developed by LeCun et al. [49], which comprises 60,000 images for training and 10,000 images for testing, with 6000 images per class. A part of this dataset can be seen in Figure 5. In the same fashion, the Arabic MNIST dataset comprises 60,000 images from digit 0 to digit 9, with each class having 6000 handwritten images.

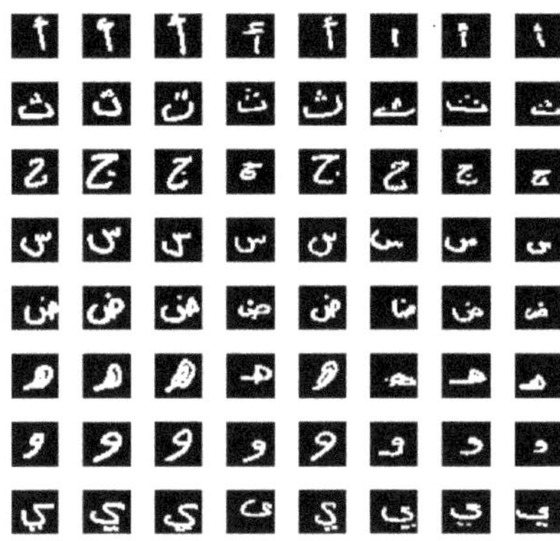

Figure 5. The Arabic MNIST dataset.

The second dataset [3] comprises 16,800 characters written by 60 participants. The images were scanned at a resolution of 300 dpi. Each image was segmented automatically using MATLAB 2016a, which automatically coordinated every image. The database was then partitioned into 13,400 images for training and 480 images for testing, with 120 images per class, which sums up to 3360 images in total.

The images were processed by converting them to grayscale using OpenCV so that the images had a single filter instead of three, then row-wise values were extracted side by side, with 784 columns. The labels for the images were used as the target values, thus generating a csv file. The Kaggle dataset already had a csv file with 1024 columns. These columns contained 0 for a black value and 1 for a white value for an image. Then, the dataset was separated into two csv files for training and testing.

4.2. Experimental Setup

The choice of datasets was the same for all the models trained (both standalone and hybrid), the datasets used are the Arabic MNIST dataset and the Arabic character dataset. All experiments were conducted using Google Colab (short for Colaboratory), which is a product from Google research that allows users to write and execute Python code with either a CPU or a GPU. As for libraries, the Keras library was employed to create the deep neural networks, the Python programming language (version 3.6.3) being used for all of them.

The proposed system was evaluated using the open-source MNIST Arabic dataset. This system calculated evaluation measures, such as accuracy, precision, recall, and F1-measure, to check the proposed system's consistency and performance. Accuracy is the ratio of the number of correct predictions to the total number of predictions. Equation (1) shows the accuracy measure. These evaluation measures are described in terms of TPs (true positives), TNs (true negatives), FNs (false negatives), and FPs (false positives).

$$Accuracy = \frac{TN + TP}{TP + TN + FP + Fn} \tag{1}$$

Precision is the ratio of true positives to true and false positives. The equation of precision is shown in Equation (2).

$$Precision = \frac{TP}{TP + FP} \tag{2}$$

The recall is the ratio of correctly identified positive examples to the total number of positive models. Equation (3) shows the recall evaluation measure.

$$Recall = \frac{TP}{TP + FN} \tag{3}$$

The F1-measure is a valuable metric in machine learning, which sums up the predictive performances to the combination of two other metrics. Equation (4) shows the F1-measure evaluation measure.

$$F1 - Measure = 2 \times \frac{Precision \times Recall}{Precision + Recall} \tag{4}$$

4.3. Results and Analysis

The task of designing a system that can automatically detect Arabic handwritten characters is very necessary, so we set out to use every means possible to achieve the best performance. To do this, we split the work into two parts: the standalone and the hybrid models. The results obtained from these experiments are reported below to show the performance evaluation for the models trained on both the Arabic MNIST and Arabic character datasets.

4.3.1. Standalone Machine-Learning Models

The standalone models are deep neural networks that carry out the entire classification task, from feature extraction to the final detection of patterns and the ultimate classification of images. The first model trained was the XGBoost model, then the random forest model was trained. Next, the CatBoost model was trained for 100 epochs, with a learning rate of 0.1. Then, the logistic regression model was trained with a cost function of the log. Finally, a Support Vector Machine (SVM) was trained, but since the SVM is a non-linear model, two different kernels were used: the RBF and the linear kernel. The results obtained by the standalone machine-learning models broke down into two tables. Table 5 shows the results for the models trained on the Arabic MNIST digit dataset, while Table 6 show the results obtained by the models trained on the Arabic character dataset.

Table 5. Standalone machine-learning models trained on the Arabic MNIST digit dataset.

Model	Precision (Macro Average)	Recall (Macro Average)	F1-Score (Macro Average)	Accuracy Score
Random Forest	0.99	0.99	0.99	0.98714
CatBoost	0.95	0.95	0.95	0.95476
SVM (Linear)	1.0	1.0	1.0	0.99628
SVM (RBF)	1.0	1.0	1.0	0.99628
AdaBoost				0.69600
XGBoost	1.0	1.0	1.0	1.00000
Logistic Regression	1.0	1.0	1.0	0.99957

The proposed system was evaluated on the Arabic MNIST digit dataset. Evaluation measures were calculated against this dataset, with precision, recall, F1-measure, and accuracy calculated for the test dataset. Machine-learning algorithms, named Random Forest, CatBoost, SVM, AdaBoost, XGBoost, and Logistic Regression, were used for the model training and testing on the Arabic MNIST digit dataset. The XGBoost results were the highest compared to the other algorithms, but Logistic Regression and Linear SVM also performed outstandingly. The results shown in Table 5 were remarkable for the dataset.

Table 6. Standalone machine-earning models trained on the Arabic character dataset.

Model	Precision (Macro Average)	Recall (Macro Average)	F1-Score (Macro Average)	Accuracy Score
Random Forest	0.64	0.64	0.63	0.6336
CatBoost	0.47	0.47	0.46	0.4643
SVM (Linear)	0.41	0.41	0.41	0.4233
SVM (RBF)	0.66	0.65	0.65	0.6529
Feed-Forward Neural Network (Adam)	0.66	0.61	0.63	0.6101
XGBoost	0.49	0.49	0.48	0.4854
Logistic Regression	0.66	0.65	0.65	0.6529

As can be seen from Table 5, above, almost all the models performed extremely well. This goes to show that the models were fine-tuned in such a way that they conformed to the dataset used. The Arabic MNIST digit dataset is a standard dataset used for artificial intelligence projects. This means that it has been standardized to avoid overfitting, which indicates that our models are valid; however, as shown in Table 6, the results obtained were average, with the highest being those for Logistic Regression and the SVM (RBF kernel), both of which achieved an accuracy of 0.65. The accuracies were as low as 0.43 for the SVM (linear kernel).

Table 6 shows the evaluated results against the Arabic character dataset for the following machine-learning algorithms. The machine-learning algorithms were Random Forest, SVM, Neural Network, XGBoost, and Logistic Regression. Based on the feature extraction in the machine learning, the results were evaluated for the Arabic character dataset. SVM (RBF) had better results than the other algorithms, but the overall results were not the best.

4.3.2. Standalone Deep-Learning Models

The second phase of the research consisted of the experiments on the standalone deep neural networks which performed the tasks of feature extraction and classification. Three deep neural networks were trained and evaluated, including the feed-forward network, the convolutional neural network, and the neural network integrated with transfer learning. These models were tested with different optimizers to see how they performed, and a summary of the results obtained is presented in Table 7.

Table 7. Standalone deep-learning models trained on the Arabic MNIST digit dataset.

Model	Precision (Macro Average)	Recall (Macro Average)	F1-Score (Macro Average)	Accuracy Score
Feed-Forward Neural Network (Adam)	1.0	1.0	1.0	0.98714
CNN	1.0	1.0	1.0	0.99063
Transfer Learning	1.0	1.0	1.0	0.99673

Figure 6 shows the loss- and accuracy-curve graphs for the feed-forward network, for which the variation was very low after the 10th iteration. The purpose was to check the variation in the iterations, so that more iterations were required. The variational changes after the 10th iteration were only minor.

Figure 7 shows the curves for both the loss and accuracy of the training and validation; the model had an almost perfect training accuracy but a more realistic validation accuracy, which goes to show that the model performed well.

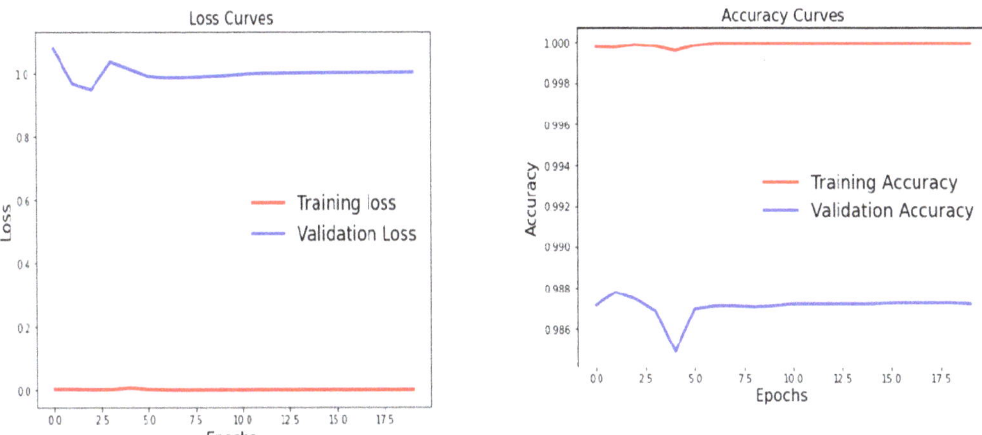

Figure 6. Graphs of the loss and accuracy curves for the feed-forward network.

Figure 7. Graphs of loss and accuracy curves for the CNN.

The loss and accuracy curves for the convolutional neural network can be seen in Figure 7. The model encountered a few problems at the beginning of the experiment, which caused it to move rapidly; however, with more epochs, the line smoothed out, which signifies the model's effectiveness. Figure 8 shows the confusion matrix of the CNN classifier. Variations in the graphs show significant changes in the loss and accuracy for the training and validation. This impact was seen due to the significant change in the dataset. During training, the changes in loss and validation for the loss and accuracy were clearly seen.

The results obtained for the experiments conducted on our two datasets were impressive. The MNIST dataset was used to train the feed-forward network, the CNN, and the transfer-learning models, which achieved 0.9871, 0.9906, and 0.9967 accuracies, respectively. These are very effective accuracies, and it can be confidently said that the models can be used. On the other hand, the Arabic character dataset was used to train only the CNN with an adagrad optimizer, and it achieved an accuracy of 0.8992, see Table 8, which is not as high as that of MNIST but still remarkably high.

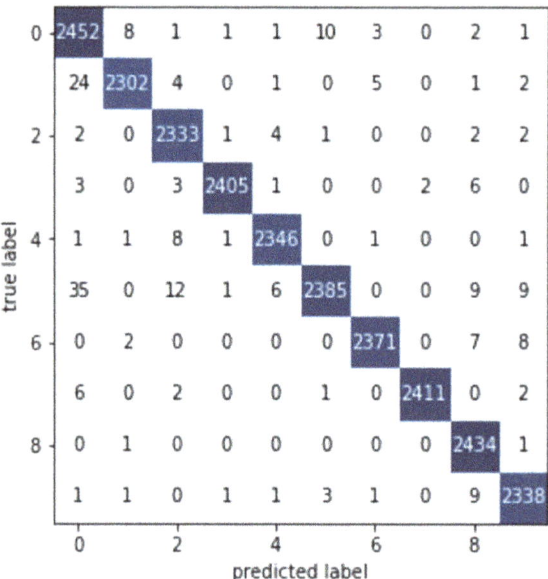

Figure 8. The confusion matrix of the CNN model.

Table 8. Standalone deep-learning models trained on the MNIST Arabic digit dataset.

Model	Precision (Macro Average)	Recall (Macro Average)	F1-Score (Macro Average)	Accuracy Score
CNN (Adagrad)	0.90	0.90	0.90	0.8992

4.3.3. Hybrid Models

The final phase of this research was the experiment on the hybrid models, which were combinations of more than one model. These hybrid networks were a combination of a convolutional neural network (CNN), which performed the task of feature extraction because of such networks' eminent suitability for the task. Then, the extracted features were passed on to various machine-learning models, which carried out the classification. The machine-learning models used were the SVM, Random Forest, AdaBoost, XGBBoost, CatBoost, and Logistic Regression. A summary of all the results obtained is shown in Tables 9 and 10, below.

Table 9. Hybrid machine-learning models trained on the Arabic MNIST digit dataset.

Model	Precision (Macro Average)	Recall (Macro Average)	F1-Score (Macro Average)	Accuracy Score
CNN + SVM	0.97	0.97	0.97	0.9710
CNN + Random Forest	0.94	0.94	0.94	0.9390
CNN + AdaBoost	0.57	0.55	0.54	0.5470
CNN + XGBoost	0.93	0.93	0.93	0.9320
CNN + Logistic Regression	0.94	0.94	0.94	0.9388

The hybrid models were purely experimental. Since we combined models that naturally are standalone, it can be seen from the MNIST dataset experiment that all the models did well, with results exceeding 0.9, with the exception of CNN + AdaBoost, which achieved a classification accuracy of 0.55. The Arabic character dataset experiments did not produce such good results, with the highest performance attained by the CNN + SVM hybrid model, which had a classification accuracy of 0.872, and the lowest by the CNN + AdaBoost, which had a very poor accuracy score of 0.1690. This result goes to show that the CNN + AdaBoost hybrid model is not suitable for this classification task because it had the lowest classification accuracy for both datasets. In the hybrid approach, the combinations were set to check the model performance, the CNN + SVM set performing better than the other varieties. CNN + AdaBoost performed very poorly as compared with the other combinations. According to this system analysis, these types of machine-learning algorithms were not best-suited to this problem. Ultimately, which combination of algorithms is best depends on the nature of the problem to be solved.

Table 10. Hybrid machine-learning models trained on the Arabic character dataset.

Model	Precision (Macro Average)	Recall (Macro Average)	F1-Score (Macro Average)	Accuracy Score
CNN + SVM	0.87	0.87	0.87	0.872
CNN + Random Forest	0.78	0.78	0.87	0.804
CNN + AdaBoost	0.14	0.17	0.13	0.1690
CNN + XGBoost	0.78	0.77	0.77	0.7990
CNN + Logistic Regression	0.83	0.83	0.83	0.8560

5. Discussion and Comparisons

The primary purpose of this study was to develop a hybrid model that recognizes Arabic handwritten characters accurately. In this section, the performances of the hybrid CNN-based architectures will be discussed from three perspectives: the basic CNN architectures and the ML classifiers utilized in the hybrid models.

The Arabic MNIST and Arabic character datasets were used to assess the hybrid models developed in the current study. Ten standalone machine-learning models using the Arabic MNIST dataset, eight standalone machine-learning models trained on the Arabic character dataset, and five hybrid models for both the Arabic MNIST and Arabic character datasets were developed. According to the results, the performances of several CNN-based models for the Arabic character dataset were considerably inferior to the performances for the Arabic MNIST [49] dataset. Therefore, the Arabic character dataset can be considered a more complicated and challenging dataset. In this study, the best performance for the Arabic MNIST dataset was 97% (Figures 9 and 10), achieved with the hybrid model that combined a CNN and an SVM.

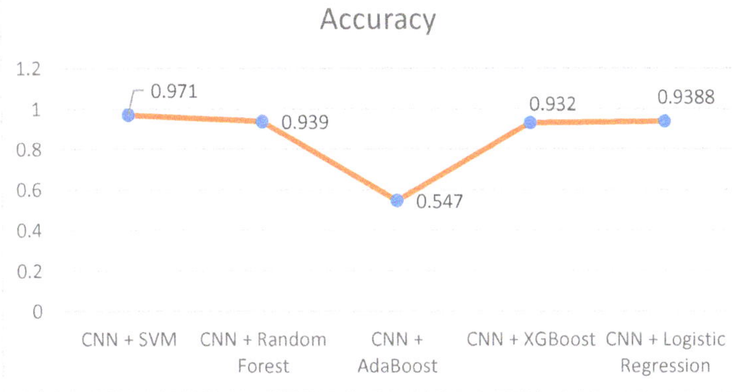

Figure 9. The accuracy of the hybrid models on the MNIST dataset.

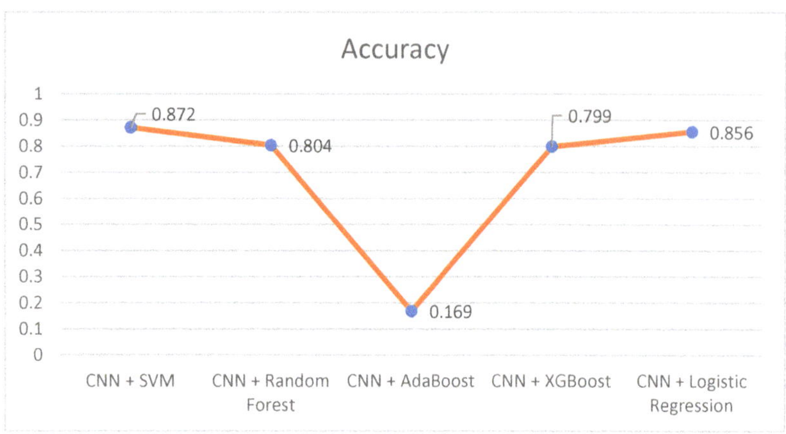

Figure 10. The accuracy of the hybrid models trained on the Arabic character dataset.

For the Arabic MNIST dataset, the Logistic Regression model and the XGBOOST model outperformed all the models in the research field in this area; the previous record of 99.71% was outperformed by our machine-learning approach. The hybrid models, such as CNN + SVM, reached near accuracy. However, hybrid models are more robust than standalone CNN models. A higher recall rate was observed for the hybrid models. The models which had the worst performances were the AdaBoost and CNN + AdaBoost models.

For the Arabic MNIST digit dataset, the machine-learning methods Random Forest, CatBoost, SVM, AdaBoost, XGBoost, and Logistic Regression were employed for model training and testing. In comparison to the other algorithms, XGBoost yielded the best results, while Logistic Regression and Linear SVM also performed quite well. Each of these models performed well. This demonstrates that the models were fine-tuned in accordance with the datasets used. The Arabic MNIST digit dataset is a typical dataset for artificial intelligence research. This implies that it has been standardized to prevent overfitting, indicating that our models are legitimate. However, on the Arabic character dataset, the results obtained were average, with Logistic Regression and the SVM (RBF kernel) achieving the best accuracy (0.65), followed by the SVM (linear kernel). The SVM accuracy dropped as low as 0.43 (with the linear kernel). The machine-learning methods examined were Random Forest, SVM, Neural Network, XGBoost, and Logistic Regression. On the basis of feature extraction in machine learning, the findings for the Arabic character dataset were reviewed, and the SVM (RBF) yielded superior results compared to the other algorithms, although the results were not the greatest overall.

In contrast, three deep neural networks were trained and assessed, including a feed-forward network, a convolutional neural network, and a neural network along with transfer learning. These models were evaluated using several optimizers to determine their performances. Experiments performed on our two datasets yielded amazing outcomes. The MNIST dataset was used to train the feed-forward network, the CNN, and the transfer-learning models, which produced corresponding accuracy levels of 0.9871, 0.9906, and 0.9967. These are very reliable accuracies; thus, it is fair to assume that the models can be used. In parallel, the Arabic character dataset was used to train the CNN solely with an adagrad optimizer, and it achieved an accuracy of 0.8992, which was not as high as that achieved for the MNIST dataset but still remarkable.

The last batch of models consisted of hybrid models. These hybrid networks incorporated a convolutional neural network (CNN), which performed the feature extraction, since such networks are superior for the task. The extracted features were then sent to several machine-learning models, which performed classification. SVM, Random Forest, AdaBoost, XGBBoost, CatBoost, and Logistic Regression were the machine-learning models used. Regarding the hybrid model approach, the model combination performances were evaluated to assess the approach. The CNN + SVM set was proven to outperform the others. However, the CNN + AdaBoost fared very badly compared to the other combinations.

5.1. Arabic MNIST Dataset

For the standalone models trained using the Arabic MNIST dataset, most of the models performed excellently, with accuracies above 95%. The best performances were observed for XGBoost (with values for precision, recall, F1-score, and accuracy of 1), Logistic Regression (with precision, recall, and F1-score values of 1.0 and an accuracy of 0.99957), the SVMs (with precision, recall, and F1-score values of 1 and an accuracy of 0.99628), followed by the transfer-learning model (with an accuracy of 0.99743) and the CNN (with an accuracy of 0.99063, and precision, recall, and F1-score values of 1). AdaBoost, on the other hand, obtained the lowest performance, with an accuracy of 0.696.

Considering the hybrid models, see Figure 9, the best classification performance was observed for the combination of CNN and SVM (with an accuracy of 0.9710, and precision, recall, and F1-score values of 0.97), followed by the CNN and Random Forest (with an accuracy of 0.9390), CNN and Logistic Regression (accuracy of 0.9388), and CNN and XGBoost (accuracy of 0.9320). The CNN and AdaBoost hybrid model obtained the weakest performance, with an accuracy of 0.5470, precision of 0.57, recall of 0.55, and an F1-score of 0.54.

5.2. Arabic Character Dataset

The standalone Convolutional Neural Network outperformed all the models trained using the Arabic character dataset. The experiments performed using this dataset recorded relatively lower accuracies than those trained using the Arabic MNIST dataset due to the large class size and low interclass difference coupled with a higher variance. Regarding the Arabic character dataset, the applied algorithms obtained weak-to-moderate performances, with accuracies ranging from 0.4233 (linear SVM) to 0.8992 (CNN), suggesting that the CNN might be the best choice for classification problems on this dataset. Hybrid models obtained similar performances, with the highest accuracy score of 0.872 for the CNN and SVM model, see Figure 10, followed by the CNN and Logistic Regression model (accuracy of 0.8560), the CNN and Random Forest model (accuracy of 0.804), and CNN and XGBoost (accuracy of 0.7990). The lowest performance was obtained by the combination of the CNN and AdaBoost algorithm, with an accuracy of 0.1690, a precision of 0.14, a recall of 0.17, and an F1-score of 0.13.

5.3. Comparison

As already discussed in Section 2, the literature contains reports of various attempts to improve Arabic handwriting recognition approaches. In the current work, the hybrid model of machine-learning and deep-learning algorithms (CNN + SVM) achieved an improved result. Tables 1 and 2 present summaries of the CNN models (standalone and hybrid) and the datasets used for the Arabic handwriting recognition experiments.

Various types of datasets, AHCD, HMBD, AIA9k, OIHAC, HACDB, and Hijja, have been used to train normal CNN models. Most of the studies have used the dataset AHCD [3], for which the highest accuracy of 99.98% was achieved [9] using AlexNet.

For the hybrid models, HACDB and AHCD were used. The accuracy reached for AHCD upon its initial publication was 94.90%, but the accuracy achieved by the hybrid model [50] was 95.07%.

In the current work, the performances of the proposed models applied on the MNIST dataset were generally better than those applied on the Arabic character dataset. Thus, one can infer that the latter dataset is more complicated than the former one. Researchers are encouraged to further investigate how performance can be improved for the Arabic character dataset. Table 11 presents a comparison of the proposed work and the other common architectures trained on the common datasets.

Table 11. Comparison with other architectures.

Authors	Model	Datasets with Accuracy Rates (%)			
		AHCD	Hijja	AIA9k	MNIST
El-Sawy et al. [3]	CNN	94.9			
Altwaijry et al. [1]	CNN	97	88		
Alrobah and Albahli [29]	Hybrid CNN models: CNN + FCL, CNN + SVM,	94.5	96.3		
	CNN + XGBoost		-		
Balaha et al. [30]	CNN	97.3	-	98.4	
Younis et al. [31]	CNN	97.6	-	94.8	
Al-Taani et al. [38]	ResNet	99.5	-	-	
Current work	CNN + SVM				97.1

6. Challenges and Open Problems

The task at hand is a very challenging one, since this is a field in which not much research has been carried out, especially in the area of artificial intelligence. Most of the challenges and problems to be discussed in this section could be mitigated or completely eradicated in the near future when more work has been completed. The challenges and open problems are discussed below:

- Inadequate dataset: For this experiment, two databases were utilized to acquire dataset images for both digits and characters; even so, the dataset was not sufficient, because one of the requirements for training both machine-learning and deep-learning models is an enormous supply of data. Lack of sufficient data leads to model overfitting.
- Cursive nature of the Arabic language: The Arabic language is cursive, meaning written characters are joined together, even in digital forms. If what is to be predicted was digitally written, it would have been much easier, but the project focused on handwriting recognition. Since every writer has a unique way of writing, it is difficult for the designed system to effectively detect handwritten characters and digits.
- Imbalanced dataset: The datasets used for the experiments were subjected to normalization by the database managers. However, the real-world problems that will be encountered might not be as clean as the ones in the dataset. For example, the images to be analyzed might be zoomed-in, unclear, or blurry, such that the system cannot be used to achieve effective results.
- Presence of special characters: The Arabic language is one that has a lot of special characters, such as dots and diacritics, making the classification problem a tedious one, in that misplacing just a single dot may completely change the meaning of the conveyed message and the presence or absence of a single stroke may change what is said completely. Hence, clear images are required for the system to make correct classifications.

The challenges and open problems are summarized in Figure 11.

The major contribution of the proposed system was to apply deep learning and conventional approaches to the crucial problem of Arabic-language character recognition. The proposed system addressed Arabic handwritten character recognition and various methods were explored to find the best solution to the problem. A deep-learning CNN approach, a conventional standalone approach, and a hybrid approach were introduced to address the crucial problem of handwritten character recognition. No approach has been applied to solve this issue prior to the system proposed here.

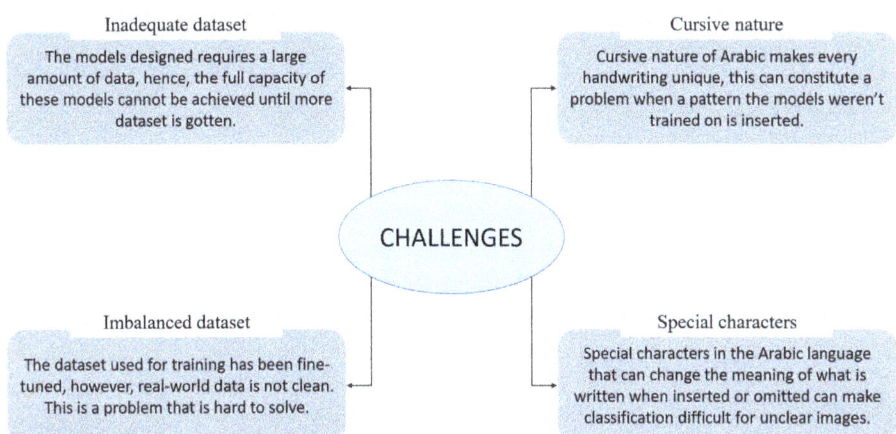

Figure 11. Challenges and open problems.

7. Conclusions and Future Studies

In this study, combining machine-learning and deep-learning algorithms to build the recognition models was worthwhile. The study involved standalone deep-learning experiments and focused attention on the idea of exploiting machine learning in classification and the advantages of deep learning in feature extraction. Several deep-learning and hybrid models were created. The best result for the standalone deep-learning models trained on the two datasets was achieved with the transfer-learning model on the MNIST dataset, which had a 0.9967 accuracy. On the other hand, the results of the hybrid models achieved good records using the MNIST dataset, with accuracies exceeding 0.9 for all the hybrid models. It should be noted that the results for the hybrid models using the Arabic character dataset were very poor, which might mean that the problem lies with the dataset itself.

To conclude, more studies are required to improve Arabic handwriting recognition and overcome the challenges it presents. Arabic handwriting might have some unique characteristics that need to be considered by researchers. What is appropriate for one language might not be appropriate for others: As mentioned in the Introduction to this study, English handwriting recognition might have received a large amount of interest in the literature. Although there is always room for improvement, the Arabic language is still in need of more attention from researchers, to improve handwriting-recognition methods and the use of appropriate datasets in experiments. It is hoped that the challenges discussed in this work will grab the attention of researchers and stimulate the proposal of more solutions and better contributions towards improvements in the field of Arabic handwriting recognition.

Author Contributions: Funding acquisition, W.A.; Methodology, W.A. and S.A.; Project administration, S.A.; Supervision, S.A.; Writing—original draft, W.A. and S.A.; Writing—review & editing, W.A. and S.A. All authors have read and agreed to the published version of the manuscript.

Funding: This research received no external funding.

Data Availability Statement: The experiment used a public dataset shared by LeCun et al. [49], and they have already published the download address for the dataset in their paper.

Acknowledgments: The researchers would like to thank the Deanship of Scientific Research, Qassim University, for funding the publication of this project.

Conflicts of Interest: The author declares that they have no conflict of interest.

References

1. Altwaijry, N.; Al-Turaiki, I. Arabic handwriting recognition system using convolutional neural network. *Neural Comput. Appl.* **2020**, *33*, 2249–2261. [CrossRef]
2. Sarkhel, R.; Das, N.; Saha, A.K.; Nasipuri, M. A multi-objective approach towards cost effective isolated handwritten Bangla character and digit recognition. *Pattern Recognit.* **2016**, *58*, 172–189. [CrossRef]
3. El-Sawy, A.; Loey, M.; El-Bakry, H. Arabic handwritten characters recognition using convolutional neural network. *WSEAS Trans. Comput. Res.* **2017**, *5*, 11–19.
4. Weldegebriel, H.T.; Liu, H.; Haq, A.U.; Bugingo, E.; Zhang, D. A new hybrid convolutional neural network and eXtreme gradient boosting classifier for recognizing handwritten Ethiopian characters. *IEEE Access* **2020**, *8*, 17804–17818. [CrossRef]
5. Babbel Magazine. Available online: https://www.babbel.com/en/magazine/ (accessed on 18 March 2022).
6. Ghadhban, H.Q.; Othman, M.; Samsudin, N.A.; Ismail, M.N.B.; Hammoodi, M.R. Survey of offline Arabic handwriting word recognition. In *International Conference on Soft Computing and Data Mining*; Springer: Cham, Switzerland, 2020; pp. 358–372.
7. Boufenar, C.; Kerboua, A.; Batouche, M. Investigation on deep learning for off-line handwritten arabic character recognition. *Cognit. Syst. Res.* **2018**, *50*, 180–195. [CrossRef]
8. Palatnik de Sousa, I. Convolutional ensembles for arabic handwritten character and digit recognition. *PeerJ Comput. Sci.* **2018**, *4*, e167. [CrossRef]
9. Elleuch, M.; Lahiani, H.; Kherallah, M. Recognizing arabic handwritten script using support vector machine classifier. In Proceedings of the 15th International Conference on Intelligent Systems Design and Applications (ISDA), Brussels, Belgium, 22–27 June 2015; pp. 551–556. [CrossRef]
10. Mudhsh, M.; Almodfer, R. Arabic handwritten alphanumeric character recognition using very deep neural network. *Information* **2017**, *8*, 105. [CrossRef]
11. Niu, X.-X.; Suen, C.Y. A novel hybrid CNN–SVM classifier for recognizing handwritten digits. *Pattern Recognit.* **2012**, *45*, 1318–1325. [CrossRef]
12. Alwaqfi, Y.M.; Mohamad, M.; Al-Taani, A.T. Generative Adversarial Network for an Improved Arabic Handwritten Characters Recognition. *Int. J. Adv. Soft. Comput. Appl.* **2022**, *14*. [CrossRef]
13. Alrobah, N.; Albahli, S. A hybrid deep model for recognizing arabic handwritten characters. *IEEE Access* **2021**, *9*, 87058–87069. [CrossRef]
14. Balaha, H.M.; Ali, H.A.; Badawy, M. Automatic recognition of handwritten Arabic characters: A comprehensive review. *Neural Comput. Appl.* **2021**, *33*, 3011–3034. [CrossRef]
15. Manwatkar, P.M.; Singh, K.R. A technical review on text recognition from images. In Proceedings of the 2015 IEEE 9th International Conference on Intelligent Systems and Control (ISCO), Coimbatore, India, 9–10 January 2015; pp. 1–5.
16. Alsanousi, W.A.; Adam, I.S.; Rashwan, M.; Abdou, S. Review about off-line handwriting Arabic text recognition. *Int. J. Comput. Sci. Mob. Comput.* **2017**, *6*, 4–14.
17. Ding, H.; Chen, K.; Yuan, Y.; Cai, M.; Sun, L.; Liang, S.; Huo, Q. A compact CNN-DBLSTM based character model for offline handwriting recognition with Tucker decomposition. In Proceedings of the 2017 14th IAPR International Conference on Document Analysis and Recognition (ICDAR), Kyoto, Japan, 9–15 November 2017; Volume 1, pp. 507–512.
18. Ji, S.; Zeng, Y. Handwritten character recognition based on CNN. In Proceedings of the 2021 3rd International Conference on Artificial Intelligence and Advanced Manufacture, Manchester, UK, 23–25 October 2021; pp. 2874–2880.
19. Alwzwazy, H.A.; Albehadili, H.M.; Alwan, Y.S.; Islam, N.E. Handwritten digit recognition using convolutional neural networks. *Int. J. Innov. Res. Comput. Commun. Eng.* **2016**, *4*, 1101–1106.
20. Ahlawat, S.; Choudhary, A.; Nayyar, A.; Singh, S.; Yoon, B. Improved handwritten digit recognition using convolutional neural networks (CNN). *Sensors* **2020**, *20*, 3344. [CrossRef] [PubMed]
21. Dutta, K.; Krishnan, P.; Mathew, M.; Jawahar, C.V. Improving CNN-RNN hybrid networks for handwriting recognition. In Proceedings of the 2018 16th international conference on frontiers in handwriting recognition (ICFHR), Niagara Falls, NY, USA, 5–8 August 2018; pp. 80–85.
22. Ahmed, R.; Gogate, M.; Tahir, A.; Dashtipour, K.; Al-Tamimi, B.; Hawalah, A.; Hussain, A. Novel deep convolutional neural network-based contextual recognition of Arabic handwritten scripts. *Entropy* **2021**, *23*, 340. [CrossRef]
23. Gan, J.; Wang, W.; Lu, K. Compressing the CNN architecture for in-air handwritten Chinese character recognition. *Pattern Recognit. Lett.* **2020**, *129*, 190–197. [CrossRef]
24. He, M.; Zhang, S.; Mao, H.; Jin, L. Recognition confidence analysis of handwritten Chinese character with CNN. In Proceedings of the 2015 13th International Conference on Document Analysis and Recognition (ICDAR), Tunis, Tunisia, 23–26 August 2015; pp. 61–65.
25. Zhong, Z.; Jin, L.; Xie, Z. High performance offline handwritten Chinese character recognition using GoogLeNet and directional feature maps. In Proceedings of the 2015 13th International Conference on Document Analysis and Recognition (ICDAR), Tunis, Tunisia, 23–26 August 2015; pp. 846–850.
26. Wu, C.; Fan, W.; He, Y.; Sun, J.; Naoi, S. Handwritten Character Recognition by Alternately Trained Relaxation Convolutional Neural Network. In Proceedings of the 2014 14th International Conference on Frontiers in Handwriting Recognition (ICFHR), Crete, Greece, 1–4 September 2014; pp. 291–296.

27. Zhong, Z.; Jin, L.; Feng, Z. Multi-font printed Chinese character recognition using multipooling convolutional neural network. In Proceedings of the 2015 13th International Conference on Document Analysis and Recognition (ICDAR), Tunis, Tunisia, 23–26 August 2015; pp. 96–100.

28. Yang, W.; Jin, L.; Xie, Z.; Feng, Z. Improved deep convolutional neural network for online handwritten Chinese character recognition using domain-specific knowledge. In Proceedings of the 2015 13th International Conference on Document Analysis and Recognition (ICDAR), Tunis, Tunisia, 23–26 August 2015; pp. 551–555.

29. Alrobah, N.; Saleh, A. Arabic handwritten recognition using deep learning: A survey. *Arab. J. Sci. Eng.* **2022**, *47*, 9943–9963. [CrossRef]

30. Balaha, H.M.; Ali, H.A.; Saraya, M.; Badawy, M. A new Arabic handwritten character recognition deep learning system (AHCR-DLS). *Neural Comput. Appl.* **2021**, *33*, 6325–6367. [CrossRef]

31. Younis, K.S. Arabic handwritten character recognition based on deep convolutional neural net-works. *Jordanian J. Comput. Inf. Technol. (JJCIT)* **2017**, *3*, 186–200.

32. Najadat, H.M.; Alshboul, A.A.; Alabed, A.F. Arabic Handwritten Characters Recognition using Convolutional Neural Network. In Proceedings of the 2019 10th International Conference on Information and Communication Systems, ICICS 2019, Irbid, Jordan, 11–13 June 2019; pp. 147–151.

33. Alyahya, H.; Ismail, M.M.B.; Al-Salman, A. Deep ensemble neural networks for recognizing isolated Arabic handwritten characters. *Accent. Trans. Image Process. Comput. Vis.* **2020**, *6*, 68–79. [CrossRef]

34. Almansari, O.A.; Hashim NN, W.N. Recognition of isolated handwritten Arabic characters. In Proceedings of the 2019 7th International Conference on Mechatronics Engineering (ICOM), Putrajaya, Malaysia, 30–31 October 2019; pp. 1–5.

35. Das, N.; Mollah, A.F.; Saha, S.; Haque, S.S. Handwritten Arabic Numeral Recognition using a Multi Layer Perceptron. *arXiv* **2010**, arXiv:1003.1891.

36. Elleuch, M.; Kherallah, M. Boosting of Deep Convolutional Architectures for Arabic Handwriting Recognition. *Int. J. Multimed. Data Eng. Manag.* **2019**, *10*, 26–45. [CrossRef]

37. Noubigh, Z.; Mezghani, A. Contribution on Arabic Handwriting Recognition Using Deep Neural Network. In *International Conference on Hybrid Intelligent Systems*; Springer: Cham, Switzerland, 2021. [CrossRef]

38. Al-Taani, A.T.; Ahmad, S.T. Recognition Of Arabic Handwritten Characters Using Residual Neural Networks. *Jordanian J. Comput. Inf. Technol. (JJCIT)* **2021**, *7*, 192–205. [CrossRef]

39. AlJarrah, M.N.; Zyout, M.M.; Duwairi, R. Arabic Handwritten Characters Recognition Using Convolutional Neural Network. In Proceedings of the 12th International Conference on Information and Communication Systems (ICICS), Valencia, Spain, 24–26 May 2021; pp. 182–188. [CrossRef]

40. Elkhayati, M.; Elkettani, Y.; Mourchid, M. Segmentation of handwritten Arabic graphemes using a directed convolutional neural network and mathematical morphology operations. *Pattern Recogn.* **2022**, *122*, 108288. [CrossRef]

41. Elleuch, M.; Tagougui, N.; Kherallah, M. Arabic handwritten characters recognition using Deep Belief Neural Networks. In Proceedings of the IEEE 12th International Multi-Conference on Systems, Signals & Devices (SSD15), Sfax, Tunisia, 16–19 March 2015; pp. 1–5. [CrossRef]

42. Kef, M.; Chergui, L.; Chikhi, S. A novel fuzzy approach for handwritten Arabic character recognition. *Pattern Anal. Appl.* **2016**, *19*, 1041–1056. [CrossRef]

43. Korichi, A.; Slatnia, S.; Aiadi, O.; Khaldi, B. A generic feature-independent pyramid multilevel model for Arabic handwriting recognition. *Multimed. Tools Appl.* **2022**, *81*, 1–21. [CrossRef]

44. Korichi, A.; Slatnia, S.; Aiadi, O.; Tagougui, N.; Kherallah, M. Arabic handwriting recognition: Between handcrafted methods and deep learning techniques. In Proceedings of the 2020 21st International Arab Conference on Information Technology (ACIT), Muscat, Oman, 21–23 December 2020; pp. 1–6.

45. Korichi, A.; Slatnia, S.; Tagougui, N.; Zouari, R.; Kherallah, M.; Aiadi, O. Recognizing Arabic Handwritten Literal Amount Using Convolutional Neural Networks. In Proceedings of the International Conference on Artificial Intelligence and its Applications, Vienna, Austria, 29–30 October 2022; pp. 153–165.

46. Albahli, S.; Nawaz, M.; Javed, A.; Irtaza, A. An improved faster-RCNN model for handwritten character recognition. *Arab. J. Sci. Eng.* **2021**, *46*, 8509–8523. [CrossRef]

47. Prokhorenkova, L.; Gusev, G.; Vorobev, A.; Dorogush, A.V.; Gulin, A. CatBoost: Unbiased boosting with categorical features. *Adv. Neural Inf. Process. Syst.* **2018**, *31*.

48. Yang, J.; Ma, J. Feed-forward neural network training using sparse representation. *Expert Syst. Appl.* **2019**, *116*, 255–264. [CrossRef]

49. LeCun, Y.; Cortes, C.; Burges, C.J. MNIST Handwritten Digit Database. **2010**, *18*. Available online: http://yann.lecun.com/exdb/mnist/ (accessed on 15 August 2022).

50. Shams, M.; Elsonbaty, A.; ElSawy, W. Arabic handwritten character recognition based on convolution neural networks and support vector machine. *Int. J. Adv. Comput. Sci. Appl.* **2020**, *11*, 144–149. [CrossRef]

Article

On a Framework for Federated Cluster Analysis

Morris Stallmann *,† and Anna Wilbik *,†

Department of Advanced Computing Sciences, Faculty of Science and Engineering, Maastricht University, 6229 EN Maastricht, The Netherlands
* Correspondence: m.stallmann@maastrichtuniversity.nl (M.S.); a.wilbik@maastrichtuniversity.nl (A.W.)
† These authors contributed equally to this work.

Abstract: Federated learning is becoming increasingly popular to enable automated learning in distributed networks of autonomous partners without sharing raw data. Many works focus on supervised learning, while the area of federated unsupervised learning, similar to federated clustering, is still less explored. In this paper, we introduce a federated clustering framework that solves three challenges: determine the number of global clusters in a federated dataset, obtain a partition of the data via a federated fuzzy c-means algorithm, and validate the clustering through a federated fuzzy Davies–Bouldin index. The complete framework is evaluated through numerical experiments on artificial and real-world datasets. The observed results are promising, as in most cases the federated clustering framework's results are consistent with its nonfederated equivalent. Moreover, we embed an alternative federated fuzzy c-means formulation into our framework and observe that our formulation is more reliable in case the data are noni.i.d., while the performance is on par in the i.i.d. case.

Keywords: federated learning; framework; cluster analysis; cluster number determination; federated fuzzy Davies–Bouldin index; federated cluster validation metric; federated fuzzy c-means

Citation: Stallmann, M.; Wilbik, A. On a Framework for Federated Cluster Analysis. *Appl. Sci.* **2022**, *12*, 10455. https://doi.org/10.3390/app122010455

Academic Editors: Pavlos S. Efraimidis, Avi Arampatzis and George Drosatos

Received: 18 September 2022
Accepted: 8 October 2022
Published: 17 October 2022

Publisher's Note: MDPI stays neutral with regard to jurisdictional claims in published maps and institutional affiliations.

Copyright: © 2022 by the authors. Licensee MDPI, Basel, Switzerland. This article is an open access article distributed under the terms and conditions of the Creative Commons Attribution (CC BY) license (https://creativecommons.org/licenses/by/4.0/).

1. Introduction

The success of machine learning (ML) can partly be attributed to the availability of good and sufficiently sized training datasets. Often, the data are stored on a central server, where ML models are trained. However, the data might initially be distributed among many clients (e.g., smartphones, companies, etc.). There are situations where gathering the data on a central server is not feasible, e.g., due to privacy regulations (such as GDPR) [1]), the amount of data, or other reasons. Federated learning (FL) is an approach that allows clients to jointly learn ML models while keeping all data local [2]. Authors describe the generic FL training process by five steps:

1. Client selection: Select clients participating in the training.
2. Broadcast: A central server initializes a global model and shares it with the clients.
3. Client computation: Each client updates the global model by applying a training protocol and shares the updates with the central server.
4. Aggregation: The central server applies an aggregation function to update the global model.
5. Model update: The updated global model is shared with the clients.

This protocol can be repeated multiple times until a convergence criterion is met. Training a model following such a process has been successfully applied to a variety of use cases, e.g., for next-word predictions on smartphones [3], vehicle image classification [4], data collaboration in the healthcare industry [5,6], on IoT data [7–9], and many more. For comprehensive surveys please refer to [2,10] or [11]. Many works focus on supervised learning while the area of federated unsupervised learning, similar to federated clustering, is less explored. Cluster analysis is widely applied across many different disciplines as diverse as medical research [12], social and behavioral sciences [13,14], strategic management [15], or marketing [16,17], just to name a few. In all of these areas of application, the data could be initially distributed and hard to centralize. Hence, they are all potential application areas for a federated cluster analysis framework. Ref. [18] described cluster analysis as seven steps that need to be performed in order to derive insights (Section 2.1). In federated clustering, only one step has been (explicitly) addressed, i.e., the clustering method [19–22]. Other important steps, such as cluster validation or determining the number of clusters, have no federated equivalent yet.

Federated clustering is an FL setting, where the goal is to group together (local) data points that are globally similar to each other. That is, data points are distributed among multiple clients and are clustered based on a global similarity measure, while all data remain local on client devices. Existing works largely focus on developing and applying a federated clustering method, i.e., partitioning the data given the number of (global) clusters K. While being an important step in the clustering framework, the lack of more comprehensive frameworks (for example, including determination of the number of clusters and cluster validation) might hinder application in practice.

We contribute to closing this gap by introducing a multistep federated (fuzzy) clustering analysis framework for the first time. In particular, we propose a federated version of the well-known (fuzzy) Davies–Bouldin index for cluster validation, show how to use it for the determination of the number of clusters, and apply a federated fuzzy c-means algorithm to solve the soft clustering problem. Even though independently developed, we note that our federated fuzzy c-means algorithm is closely related to other works in the area of federated clustering [19–22]. It combines local fuzzy c-means on the client side with k-means on the global server side. Each idea in itself is not new, but the combination is, and we observe that our formulation is more reliable than other federated fuzzy clustering methods in case the data are non-i.i.d. To the best of our knowledge, there exists no cluster validation index for federated cluster validation yet. Moreover, no work addressing the problem of determining the number of clusters in a federated setting is known to us.

In the remainder of this section, we demonstrate the need for a federated clustering framework. In subsequent sections, we review relevant works from nonfederated and federated cluster research in Section 2.1. In Section 2.2, we introduce the individual pieces of our framework before fitting them together. Section 3 contains an experimental evaluation on real-world and artificial data to demonstrate the framework's effectiveness and uncover shortcomings. Finally, Section 4 concludes this work.

1.1. Motivational Example

Previous works in federated clustering mostly assume application scenarios to be given. However, it is not necessarily obvious that sophisticated federated cluster analysis algorithms are indeed required and, for example, exchanging locally optimal cluster analysis results is not sufficient. This motivational example is designed to close the gap. In the following example, locally optimal results obtained by nonfederated fuzzy c-means do not reveal the global cluster structure. We illustrate this example by outlining a potential practical application.

Imagine a multinational company with several local markets selling similar consumer goods in all markets. Each local market has data about their customers (e.g., age, place of residency, sold good, etc.) and applies (fuzzy) clustering algorithms to generate customer segments. The cluster analysis insights are utilized to steer marketing activities. The company wishes to derive global clusters to better understand their global customer base and identify unlocked potential in the local markets. Due to strict privacy regulations, the company is not allowed to gather all data in a central database (e.g., European customer data are not allowed to be transferred to most countries outside of Europe). The company could ask each local market to share their local cluster centers, but this approach disregards that clusters might only become apparent when the data are combined. Such a situation exists, as we will show. For the purpose of this example, we spare the details of the dataset creation, because a detailed explanation can be found in Section 3.2.1. It is enough to know that there are five global clusters in the dataset, because of how it was created. Four of the clusters have relatively high cardinality, and the fifth cluster has fewer points. We verify that the correct number of clusters is detected in the centralized dataset. To achieve this, the (nonfederated) fuzzy c-means algorithm (Section 2.1.2) is applied with multiple (potential) number of clusters K, and the result is evaluated with the fuzzy Davies–Bouldin index (Section 2.1.3). The best fuzzy Davies–Bouldin score is achieved with $K = 5$, and we can conclude that the correct number of clusters can be found in the centralized case.

This example is designed to prove that global structure in federated data can hide behind locally optimal solutions. To create a federated setting, the data are distributed among three clients in a certain way: each client receives data from two of the four bigger clusters and a few points from the smaller cluster (see Figure 1 for a visualization). Next, we calculate the number of local clusters for all clients using the same method as before (nonfederated fuzzy c-means in combination with nonfederated Davies–Bouldin). For each client now, the best number of clusters is 2 (as it gives the best DB index score) and none identify the smaller cluster present in their data. That calls for a federated cluster analysis (federated clustering method and federated cluster validation metric) that is able to detect the global structure.

In the multinational company example, the smaller cluster could represent a group of customers that each local market falsely assigned to their bigger clusters. As a consequence, the company targets those customers with inappropriate marketing activities. Exploiting insights from the global cluster analysis could lead to new, more targeted marketing strategies and unlock (previously hidden) potential.

This work is concerned with introducing the federated fuzzy Davies–Bouldin (Fed-FuzzDB) index as a federated cluster validation metric. It can be leveraged to identify all five clusters in the data without the need for sharing the raw data, as we will see in Section 3.2.1. To the best of our knowledge, there exists no other federated cluster validation metric in the literature. Another, equally important, challenge is to correctly identify the global cluster structure given the number of clusters. Our framework applies a federated fuzzy c-means algorithm with k-means averaging (FedFCM, Section 2.2.3) to address this challenge. We will see that the combination of FedFuzzDB and FedFCM leads to promising results, but note that the framework could also be used with other clustering algorithms.

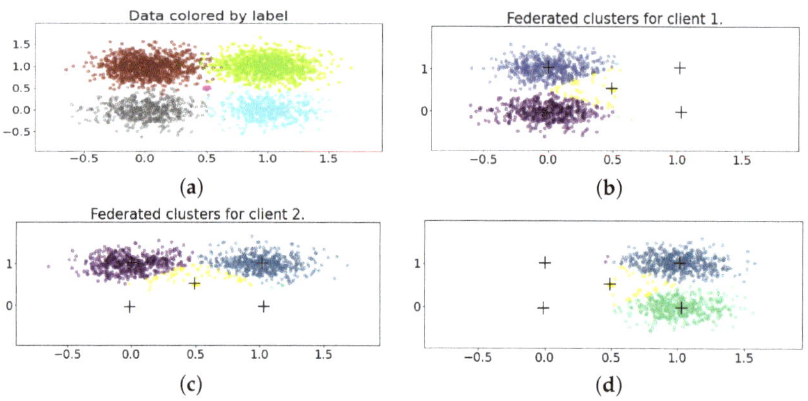

Figure 1. Motivational example. (**a**) The centralized dataset. Colors correspond to ground truth partitions. (**b–d**) The distributed dataset. Crosses denote the clustering result of the federated clustering framework. Original cluster centers are recovered even though no client alone was able to do so.

2. Materials and Methods

Our overall framework consists of federated versions of the fuzzy c-means with k-means averaging and a federated version of the (fuzzy) Davies–Bouldin index. Therefore, we revisit these well-known concepts in the next subsections (Sections 2.1.1–2.1.3). Moreover, we provide a brief overview of works about federated clustering (Section 2.1.4). Finally, we introduce our new framework in Section 2.2 and corresponding subsections.

2.1. Background: Related Work

2.1.1. Nonfederated k-Means Clustering

Let X be a given dataset. The objective of the k-means clustering is to find cluster centers $c = (c_1, c_2, \ldots, c_K)$ (and corresponding assignments) such that the following expression is minimized (see, e.g., [23]):

$$\hat{J}(c) = \sum_{k=1}^{K} \sum_{x \in X} I(x,k) ||x - c_k||^2 \tag{1}$$

$$I(x,k) = \begin{cases} 1, & x \text{ is assigned to cluster } k, \\ 0, & \text{otherwise.} \end{cases}$$

Each data point x is assigned to its closest cluster center, where closeness is defined by Euclidean distance. The assignment is given by function $a : X \times \{c_1, \ldots, c_K\} \to \{1, \ldots, K\}$:

$$a(x, \{c_k\}_{k=1}^{K}) := \underset{1 \leq k \leq K}{\arg\min} ||c_k - x||^2 \tag{2}$$

A widely used iterative algorithm for finding such a clustering works is as follows [23]:

1. Initialize cluster centers $c_1, \ldots, c_K$.
2. Assign clusters for all $x \in X$ according to $a(x, \{c_k\}_{k=1}^{K})$ (Equation (2)).
3. Recalculate cluster centers c_k by solving the following problem, i.e., calculating the mean of points assigned to the same cluster:

$$c_k = \min_{m \in \mathbb{R}^d} \sum_{x \in X : a(x) = k} ||x - m||^2, k = 1, \ldots, K.$$

4. Check if the convergence criterion is met, i.e., whether the assignment did not change (much) compared to the previous iteration. Let $a_i^{(t-1)}$ be the assignment vector of the previous iteration for data point $x_i \in X$, i.e., the k-th entry is 1 if $a(x_i, \{c_k\}) = k$, and zero otherwise. Let $a_i^{(t)}$ be the assignment of the current iteration. Further, let $A^{(t-1)}$ and $A^{(t)}$ be the assignment matrices, where the i-th row equals $a_i^{(t-1)}$ and $a_i^{(t)}$, respectively. Then, the algorithm converges if the difference between the two assignment matrices is smaller than some predefined ϵ:

$$||A^{(t-1)} - A^{(t)}||^2 < \epsilon. \tag{3}$$

If the algorithm did not converge yet, move back to step 2. If it did converge, terminate. $\hat{J}(c)$ is monotonously decreasing with each iteration, but it is known that the algorithm might become stuck in a local minimum. In fact, it does not offer any performance guarantee, and [24] argues that it often fails due to its sensitivity to the initialization method. In [24], the still-popular initialization method k-means++ is introduced. It subsequently chooses random cluster centers that are likely to be far from already chosen centers. In our experiments, we use the scikit-learn implementation that applies the k-means++ initialization method, too [25].

2.1.2. Nonfederated Fuzzy c-Means Clustering

Fuzzy c-means is a well-known soft clustering method that assigns a membership index u_{ij} for clusters $j = 1, \ldots, K$ to data points $x_i \in X$ such that $\sum_{j}^{K} u_{ij} = 1 \; \forall i = 1, \ldots, N$. The term soft clustering refers to the fact that points are allowed to belong to more than one cluster. In contrast, a hard clustering method such as k-means assigns each point to exactly one cluster.

For given data X, the objective is to find cluster centers c_j and membership matrix $U = [u_{ij}]$ such that the following expression is minimized [26]:

$$J_m(U,c) = \sum_{i=1}^{N} \sum_{j=1}^{K} (u_{ij})^m \|x_i - c_j\|^2, \tag{4}$$

$$u_{ij} := \frac{1}{\sum_{k=1}^{K} \left(\frac{\|x_i - c_j\|^2}{\|x_i - c_k\|^2}\right)^{\frac{2}{m-1}}}, \tag{5}$$

where $\|y\| := \sqrt{\sum_{l=1}^{n} y_l^2}$. It is closely related to the k-means clustering, and the main difference lies in the assignment matrices A in k-means versus U in fuzzy c-means.

Parameter $m > 1$ controls how fuzzy the cluster assignments should be. The greater the m, the more fuzziness in the assignment, i.e., points are assigned to more clusters with smaller values. A common choice that we also employ in all of our experiments is $m = 2$.

A widely used algorithm to find a solution to the optimization problem was introduced by [26] and follows four basic steps:

1. Initialize matrix $U := U^0$.
2. In iteration t, (re)calculate cluster centers c_j according to:

$$c_j = \frac{\sum_i u_{ij}^m x_i}{\sum_i u_{ij}^m} \tag{6}$$

3. Update membership matrix U^{t+1} according to Equation (5).
4. Check if the convergence criterion is met: $\|U^{t+1} - U^t\| \leq \epsilon$ for some predefined ϵ, i.e., did the memberships change by at most ϵ? If it was not met, return to step 2 after setting $U^t = U^{t+1}$. Terminate if it was met.

The time-complexity of the algorithm is quadratic in the number of clusters K, and methods to reduce the complexity have been proposed [27]. Similar to k-means, other shortcomings of the algorithm are sensitivity to the cluster initialization and sensitivity to noise, as noted in [28]. Those challenges have been addressed by subsequent works, but each auxiliary method comes with its own shortcomings [28]. Clustering in high-dimensional spaces is another well-known challenge for clustering algorithms in general [29], and fuzzy c-means in particular [30], due to high sparsity in high-dimensional spaces. Authors of [30] show that fuzzy c-means centers are likely to converge to the center of mass of the whole dataset in high-dimensional spaces. It remains up to the practitioners to decide on a suitable method for their specific problems.

For the introduction of federated fuzzy c-means, we focus on the original formulation and extend it to the federated setting.

2.1.3. Davies–Bouldin Index

The Davies–Bouldin index was introduced in [31] as a method to measure cluster validity. One of its advantages is that it only requires distances between a "vector characteristic of a cluster" (i.e., cluster center) and the vectors belonging to the cluster, as opposed to pairwise distances between all vectors in the dataset, as in other cluster validation methods. That makes it also particularly interesting for the federated clustering setting, where a pairwise distance matrix is hard to obtain, but distances to the cluster center can be calculated locally, shared, and averaged by the central server.

Informally speaking, the validation measures how well "cluster spread" is balanced against "dissimilarity between cluster centers". A good clustering is achieved with low spread and high cluster center dissimilarity, but these goals are potentially conflicting. Formally, in [31], the nonfederated measure for a hard clustering is defined as:

$$\bar{R} := \frac{1}{K} \sum_{i=1}^{K} R_i \tag{7}$$

$$R_i := \max_{i \neq j} R_{ij} \tag{8}$$

$$R_{ij} := \frac{S_i + S_j}{M_{ij}} \tag{9}$$

$$S_i := \left(\frac{1}{T_i} \sum_{j=1}^{T_i} ||x_j - c_i||^q\right)^{\frac{1}{q}} \text{ "cluster spread"} \tag{10}$$

$$M_{ij} := \left(\sum_{k=1}^{D} |c_i[k] - cj[k]|^p\right)^{\frac{1}{p}} \text{ "center distances",} \tag{11}$$

where c_i is the characteristic vector (read: center) of cluster i, and D is the dimension of the data. Note that R_{ij} is big when two cluster centers are close to each other (small M_{ij}), or the "spread" of the clusters is big (big S_i). Additionally, the cluster spread is usually smaller if we have many clusters. However, this often comes at the expense of closer centers. Roughly speaking, the index measures how well those two characteristics are balanced, and a smaller value indicates a better cluster result. In our experiments, we chose $p = 2$ and $q = 1$ for our computations.

A soft version of the Davies–Bouldin index for fuzzy clustering was introduced in [32]. In soft clustering, every point can belong to every cluster, but in Equation (10), only points belonging to the same cluster are considered. Hence, the "spread" of a cluster must be defined differently. Authors of [32] propose the following adaptation:

$$S_i^f := U_i\left(\frac{1}{N} \sum_{j=1}^{N} ||x_j - c_i||^q\right)^{\frac{1}{q}}, \tag{12}$$

$$U_i := \frac{1}{N} \sum_{j=1}^{N} u_{ij}. \tag{13}$$

Each $x \in X$ can belong to each cluster i. As a consequence, the spread of each cluster needs to be calculated by considering the whole dataset and is then multiplied by the average assignment for cluster i, i.e., U_i. The calculation of the index $\bar{R}$ proceeds as outlined above, with S_i^f instead of S_i.

2.1.4. Federated Clustering

Due to the similarity in terminology, we start by contrasting clustered federation with federated clustering. Clustered federation is concerned with identifying clusters of clients or model updates that are suitable to be grouped for a focused update of global supervised FL models. It has been proven to be effective when addressing issues caused by non-i.i.d. data among clients [33–36].

In contrast, federated clustering is concerned with identifying global clusters in distributed data without sharing the data and, to the best of our knowledge, has not been explored as much. In [19], the k-means algorithm was extended to the federated setting. Client devices execute the k-means algorithm and share cluster centers with the central server. Authors propose a global averaging function that calculates a weighted mean of local cluster centers in order to update global cluster centers. The weights are given by the number of local data points assigned to the clusters. Further, the federated fuzzy clustering equivalent of that approach was introduced in [20] and similarly in [21]. In this approach, the clients execute the fuzzy c-means algorithm and share the results. Then, the fuzzy assignment vectors are used as weights instead of number of data points given by the hard assignments. In their experimental sections, both works focus on scenarios where the data are uniformly distributed among the clients. Ref. [20] found that the federated clustering result was consistent with centralized clustering result, the algorithm converged quickly, and the clustering result was not impacted much by the number of clients participating. Additionally, Ref. [21] observed that even if clients became unavailable during the federation, the algorithm still found good results. However, these findings are limited to scenarios where the data are uniformly distributed among clients. Finally, we acknowledge that [20] also introduced a formulation for vertical FL, which is beyond the scope of this work.

A different approach on averaging the local cluster centers to obtain global cluster centers was taken in [22] in the context of one-shot learning. The clients performed k-means clustering. On the central server, the global cluster centers were computed by applying the k-means algorithm again to the shared local cluster centers. Besides numerical experiments, they also provided proof that the result was similar to an "oracle clustering" (e.g., clustering on the assumed centralixed dataset). The federated k-means algorithm by [19] appears to be the first work in the area of federated clustering. All other papers were published around the same time and appear to be independent of each other.

In [18], the cluster analysis framework is described in terms of seven steps: selecting training examples for the cluster analysis, selecting variables for the cluster analysis, preprocessing (e.g., standardizing) the data, selecting a (dis)similarity measure, selecting a clustering method, determining the number of clusters, and interpreting and validating the results. Note that the aforementioned works are mostly concerned with the clustering method and (implicitly) with the dissimilarity measure in a federated setting. With this work, we aim to also contribute to determining the number of clusters and cluster validation in a federated setting; however, similar to the other works, we assume the experimental datasets to be preprocessed and prepared for analysis.

2.2. The Federated Fuzzy Clustering Framework

In this section, we build upon the previous section and introduce the federated versions of the fuzzy c-means algorithm (Section 2.2.1) and fuzzy Davies–Bouldin index (Section 2.2.2). In Section 2.2.3 the pieces are assembled to form a cluster analysis framework performing three steps: determine the number of clusters K, derive a clustering for K, and validate the clustering through a federated cluster validation metric.

2.2.1. Federated Fuzzy c-Means with k-Means Averaging

Our proposed federated fuzzy c-means algorithm (FedFCM) is an extension of the iterative fuzzy c-means algorithm to the federated learning setting similar to [20,21], but with a different take on the global cluster center calculation. The global cluster calculation is similar to the one proposed in [22], where it is applied in the context of federated one-shot k-means. This idea was first mentioned and discussed in our preliminary work [37].

In the federated scenario, the data are not stored in a centralized database, but distributed among multiple clients. The goal is to learn a global clustering that is similar to the clustering of the centralized data while the data stay private. The general procedure is as follows: Each client runs a number of fuzzy c-means iterations locally, and sends the resulting cluster centers to a central server. The central server is responsible for calculating meaningful global clusters from the local learners' results. After calculating the global centers, they are shared with the clients that use them to recalculate their local centers, which in turn are shared with the central server, and so forth. That procedure is repeated until the global centers remain stable.

The creation of a global model from clients' local model updates was first introduced by [38] and is known as federated averaging (FedAvg). Our averaging method is a k-means averaging that was independently developed and applied in the context of federated one-shot clustering in [22].

Let data X be distributed among P parties (clients), i.e., $X = \bigcup_{l=1}^{P} X^{(l)}$. The protocol reads as follows:

1. The central server initializes K global cluster centers $c_1, \ldots, c_K$.
2. The central server shares $c_1, \ldots, c_K$ with the clients.
3. Client l calculates membership matrix U_l^{t+1} according to Equation (5) and generates local cluster centers $c_1^{(l)}, \ldots, c_K^{(l)}$ according to Equation (6) ($l = 1, \ldots, P$) and repeats until local convergence.
4. Client l shares $c_1^{(l)}, \ldots, c_K^{(l)}$ for $j = 1, \ldots, K$ and $l = 1, \ldots, P$.
5. The central server updates $c_1, \ldots, c_K$ by applying an averaging function $avg([c_1^{(l)}, \ldots, c_K^{(l)}]_{l=1}^P)$.
6. The central server checks a convergence criterion. If not converged, go back to step 2.

Since the central server has only access to the local cluster centers $c_k^{(l)}$, the previous convergence criterion can not be applied. As an alternative, we check whether the cluster centers changed by less than ϵ between two iterations. Let c_{k_t} be the global cluster center k after time step t. Then, the convergence criterion can be formulated as follows: $\sum_{k=1}^K ||c_{k_t} - c_{k_{t+1}}|| \leq \epsilon$. Note that this new criterion might lead to different cluster centers than in the previous formulation. It might let the centers move closer to the center of mass even though the assignments might have stabilized already.

In order to find meaningful global clusters, it is essential to find a good averaging function $avg(\cdot)$ used in step 5 of the framework outlined above.

To address this challenge, we apply a k-means averaging function, similar to the one in [22]:

$$avg : \mathbb{R}^{P \times d \times K} \to \mathbb{R}^{d \times K}$$
$$avg([c_k^{(l)}]_{k,l}) := kmeans([c_k^{(l)}]_{k,l}), \tag{14}$$
$$= [c_k]_k,$$

where $kmeans(\cdot)$ denotes a function that applies the k-means clustering algorithm and outputs the k cluster centers it converged to. This averaging function applies the k-means algorithm to all reported local cluster centers to find new global cluster centers. It does introduce increased complexity compared to federated fuzzy c-means in [20,21], but we observed robust results in preliminary experiments [37]. At the same time, sharing only the local cluster centers (as opposed to local centers and assignment vectors such as in [19–21]) increases the privacy of the data.

We know that both the fuzzy c-means and k-means algorithm converge (even though possibly to local optima) when applied separately. Convergence means that the centers and assignment vectors are guaranteed to stabilize after finitely many iterations. Our federated algorithm converged if the global cluster centers stabilized. For that to happen, the local cluster centers must have stabilized. The local cluster center, in turn, stabilizes if the global centers do not change much between two iterations. We want to provide intuition on why we observe such a behavior in our experiments. Each clients starts with calculating local cluster centers in the first iteration and reports them back to the central server. The central server essentially groups together all centers that are close to each other, calculates the average of close centers, and reports those averages back to the clients as their new centers. Due to this averaging, it is likely that for any previous local center there is a new nearby global center (which is a function of that previous center). The client updates the global center with its local data. Since it is close to a previously reported center, the new local centers do not deviate much from the global center, and the update is small. If a new center has low cluster membership, the update is naturally small. With small updates, however, we know the k-means algorithm to converge. Usually, the updates can be quite big after the first global round, but are small thereafter, which is consistent with the behavior of k-means one-shot learning with k-means averaging [22]. Even though this is not a formal convergence proof, we hope to provide insights into how the algorithm is expected to behave.

2.2.2. Federated Fuzzy Davies–Bouldin Index

As described in the introduction (Section 1.1, Figure 1), there is a need for federated cluster validation metrics. It is not enough to calculate metrics such as the Davies–Bouldin index locally and draw conclusions from there. Therefore, we formulate a federated version of the index. In the federated setting, the global server does not have access to the clients' data. That means that it cannot carry out the calculation of the soft Davies–Bouldin index $\bar{R}$ directly. Specifically, the central server can not directly calculate the cluster spread S_i^f (Equation (12)) that requires the calculation of distances between cluster centers c_i and data points x_j. Through a simple transformation we see that sharing the data points is not required:

$$S_i^f = U_i \left(\frac{1}{N} \sum_{j=1}^{N} ||x_j - c_i||^q \right)^{\frac{1}{q}} \tag{15}$$

$$= U_i \left(\frac{1}{\sum_{l=1}^{P} N_l} \sum_{l=1}^{P} \sum_{j=1}^{N_l} ||x_j^{(l)} - c_i||^q \right)^{\frac{1}{q}} \tag{16}$$

$$U_i = \frac{1}{\sum_{l=1}^{P} N_l} \sum_{l=1}^{P} \sum_{j=1}^{N_l} u_{ij}^{(l)} \tag{17}$$

Hence, for the calculation of S_i^f, each client l needs to share its number of data points N_l, its local local cluster spread $\sum_{j=1}^{N_l} ||x_j^{(l)} - c_i||^q$, and its local average assignment vectors $\sum_{j=1}^{N_l} u_{ij}^{(l)}$ for $i = 1, \ldots, K$. With that information, the global server can calculate S_i^f. Since the global server calculates (and knows) the cluster centers, it can also calculate M_{ij}, i.e., the distances between centers c_i and c_j (Equation (11)), and, finally, R_{ij} for all (i, j), R_i and the index $\bar{R}$. Note that the federated and nonfederated versions of the Davies–Bouldin index produce the same result given $X = \bigcup_{l=1}^{N} X^{(l)}$, and the nonfederated and federated cluster centers are the same. Generally, the first assumption holds in our experiments while the second one is the subject of study and cannot be guaranteed. In fact, due to different convergence criteria in the nonfederated and federated fuzzy c-means algorithms, the centers are often different. Generally, however, we expect federated clustering and nonfederated algorithms to converge to similar centers.

2.2.3. The Complete Framework

Our proposed framework for federated clustering addresses three core challenges: Estimate the number of clusters in the federated dataset, obtain a cluster result (i.e., centers and a data partitioning) that is similar (or not worse) to the one on the same but centralized dataset, and assess the federated cluster result via a federated validation metric. Note that the challenges are closely related. In order to compare two clustering results (for example, with different numbers of clusters), there must be an evaluation metric. This evaluation metric is the FedFuzzDB index. To obtain the federated clustering result, a federated clustering method must be applied. In our case, this is FedFCM (and for comparison federated fuzzy c-means with federated averaging). The overall framework applies the following steps:

1. Decide on a range for number of clusters K to test: $[K_{min}, K_{max}]$.
2. For each $K \in [K_{min}, K_{max}]$:

    (a) Obtain a clustering with FedFCM as described in Section 2.2.1.
    (b) Calculate the FedFuzzDB index of that clustering as described in Section 2.2.2 and store the result.

3. Choose $K \in [K_{min}, K_{max}]$ with the minimum FedFuzzDB index as the number of clusters or apply the elbow method (see, e.g., [39]).

The initial guess for $[K_{min}, K_{max}]$ is not subject to a more principled study in this work, but we acknowledge that choosing a good range is crucial. It is known that the Davies–Bouldin index sometimes favors higher number of clusters [40]. Therefore, introducing a tight upper bound can be important. In a federated setting, this is even harder, as some clusters might not even be present in any client's data, but only form when the data are combined (Sections 1.1 and 3.2.1). As a rule of thumb, we note that the minimum number of global clusters is given by the minimum number any client could identify on its own (note that two local clusters might turn out to belong to the same cluster in the global dataset). The maximum is (roughly) given by $\sum_{l=1}^{P} K_l + forming - overlapping$. K_l is the number of clusters locally in client l, $forming$ is the number of clusters that only form when the data are combined, and $overlapping$ is the number of clusters that overlap. $forming$ and $overlapping$ are the hardest to estimate, even with knowledge from an initial local-only clustering. As a very rough rule of thumb, we apply $K_{max} \approx \min_l K^{(l)} + \max_l K^{(l)}$ in our real-world experiment.

Before continuing with experiments, we note that there is a potential privacy risk and suggest a simple prevention mechanism. Let us summarize the local information that is required to be shared with the central server for the overall framework. For the execution of federated fuzzy c-means with k-means averaging, only the local cluster centers need to be shared, which is already a privacy improvement over some of the existing methods. For the calculation of the FedFuzzDB index, however, each client l needs to share more information: its number of total data points N_l, the total spread $\sum_{j=1}^{N_l} ||x_j^{(l)} - c_i||^q$, and the local average assignment vectors $\sum_{j=1}^{N_l} u_{ij}^{(l)}$ for all clusters $i = 1, \ldots, K$. As noted in [21], this information can be used to formulate a system of nonlinear equations where data $x_i^{(l)}$ are the only unknowns ($i = 1, \ldots, N_l, l = 1, \ldots, P$). While not necessarily easy to solve, this imposes a privacy risk. In [21], the server does not know N_l, which is an effective prevention, as they explain. Hence, if we hide N_l from the central server, we prevent the privacy risk. Luckily, the calculation of the FedFuzzDB index only requires $\sum_{l=1}^{P} N_l$. If we outsource the calculation of $\sum_{l=1}^{P} N_l$ to an intermediate server, we can circumvent the risk. Another option is to perform the clustering and the validation on different servers that cannot communicate with each other, as the system cannot be solved for X without the cluster center information, either. However, that would require that the distances between clusters i and j (M_{ij}) need to submitted to the cluster validation server, e.g., by one of the clients. All of that might not be required if the central server can be completely trusted (for example, when local markets cooperate under the orchestration of the parent company). Usually, it is assumed that the central server only keeps updates for immediate aggregation. Hence, the cluster center information is supposed to be not available anymore when FedFuzzDB is calculated.

3. Results

Our framework is evaluated on three different groups of data. Firstly, we handcraft a non-i.i.d. clustering scenario and apply the complete clustering framework to it. The emphasis in this experiment is on demonstrating and motivating the use of the federated clustering framework to obtain a global cluster structure without sharing the raw data. Secondly, we create federated scenarios from 100 well-known cluster benchmark datasets from the literature [41] by uniformly distributing the data to clients. This set of experiments allows us to study how the framework performs with data of different dimensionality and different spread, and how it behaves with increasing number of clients, but fixed number of total data points. We will see that high sparsity is harmful and big cluster overlap is harmful in the federated and nonfederated setting. Thirdly, we apply the complete framework to more real-world-inspired data and demonstrate how to use it in practice. We will see that the federated and nonfederated results are mostly consistent, and the federated clustering is even slightly better in terms of FedFuzzDB.

Moreover, we compare our federated fuzzy c-means with k-means averaging to federated fuzzy c-means, as introduced in [20,21]. We note that scenarios where the data are nonuniformly distributed among clients are not investigated by these works. In fact, we observe that the method does not converge reliably in such scenarios (see Section 3.2.1 and our preliminary work [37]). Further, note that we do not compare the method to hard clustering methods such as [19] or [22], because we only introduced the fuzzy version of the federated Davies–Bouldin index.

Before describing each dataset in more detail and reporting the results, we introduce the evaluation methods for our experiments.

3.1. Evaluation Method

There are two main questions we want to answer with our evaluation.

1. How reliably can the framework detect the correct number of clusters in federated datasets?
2. How good is the federated clustering result, and how does it compare to an assumed central clustering result?

To answer the first question, we first need to define the "correct" number of clusters. In the first two sets of experiments (Sections 3.2.1 and 3.2.2), we know the ground truth number of clusters, because the data are artificially created by sampling from Gaussian distributions. Given the ground truth number of clusters K_{true} and the detected number of clusters K_{det}^i in experiment i, we simply count how often the correct number of clusters was found and report the percentage of correct numbers:

$$p_{correct} = \frac{\sum_{i=1}^{N_{exp}} \mathbb{1}_{\{K_{det}^i = K_{true}\}}}{N_{exp}}, \tag{18}$$

where N_{exp} is the total number of experiments. Moreover, we want to study how federated clustering compares to nonfederated clustering. Therefore, we report $p_{correct}$ for the federated dataset and for the same centralized dataset.

The second challenge is to evaluate the clustering result itself. On the one hand, we use the (federated) Davies–Bouldin index introduced in this work. Moreover, we calculate a "knowledge gap" metric whenever ground truth cluster centers are known:

$$gap := \sum_{k=1}^{K} \sqrt{\sum_{i}^{D} \frac{(\tilde{c}_k[i] - c_k[i])^2}{Var(x[i])}}, \tag{19}$$

between two sets of cluster centers $\tilde{c} = (\tilde{c}_1, \ldots, \tilde{c}_K)$, and $c = (c_1, \ldots, c_K)$ and $Var(x[i])$ denotes the variance in the i-th dimension. This gives us a normalized measure of the distance between the cluster centers and another indication of whether the algorithm converged to a meaningful result.

3.2. Experiments

As our parameter setup, we chose $\epsilon = 0.001$ in the centralized case, for local convergence, and for global convergence. As noted before, we chose $m = 2$, $p = 2$, and $q = 1$ for all experiments. Code to reproduce the results is available (https://github.com/stallmo/federated_clustering).

3.2.1. Framework Demonstration on Artificial Data

We revisit the motivational example from the introduction (Section 1.1). In this example, we start with a centrally created dataset by drawing from five Gaussian distributions: 500 examples each drawn from distributions centered at $\mu_1 = (0,0), \mu_3 = (1,1)$, 1000 examples each drawn from distributions centered at $\mu_2 = (0,1), \mu_4 = (1,0)$ and standard deviation $\sigma_1 = 0.2$, 120 examples drawn from a distribution centered at $\mu_5 = (0.5, 0.5)$ and

$\sigma_2 = 0.01$ (see also Figure 1). Hence, there are five ground truth centers. First, we verify that the five ground truth centers can indeed be found in a nonfederated scenario. We obtain clustering results for $K = 2, \ldots, 7$ and calculate the (nonfederated) Davies–Bouldin index for each result. As expected, the index is smallest for $K = 5$ (Table 1). Second, the data are distributed to three clients such that all clients have data from three clusters in total: two of the four bigger clusters and a few points from the smaller cluster. In particular, each client receives 40 points from distribution (μ_5, σ_2). Client 1 receives 500 points each from distributions (μ_1, σ_1) and (μ_2, σ_1). Client 2 receives 500 points each from (μ_2, σ_1) and (μ_4, σ_1). Client 3 receives 500 points each from (μ_3, σ_1) and (μ_4, σ_1). Next, each client applies the (nonfederated) fuzzy c-means separately on their local data for $K \in \{2, 3, 4, 5\}$ and calculates the (nonfederated) fuzzy Davies–Bouldin index. In this experimental setup, all clients would conclude that they have only two clusters, as $K = 2$ results in the smallest Davies–Bouldin index (Table 1). The clients only detect the two bigger clusters in their data and disregard the smaller cluster. When applying the federated clustering framework outlined in Section 2.2, the correct number of clusters $K = 5$ is found. Please refer to Table 1 for an overview of the results. This experiment shows that the framework is capable of identifying global cluster structure even though it is hidden behind local optima without sharing the raw data. For comparison, we repeat the same experiment with the federated fuzzy c-means formulation that applies federated averaging, as introduced in [20,21] (see also Section 2.1). Note that the setting is non-i.i.d. in the sense that not all clients have data from the same clusters and that such a situation was not part of the analysis in [20,21]. For an overview of the results, please refer to Table 2. We observe that the federated averaging formulation struggles to identify the ground truth centers in this setting. The Davies–Bouldin index is generally higher with federated averaging than with k-means averaging. This indicates that the clustering can be considered less meaningful. The same is indicated by the higher ground truth gap, i.e., the ground truth centers could not be found. Consequently, this also leads to a wrong estimate for the number of global cluster centers. All in all, the results with federated averaging appear to be less reliable than k-means averaging on non-i.i.d. data. However, as we will see in the next section, the results on i.i.d. data are similar and, therefore, consistent with [20,21].

The drop in performance can be explained by the lack of a "grouping mechanism". The grouping mechanism must identify a group of local centers that belong to the same global center and, hence, are used to update that global center. In the case that each client has data from the same clusters (thus, finds and reports similar centers locally), that matching is (implicitly) given. Since all clients have points from the same clusters, all local updates will move in the same direction and there is no ambiguity. With widely different data locally, the local updates will also be very different and there must be a mechanism to deal with the ambiguity, e.g., a grouping of local updates. Using k-means as averaging function directly provides such a mechanism and, as a consequence, produces more reliable results in non-i.i.d. settings.

Table 1. Local and federated clustering results on the motivational dataset with k-means averaging. The best result per experiment (column) is bold.

Fuzzy Davies–Bouldin	Central (Nonfederated)	Federated	Local Client 1	Local Client 2	Local Client 3
$K = 2$	0.8707	0.8179	**0.6426**	**0.6437**	**0.6381**
$K = 3$	0.5687	0.6055	0.7289	0.6991	0.7704
$K = 4$	0.4869	0.4951	1.0637	1.0706	1.1248
$K = 5$	**0.4348**	**0.4289**	0.9260	0.8927	0.9496
$K = 6$	0.5440	0.6202	—	—	—
$K = 7$	0.6707	0.5072	—	—	—
$K = 8$	0.5680	0.6221	—	—	—

Table 2. Local and federated clustering results on the motivational dataset with federated averaging. The best result per experiment (column) is bold

Fuzzy Davies–Bouldin	Central (Nonfederated)	Federated	Local Client 1	Local Client 2	Local Client 3
$K = 2$	0.8707	1.0047	1.0356	0.8620	0.9685
$K = 3$	0.5687	**0.9143**	**0.8880**	**0.8370**	**0.8647**
$K = 4$	0.4869	1.1683	1.1924	0.9221	1.1662
$K = 5$	**0.4348**	2.9158	2.8867	2.4132	2.9538
$K = 6$	0.5440	1.5063	—	—	—
$K = 7$	0.6707	0.9248	—	—	—
$K = 8$	0.5680	1.2115	—	—	—

3.2.2. Evaluation on Benchmark Data

We test our framework on cluster benchmark sets from an online repository (http://cs.uef.fi/sipu/datasets/, accessed on 17 March 2022) [41]. In particular, the G2 sets were introduced in [42] and each set was generated by drawing 2048 samples from two Gaussian distributions with different means, i.e., each set contains two ground truth centers. The Gaussians are centered at $\mu_1 = (500, 500, \dots)$ and $\mu_2 = (600, 600, \dots)$ with standard deviations $\sigma \in \{10, 20, \dots, 100\}$ and dimension $D \in \{2, 4, 8, \dots, 1024\}$. In total, there are 100 sets with varying dimension and standard deviation. In order to evaluate the federated clustering framework, we randomly (but uniformly) distribute the points among $P \in \{2, 5, 10\}$ clients. Note that the number of samples is fixed to be 2048 such that with an increasing number of clients, each clients has fewer samples. Each experiment is repeated 20 times and the averages over the runs are reported.

The first step is to determine the correct number of clusters for each G2 set. We follow the procedure described in Section 2.2.3 for $K \in [2, \dots, 6]$ and choose the minimum as the framework's guess. An overview of the results can be found in Table 3 and Figures 2 and 3 for the framework with k-means averaging. Moreover, the complete framework is evaluated with the federated averaging method for comparison. Results are summarized in Table 3 as well. For the calculation of the correct number of cluster guesses metric in Table 3, we only consider datasets with clustering results with a federated fuzzy Davies–Bouldin index below 1.3. Through exploratory analysis we found that a higher Davies–Bouldin index often shows that the algorithm could not converge (which we also discuss later in this section). In such cases, the framework's cluster number guess is meaningless, because the clustering itself is not meaningful. Note that the value of 1.3 coincides with the 75% quantile in all scenarios (central, k-means averaging, and federated averaging) such that the number of considered datasets is similar for all evaluations.

First, we observe that in the nonfederated, central clustering case, the correct number of clusters can be found in 92.2% of all cases with an federated fuzzy DB index below 1.3. This detection rate is slightly lower in the federated case. Generally, it decreases with an increasing number of clients (while keeping the number of data points fixed). This effect is independent of the clustering method. The effect can be explained by sparsity, as we discuss at the end of this section. High sparsity leads to decreased cluster algorithm performance and, as a consequence, to less meaningful number of cluster detection.

Table 3. Correct cluster guesses with different numbers of clients (DB index below 1.3).

Correct	Central	Two Clients	Five Clients	Ten Clients
k-means avg.	92.9%	91.4%	88.0%	87.2%
Federated avg.		90.4%	89.5%	88.9%

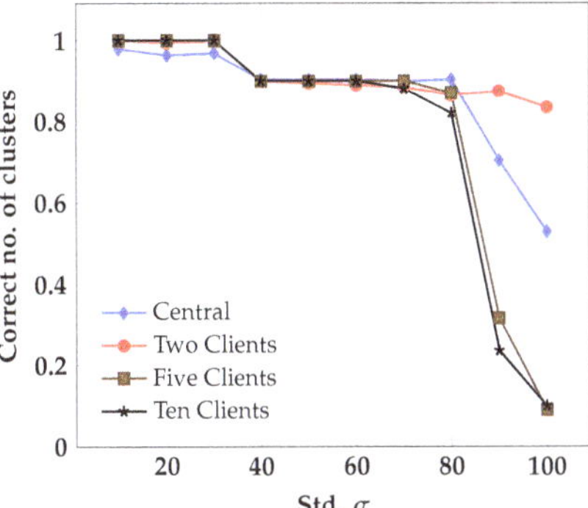

Figure 2. Correct number of cluster guesses on all 100 G2 sets per standard deviation σ. The values are averaged over all dimensions and runs. We observe a decline of correct number of cluster guesses with increasing σ. The figure shows results of the k-means averaging, but they are similar with federated averaging.

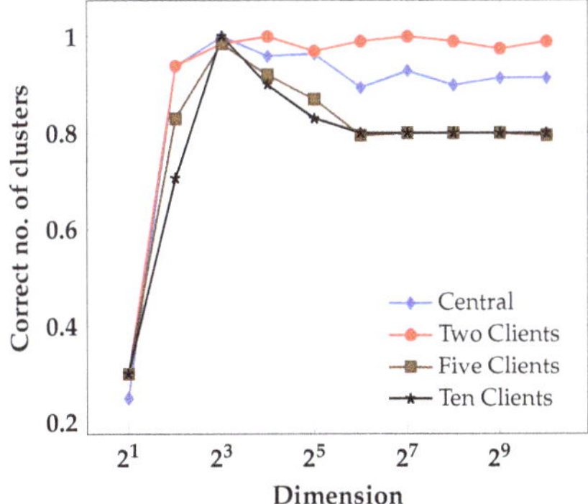

Figure 3. Correct number of cluster guesses on all 100 G2 sets per dimension. The values are averaged over all values of σ and runs. We observe few correct guesses in the two dimensional case, a peak for $D = 8$, then a decline that stabilizes after $D = 64$. The overall trend is similar for all number of clients. The figure shows results of the k-means averaging, but they are similar with federated averaging.

We want to demonstrate the effect of sparsity by taking a closer look at the full G2 set for the k-means averaging. The performance generally worsens with increasing σ (Figure 3; averaged over all D), independent of the number of clients. For $\sigma \in \{10, 20, 30\}$, the detection rate is close to one. For $\sigma \in \{40, 50, 60, 70, 80\}$, the detection rate is between 0.8 and 0.9 with lower numbers for higher number of clients. Finally, there is a noticeable performance drop when $\sigma \in \{90, 100\}$ with the steepest decline when $P \in \{5, 10\}$, where in the majority of cases, the correct number of cases is not detected. Moreover, we also observe detection rates varying across dimensions (Figure 3; averaged over all σ). For $D = 2$, the correct number of clusters is only detected in 0.25 to 0.3 of all cases. Then, it peaks with a detection rate close to 1 for $D = 8$ before decreasing and stabilizing at $D = 64$ and being constant thereafter. The trend is similar for all P, with small exceptions for $P = 2$. However, the level of detection rate is smallest for $P \in \{5, 10\}$, with the exception of $D = 2$, where it is even slightly higher than in the central case.

Second, we report the results of the clustering itself in terms of (federated) fuzzy Davies–Boulding index and knowledge gap . Overall, we see that the results are mostly similar in the central case and in the federated scenarios for $P \in \{2, 5, 10\}$ for either clustering method: The 0.25, 0.5, and 0.75 quantiles and the minimum values for both metrics are similar. However, the maximum value for the federated fuzzy Davies–Bouldin index shows some variation (Tables 4 and 5). Similarly, the knowledge gap statistics are consistent, but not the same (Tables 6 and 7). Hence, in some cases, FedFCM converges to different centers. As we will explain, the differences mostly occur due to sparsity. It is important to note that those are the cases where FedFCM did not find a good clustering (high knowledge gap and high FedFuzzDB) in either the central case or in the federated settings. Overall, we conclude that the clustering results on i.i.d. data are similar in the central case and with both federated clustering methods.

Table 4. Statistics of the (federated) fuzzy Davies–Bouldin index on 100 G2 test sets with k-means averaging.

Fuzzy Davies–Bouldin ($K = 2$)	Central	Two Clients	Five Clients	Ten Clients
25% Quantile	0.6762	0.6771	0.6766	0.6766
50% Quantile	0.8640	0.8638	0.8627	0.8627
75% Quantile	1.3092	1.3073	1.2961	1.2944
Minimum	0.5460	0.5459	0.5459	0.5458
Maximum	56,784.5518	57.6910	23.6987	20.0192

Table 5. Statistics of the (federated) fuzzy Davies–Bouldin index on 100 G2 test sets with federated averaging.

Fuzzy Davies–Bouldin ($K = 2$)	Central	Two Clients	Five Clients	Ten Clients
25% Quantile	0.6762	0.6767	0.6766	0.6766
50% Quantile	0.8640	0.8627	0.8623	0.8620
75% Quantile	1.3092	1.2997	1.2990	1.2934
Minimum	0.5460	0.5460	0.5459	0.5459
Maximum	56,784.55	31,307.76	54,809.42	9021.3485

Table 6. Statistics of the knowledge gap on the 100 G2 test sets with k-means averaging.

Knowledge Gap ($K = 2$)	Central	Two Clients	Five Clients	Ten Clients
25% Quantile	0.2850	0.2724	0.2700	0.2587
50% Quantile	0.8904	0.8897	0.8900	0.8898
75% Quantile	5.9544	5.9272	5.8335	5.4764
Minimum	0.0286	0.0278	0.0265	0.0253
Maximum	77.6472	77.540	77.7092	76.0134

Table 7. Statistics of the knowledge gap on the 100 G2 test sets with federated averaging.

Knowledge Gap ($K = 2$)	Central	Two Clients	Five Clients	Ten Clients
25% Quantile	0.2850	0.2729	0.2693	0.2679
50% Quantile	0.8904	0.8888	0.8902	0.8899
75% Quantile	5.9544	5.5436	5.4888	5.4320
Minimum	0.0286	0.0285	0.0283	0.0281
Maximum	77.6472	77.6472	77.6472	77.6472

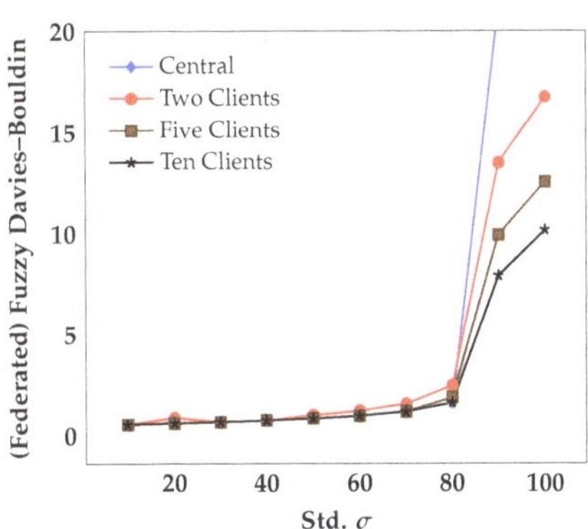

Figure 4. The federated Davies–Bouldin index on 100 G2 sets per standard deviation σ. The values are averaged over all values of D and runs. As expected, we see an increasing index with higher σ. However, the index suggests that it is hard to find good clusterings with FedFCM for $\sigma \geq 80$. The values outside this plot are 434.4732 and 2463.9748 (central).

With this in mind, we enter the discussion of the results and offer an explanation for some of the observations. First, we want to understand why the cluster number detection rate is lower for higher values of σ. Recall that the Davies–Bouldin index is the ratio of "cluster spread" and "center closeness". Hence, the index is high for clusters that naturally have a high spread, i.e., high σ as depicted in Figure 4 (while keeping the center distances fixed, as in our experiments). In such cases, the index could be reduced by introducing a new cluster, because the gain through the lower spread is relatively big. This behavior is intensified by the poor performance of fuzzy c-means on sparse data. In such cases, (local) fuzzy c-means centers (regardless of K) tend to converge to the center of mass of the whole dataset [30]. Hence, the global centers are also all close to the center of mass, and, thus, to each other. This leads to favoring a higher number of clusters. With fewer data points per client (i.e., more clients in our experiments), the data become even more sparse and the effect more severe. Overall, we attribute the lower detection rate to poor performance of FedFuzzDB and FedFCM in sparse spaces. Note that this is a shortcoming of the nonfederated equivalents as well. Second, we want to understand why the cluster number detection rate is so much lower in the two-dimensional case (Figure 3). As opposed to the high-dimensional case, in two dimensions, we are faced with a very dense space and a significant cluster overlap even for smaller values of σ. Through visual inspection, we found that in some cases it is even questionable whether there exist two clusters, because of the high overlap. Even though FedFCM identifies cluster centers correctly for $K = 2$ (small knowledge gap), the FedFuzzDB can be reduced by introducing more clusters because of the high spread, even for smaller values of σ. We attribute the low detection rate in the two-dimensional G2 sets to the FedFuzzDB index and its bias towards more clusters in data with high overlap. Again, the federated and nonfederated versions both suffer from this effect alike.

In summary, with our experiments, we show that the federated and nonfederated fuzzy Davies–Bouldin in interplay with the federated and nonfederated fuzzy c-means algorithms behave similarly in most tested situations. We tested the behavior on data with big cluster overlap (high value of σ in low-dimensional spaces) or sparsity. Sparsity was introduced through a big spread in high-dimensional spaces or an increased number of clients with a fixed dataset size. While generally reliable, the federated and nonfederated cluster algorithms struggle with extreme overlap or extreme sparsity. The federated and nonfederated indices favor a higher number of clusters in such situations. Overall, we see promising results and a good consistency between the federated and nonfederated settings.

3.2.3. Evaluation on Real-World Data

In the last two sections it was investigated how the framework behaves on artificial data with well-controlled properties. With this final experiment, we want to evaluate the framework on more real-world-inspired data and demonstrate how it could be used in practice. The data for this experiment were first introduced in [43] and can be accessed through the UCI machine learning repository (https://archive.ics.uci.edu/ml/datasets/Bank+Marketing, accessed on 27 March 2022) and are about customers of a bank. In particular, we are interested in the customers' recorded job, age, balance, and education. Based on the job information, we split the data to create a federated setting. Each client has data of only one job group. For example, one client has all records of students, another has all data of retired persons and another has all records of managers, etc. In total, there are 11 job categories such that there are 11 clients in the federated setting. Based on the remaining columns (age, balance, education), we want to form groups of similar customers using (federated) fuzzy c-means following the framework introduced in Section 2.2.3: determine the number of global clusters, derive a soft partitioning of the data, and validate the clustering. For comparison, we also compute the partitioning on the full, but centralized, dataset as well as the local-only datasets. Before applying the framework, we preprocess the data: we translate education into numerical values (primary: 1, secondary: 2, tertiary: 3), roughly estimating the time spent in school/university, rows with unknown values are dropped, and each column is standardized to have zero mean and standard deviation of 1. In total, we are left with 43,193 rows in the dataset. Each client holds data of only one job group: job group "management" has 9216 examples, "technician" 7355, "entrepreneur" 1411, "retired" 2145, "admin" 5000, "services" 4004, "blue-collar" 9278, "self-employed" 1450, "unemployed" 1274, "housemaid" 1195, and "student" 775 examples.

First, we need to determine the number of clusters by executing the first step of our framework. We set the minimum of clusters $K_{min} = 3$. Each client has at least 3 clusters in its local-only data. We draw that conclusion from calculating the (nonfederated) fuzzy Davies–Bouldin index and applying the elbow method (see Figure 5 for examples). The maximum number of clusters is set to $K_{max} = 9$, because it provides a buffer for the identification of forming clusters. One of the clients (entrepreneur) reports that it has six clusters and we choose $K_{max} = K_{min} + \max_l K^{(l)} = 9$ according to our rule of thumb (Section 2.2.3). For each $K \in [K_{min}, K_{max}]$, the partitioning using the federated fuzzy c-means algorithm with k-means averaging and the FedFuzzDB index is calculated. The results can be found in Figure 6. The elbow method suggests the number of global clusters to be four, five, or six. Additionally, the figure contains the results of the same analysis in the nonfederated setting. Similarly, the method suggests that there are four or six in the centralized data, showing good consistency. As common in practice, the index only gives a good indication on the number of clusters and the practitioner is left to make the final call. Notably, the nonfederated and federated index values are not the same. Generally, the FedFuzzDB index is slightly lower than the nonfederated index. That implies that the federated method returns better centers (as measured by the Davies–Bouldin index) than its nonfederated counterpart.

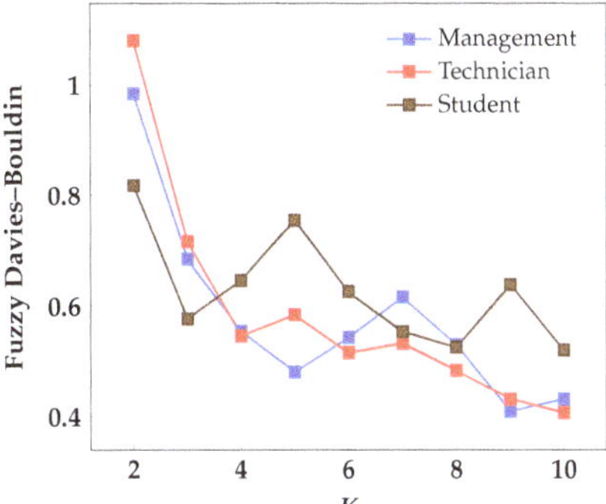

Figure 5. Local-only Davies–Bouldin index for different (but not all) clients. According to the index and the elbow method, each client has a different number clusters locally (management, $K = 5$; technician, $K = 4$; student, $K = 3$).

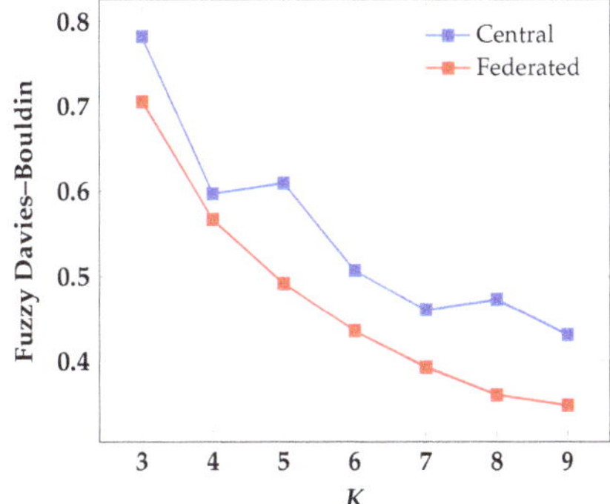

Figure 6. Federated fuzzy Davies–Bouldin index for different number of clusters.

Second, we want to understand why the federated index is lower than the nonfederated index. Therefore, we compare the federated and nonfederated cluster centers for $K = 4$. We note that three out of the four centers are almost identical. The three centers all have similar values in feature direction "balance" and are very different in the other dimensions "education" and "age". Intuitively, this makes sense because the vast majority of data have relatively low balance, and the other dimensions are key discriminators. The fourth center is the center of wealthy customers (very high account balance) in the federated and the nonfederated settings. However, in the federated setting, the center has a higher value for "balance" (3.1) compared to the central clustering (2.4). For a visualization of the centers, please refer to Figure 7. Hence, the center is further from the other centers, which is is the reason for a lower FedFuzzDB index. In the central clustering case, the center does not move as far in the balance direction, because the mass of all points has a value close to zero. Recall that in fuzzy clustering all points are considered for the calculation of the center. Hence, many points (even though with low weight for further points) still have a noticeable effect. The key difference is that the federated clustering algorithm computes global centers based on the local cluster centers, which changes the relative importance in this case. To illustrate this, seven of 40 local-only cluster centers have a balance of >2.4, which is 15.9% of all local-only cluster centers. In contrast, in the central dataset, only 3.9% of all points have a balance of >2.4. This leads to higher cluster center dissimilarity M_{ij}.

The lower FedFuzzDB index lets us conclude that the effect on the spread (and, hence, the assignments u_{ij}) is small.

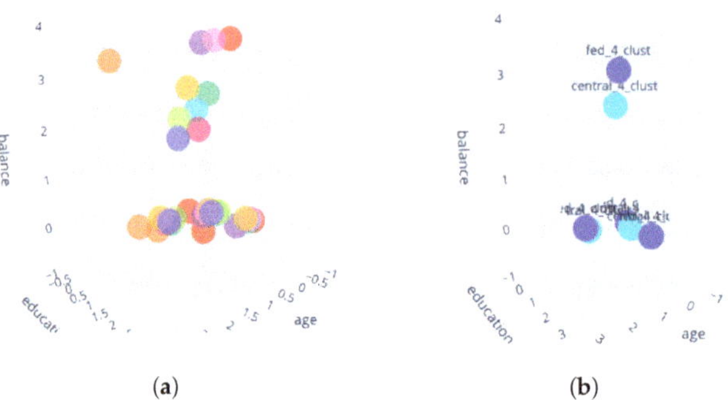

(a)   (b)

Figure 7. (**a**) Local-only clustering results for $K = 4$ of the clients in one plot. Each color corresponds to one client. (**b**) Federated and centralized cluster centers for $K = 4$. The darker points are the federated cluster centers and the brighter points are the central cluster centers.

In summary, we demonstrate how our cluster analysis framework can be applied to gain insights from real-world datasets. We see that there is a good consistency between the federated and the nonfederated cluster results, and the federated algorithm produces even better results in terms of (federated) fuzzy Davies–Bouldin index.

4. Discussion

In this work, we introduce a federated clustering framework (Section 2.2) that solves three challenges: determine the number of global clusters in a federated dataset, obtain a global clustering via a federated fuzzy c-means algorithm (Section 2.2.1), and validate the clustering through a federated cluster validation metric (Section 2.2.2). To the best of our knowledge, there exists no other similar federated cluster analysis framework. Instead, previous works mostly focus on the clustering method itself. The lack does not stem from the lack of necessity, as we show with our motivational example (Section 1.1): There exist situations where global clusters remain hidden behind local clients' optima.

The complete framework is evaluated through numerical experiments on different datasets (Section 3). We find that the framework identifies global cluster structures (correct number of clusters and data partitions) that are hidden in non-i.i.d. data (Section 3.2.1). We also find that the framework performs reliably if the data have certain properties, but fails if they do not (Section 3.2.2). In particular, it struggles with sparse data as well as with high cluster overlap. This is consistent with the equivalent nonfederated setting. In our last set of experiments, we outline how the framework can be applied in practice. It shows a good consistency with nonfederated clustering, and can even find better data partitions than in the centralized case (as measured by the Davies–Bouldin index).

Lastly, we see multiple interesting research directions for future works. One direction is to better understand the theoretical properties of the federated fuzzy c-means algorithm with k-means averaging. Moreover, the calculation of the federated fuzzy Davies–Bouldin index potentially creates a privacy risk. We suggest simple prevention mechanisms, but an in-depth analysis could lead to more sophisticated mechanisms. Furthermore, the cluster determination method still needs an initial range for the number of clusters, which can be hard to obtain. We provide a rule of thumb, but a better understanding of when and how federated clusters form could help to make this initial guess more accurate or even automate the choice. Moreover, FedFuzzDB can be extended to a federated crisp clustering or can be applied in combination with other clustering algorithms. Finally, the framework can be extended to include more steps in cluster analysis, such as federated preprocessing or feature selection.

Overall, we propose the first federated cluster validation metric, a new federated clustering approach based on existing works in the field, propose a comprehensive federated cluster analysis framework, and demonstrate how it can be applied. In comprehensive experiments, we observe promising results and identify shortcomings. Topics such as theoretical properties of the clustering algorithm and privacy evaluation of the framework have only been briefly discussed and can be addressed in more detail in future works.

Author Contributions: Conceptualization, methodology, validation, analysis: M.S. and A.W.; software: M.S.; writing—original draft preparation: M.S.; writing—review and editing: A.W.; visualization: M.S.; supervision: A.W. All authors have read and agreed to the published version of the manuscript.

Funding: This research received no external funding.

Institutional Review Board Statement: Not applicable.

Informed Consent Statement: Not applicable.

Data Availability Statement: The analyzed data can be retrieved from online repositories as referenced in the article. Online repositories are referenced in the paper.

Conflicts of Interest: The authors declare no conflict of interest.

References

1. EU. Regulation (EU) 2016/679 of the European Parliament and of the Council of 27 April 2016 on the protection of natural persons with regard to the processing of personal data (...) (General Data Protection Regulation). *Off. J. Eur. Union* **2016**, *119*, 1–88.
2. Kairouz, P.; McMahan, H.B. Advances and Open Problems in Federated Learning. *Found. Trends® Mach. Learn.* **2021**, *14*, 1–210. [CrossRef]
3. Hard, A.; Rao, K.; Mathews, R.; Ramaswamy, S.; Beaufays, F.; Augenstein, S.; Eichner, H.; Kiddon, C.; Ramage, D. Federated learning for mobile keyboard prediction. *arXiv* **2018**, arXiv:1811.03604.
4. Ye, D.; Yu, R.; Pan, M.; Han, Z. Federated Learning in Vehicular Edge Computing: A Selective Model Aggregation Approach. *IEEE Access* **2020**, *8*, 23920–23935. [CrossRef]
5. Deist, T.M.; Jochems, A.; van Soest, J.; Nalbantov, G.; Oberije, C.; Walsh, S.; Eble, M.; Bulens, P.; Coucke, P.; Dries, W.; et al. Infrastructure and distributed learning methodology for privacy-preserving multi-centric rapid learning health care: EuroCAT. *Clin. Transl. Radiat. Oncol.* **2017**, *4*, 24–31. [CrossRef]
6. Brisimi, T.S.; Chen, R.; Mela, T.; Olshevsky, A.; Paschalidis, I.C.; Shi, W. Federated learning of predictive models from federated Electronic Health Records. *Int. J. Med. Inform.* **2018**, *112*, 59–67. [CrossRef]
7. Grefen, P.; Ludwig, H.; Tata, S.; Dijkman, R.; Baracaldo, N.; Wilbik, A.; D'hondt, T. Complex collaborative physical process management: A position on the trinity of BPM, IoT and DA. In *IFIP Advances in Information and Communication Technology, Proceedings of the Working Conference on Virtual Enterprises, Cardiff, UK, 17–19 September 2018*; Springer: Cham, Switzerland, 2018; pp. 244–253.
8. Duan, M.; Liu, D.; Chen, X.; Tan, Y.; Ren, J.; Qiao, L.; Liang, L. Astraea: Self-Balancing Federated Learning for Improving Classification Accuracy of Mobile Deep Learning Applications. In Proceedings of the 2019 IEEE 37th International Conference on Computer Design (ICCD), Abu Dhabi, United Arab Emirates, 17–20 November 2019; pp. 246–254. [CrossRef]
9. Wang, X.; Han, Y.; Wang, C.; Zhao, Q.; Chen, X.; Chen, M. In-Edge AI: Intelligentizing Mobile Edge Computing, Caching and Communication by Federated Learning. *IEEE Netw.* **2019**, *33*, 156–165. [CrossRef]
10. Yin, X.; Zhu, Y.; Hu, J. A Comprehensive Survey of Privacy-Preserving Federated Learning: A Taxonomy, Review, and Future Directions. *ACM Comput. Surv.* **2021**, *54*, 1–36. [CrossRef]
11. Khan, L.U.; Saad, W.; Han, Z.; Hossain, E.; Hong, C.S. Federated Learning for Internet of Things: Recent Advances, Taxonomy, and Open Challenges. *IEEE Commun. Surv. Tutor.* **2021**, *23*, 1759–1799. [CrossRef]
12. McLachlan, G. Cluster analysis and related techniques in medical research. *Stat. Methods Med. Res.* **1992**, *1*, 27–48. [CrossRef]
13. Maione, C.; Nelson, D.R.; Barbosa, R.M. Research on social data by means of cluster analysis. *Appl. Comput. Inform.* **2019**, *15*, 153–162. [CrossRef]
14. Bolin, J.H.; Edwards, J.M.; Finch, W.H.; Cassady, J.C. Applications of cluster analysis to the creation of perfectionism profiles: A comparison of two clustering approaches. *Front. Psychol.* **2014**, *5*, 343. [CrossRef] [PubMed]
15. Ketchen, D.J.; Shook, C.L. The application of cluster analysis in strategic management research: An analysis and critique. *Strateg. Manag. J.* **1996**, *17*, 441–458. [CrossRef]
16. Punj, G.; Stewart, D.W. Cluster analysis in marketing research: Review and suggestions for application. *J. Mark. Res.* **1983**, *20*, 134–148. [CrossRef]
17. Hudson, S.; Ritchie, B. Understanding the domestic market using cluster analysis: A case study of the marketing efforts of Travel Alberta. *J. Vacat. Mark.* **2002**, *8*, 263–276. [CrossRef]

18. Milligan, G.W.; Cooper, M.C. Methodology review: Clustering methods. *Appl. Psychol. Meas.* **1987**, *11*, 329–354. [CrossRef]
19. Kumar, H.H.; Karthik, V.R.; Nair, M.K. Federated K-Means Clustering: A Novel Edge AI Based Approach for Privacy Preservation. In Proceedings of the 2020 IEEE International Conference on Cloud Computing in Emerging Markets (CCEM), Bengaluru, India, 6–7 November 2020; pp. 52–56. [CrossRef]
20. Pedrycz, W. Federated FCM: Clustering Under Privacy Requirements. *IEEE Trans. Fuzzy Syst.* **2022**, *30*, 3384–3388. [CrossRef]
21. Bárcena, J.L.C.; Marcelloni, F.; Renda, A.; Bechini, A.; Ducange, P. A Federated Fuzzy *c*-means Clustering Algorithm. In Proceedings of the International Workshop on Fuzzy Logic and Applications (WILF 2021), Vietri sul Mare, Italy, 20–22 December 2021.
22. Dennis, D.K.; Li, T.; Smith, V. Heterogeneity for the Win: One-Shot Federated Clustering. In Proceedings of the 38th International Conference on Machine Learning, PMLR 2021, Virtual, 18–24 July 2021; Meila, M., Zhang, T., Eds.; 2021; Volume 139, pp. 2611–2620.
23. Hastie, T.; Tibshirani, R.; Friedman, J. *The Elements of Statistical Learning—Data Mining, Inference and Prediction*; Springer: New York, NY, USA, 2017.
24. Arthur, D.; Vassilvitskii, S. K-Means++: The Advantages of Careful Seeding. In Proceedings of the Eighteenth Annual ACM-SIAM Symposium on Discrete Algorithms; Society for Industrial and Applied Mathematics, SODA '07, New Orleans, LA, USA, 7–9 January 2007; pp. 1027–1035.
25. Pedregosa, F.; Varoquaux, G.; Gramfort, A.; Michel, V.; Thirion, B.; Grisel, O.; Blondel, M.; Prettenhofer, P.; Weiss, R.; Dubourg, V.; et al. Scikit-learn: Machine Learning in Python. *J. Mach. Learn. Res.* **2011**, *12*, 2825–2830.
26. Bezdek, J.C.; Ehrlich, R.; Full, W. FCM: The fuzzy c-means clustering algorithm. *Comput. Geosci.* **1984**, *10*, 191–203. [CrossRef]
27. Kolen, J.; Hutcheson, T. Reducing the time complexity of the fuzzy c-means algorithm. *IEEE Trans. Fuzzy Syst.* **2002**, *10*, 263–267. [CrossRef]
28. Suganya, R.; Shanthi, R. Fuzzy C- Means Algorithm—A Review. *Int. J. Sci. Res. Publ.* **2012**, *2*, 1–3.
29. Steinbach, M.; Ertöz, L.; Kumar, V. The challenges of clustering high dimensional data. In *New Directions in Statistical Physics*; Springer: Berlin/Heidelberg, Germany, 2004; pp. 273–309.
30. Winkler, R.; Klawonn, F.; Kruse, R. Fuzzy C-Means in High Dimensional Spaces. *Int. J. Fuzzy Syst. Appl.* **2011**, *1*, 1–16. [CrossRef]
31. Davies, D.; Bouldin, D. A Cluster Separation Measure. *IEEE Trans. Pattern Anal. Mach. Intell.* **1979**, *PAMI-1*, 224–227. [CrossRef]
32. Vergani, A.A.; Binaghi, E. A Soft Davies–Bouldin Separation Measure. In Proceedings of the 2018 IEEE International Conference on Fuzzy Systems (FUZZ-IEEE), Rio de Janeiro, Brazil, 8–13 July 2018; pp. 1–8. [CrossRef]
33. Ghosh, A.; Chung, J.; Yin, D.; Ramchandran, K. An Efficient Framework for Clustered Federated Learning. In *Advances in Neural Information Processing Systems*; Larochelle, H., Ranzato, M., Hadsell, R., Balcan, M.F., Lin, H., Eds.; Curran Associates, Inc.: Red Hook, NY, USA, 2020; Volume 33, pp. 19586–19597.
34. Sattler, F.; Muller, K.R.; Samek, W. Clustered Federated Learning: Model-Agnostic Distributed Multitask Optimization Under Privacy Constraints. *IEEE Trans. Neural Netw. Learn. Syst.* **2021**, *32*, 3710–3722. [CrossRef]
35. Kim, Y.; Hakim, E.A.; Haraldson, J.; Eriksson, H.; da Silva, J.M.B.; Fischione, C. Dynamic Clustering in Federated Learning. In Proceedings of the ICC 2021—IEEE International Conference on Communications, Montreal, QC, Canada, 14–23 June 2021; pp. 1–6. [CrossRef]
36. Xie, M.; Long, G.; Shen, T.; Zhou, T.; Wang, X.; Jiang, J.; Zhang, C. Multi-center federated learning. *arXiv* **2021**, arXiv:2108.08647.
37. Stallmann, M.; Wilbik, A. Towards Federated Clustering: A Federated Fuzzy *c*-Means Algorithm (FFCM). In Proceedings of the International Workshop on Trustable, Verifiable and Auditable Federated Learning in Conjunction with AAAI 2022 (FL-AAAI-22), Vancouver, BC, Canada, 1 March 2022.
38. McMahan, B.; Moore, E.; Ramage, D.; Hampson, S.; y Arcas, B.A. Communication-efficient learning of deep networks from decentralized data. In Proceedings of the Artificial Intelligence and Statistics, PMLR, 2017; pp. 1273–1282.
39. Bholowalia, P.; Kumar, A. EBK-means: A clustering technique based on elbow method and k-means in WSN. *Int. J. Comput. Appl.* **2014**, *105*, 17–24.
40. Milligan, G.W.; Cooper, M.C. An examination of procedures for determining the number of clusters in a data set. *Psychometrika* **1985**, *50*, 159–179. [CrossRef]
41. Fränti, P.; Sieranoja, S. Clustering Basic Benchmark. 2018. Available online: http://cs.uef.fi/sipu/datasets/ (accessed on 27 March 2022).
42. Mariescu-Istodor, P.F.R.; Zhong, C. XNN graph. *LNCS* **2016**, *10029*, 207–217.
43. Moro, S.; Cortez, P.; Rita, P. A data-driven approach to predict the success of bank telemarketing. *Decis. Support Syst.* **2014**, *62*, 22–31. [CrossRef]

Article

A Message Passing Approach to Biomedical Relation Classification for Drug–Drug Interactions

Dimitrios Zaikis *,†, **Christina Karalka** † and **Ioannis Vlahavas** *

School of Informatics, Aristotle University of Thessaloniki, 541 24 Thessaloniki, Greece
* Correspondence: dimitriz@csd.auth.gr (D.Z.); vlahavas@csd.auth.gr (I.V.)
† These authors contributed equally to this work.

Featured Application: With this contribution, we aim to aid the drug development process as well as the identification of possible adverse drug events due to simultaneous drug use.

Abstract: The task of extracting drug entities and possible interactions between drug pairings is known as Drug–Drug Interaction (DDI) extraction. Computer-assisted DDI extraction with Machine Learning techniques can help streamline this expensive and time-consuming process during the drug development cycle. Over the years, a variety of both traditional and Neural Network-based techniques for the extraction of DDIs have been proposed. Despite the introduction of several successful strategies, obtaining high classification accuracy is still an area where further progress can be made. In this work, we present a novel Knowledge Graph (KG) based approach that utilizes a unique graph structure in combination with a Transformer-based Language Model and Graph Neural Networks to classify DDIs from biomedical literature. The KG is constructed to model the knowledge of the DDI Extraction 2013 benchmark dataset, without the inclusion of additional external information sources. Each drug pair is classified based on the context of the sentence it was found in, by utilizing transfer knowledge in the form of semantic representations from domain-adapted BioBERT weights that serve as the initial KG states. The proposed approach was evaluated on the DDI classification task of the same dataset and achieved a F1-score of 79.14% on the four positive classes, outperforming the current state-of-the-art approach.

Keywords: Drug–Drug Interactions; transformers; graph neural networks; language models; relation classification; domain-adaption

Citation: Zaikis, D.; Karalka, C.; Vlahavas, I. A Message Passing Approach to Biomedical Relation Classification for Drug–Drug Interactions. *Appl. Sci.* **2022**, *12*, 10987. https://doi.org/10.3390/app122110987

Academic Editors: Pavlos S. Efraimidis, Avi Arampatzis and George Drosatos

Received: 30 September 2022
Accepted: 27 October 2022
Published: 30 October 2022

Publisher's Note: MDPI stays neutral with regard to jurisdictional claims in published maps and institutional affiliations.

Copyright: © 2022 by the authors. Licensee MDPI, Basel, Switzerland. This article is an open access article distributed under the terms and conditions of the Creative Commons Attribution (CC BY) license (https://creativecommons.org/licenses/by/4.0/).

1. Introduction

Drug–Drug Interactions (DDI) refer to the pharmacological action between drugs that can occur during polypharmacy, and the co-administration of more than one drug can potentially lead to harmful adverse drug reactions that have a significant impact on public health. Most DDIs are discovered during the various drug development stages or during Phase IV clinical trials conducted on already publicly available drugs [1]. The dissemination of these findings are reported at an exponential rate, rendering the task of manually finding the most relevant information very difficult and time-consuming [2]. However, the heterogeneity of the available data regarding DDIs presents new challenges in their exploration, analysis and manageability. The identification and retrieval of documented drug interactions requires gathering and analyzing data from multiple data sources, especially in the early stages of drug development.

Moreover, as the practice of medicine and scientific research increasingly produces and depends on data, addressing these issues becomes a necessity. Therefore, the automatic extraction of DDIs from biomedical literature is important in order to accelerate this time-consuming and strenuous process. Vast amounts of relevant knowledge can be extracted from various types of information sources such as scientific literature, electronic health records, online databases and many more [3]. However, these sources contain textual information that is very diverse in terms of type, format, level of detail and differ in terms of expressiveness and semantics. Additionally, the possibility of conflicting or outdated information presents among the various sources adds to the overall complexity regarding the collection, storage and analysis of the data and consequently the extraction and exploitation of the hidden wealth of information.

DDI extraction from textual corpora is a traditional Relationship Extraction (RE) task in Machine Learning (ML) that aims to classify the interaction between drug entities [4] into specific predefined categories. Related DDI extraction studies vary based on the underlying task they aim to tackle and could be divided into pattern-based, traditional machine learning-based and deep learning-based [5]. The DDI classification task focuses on classifying the interactions between drug pairs by using gold entities with Relationship Classification (RC) techniques and are evaluated on the DDI Extraction 2013 corpus, which is considered as the benchmark dataset [6]. Similar to all underlying extraction tasks, the Deep Learning-based (DL) methods achieve the best performance and advance the state-of-the-art research in this field. Early DL-based approaches mainly utilized Convolutional Neural Networks (CNN) and Recurrent Neural Networks (RNN) as their base architectures to learn better task-specific representations with the use of contextualized information incorporated into their Neural Network-based architectures.

Liu et al. [7] introduced the first CNN-based approach for the DDI task, focusing on local sentence information without defining additional features generated by Natural Language Processing (NLP) toolkits. They applied a convolutional layer that takes the input from a look-up table constructed from word and position embeddings, leveraging the neural networks ability to automatically learn features. A max pooling layer then extracts the most important feature from each feature vector before finally classifying the interactions into one of the five classes using a softmax layer. The reported results show that the position embeddings improve the classification performance but face challenges due to the different position distribution on the test set.

Similarly, Quan et al. [8] integrated multiple word embeddings in their proposed MCCNN model, to tackle the vocabulary gap, the integration of semantic information and the manual feature selection in the DDI extraction task. The proposed approach implemented a multi-channel CNN model and fused multiple versions of word embeddings that were trained on biomedical domain corpora. However, the systems performance depends greatly on the CNN's window size, leading to errors in long sentences where the relevant drug mentions are either very close or very far from each other. In an attempt to capture long distance dependencies, Liu et al. [9] utilized syntactic features in the form of dependency parsing trees and word syntax-based embeddings in their proposed DCNN approach. Due to the small number of correctly parsed long sentences, a threshold was implemented where sentences with a length smaller than the threshold were classified by the DCNN, while the rest by a CNN. Similarly, Zhao et al. [10] utilized dependency features in combination with Part-of-Speech (PoS) and position embeddings with an auto-encoder to transfer sparse bag-of-words feature vectors to dense real value feature vectors. The proposed SCNN approach additionally implemented a rule-based negative instance filtering, leading to limited generalization ability.

To alleviate the limitations of CNN-based approaches, various DDI extraction studies employed RNN-based networks that capture long sequences using an internal memory mechanism, such as Long Short Term Memory (LSTM) and Gated Recurrent Units (GRU) networks. Accordingly, Wang et al. [11] presented a three channel bidirectional LSTM (BiLSTM) architecture to capture distance and dependency-based features with their DL-STM model. To account for the imbalanced class distribution of the DDI corpus, negative instance filtering and training set sampling were employed. However, the reported results indicate that the lengths of the instances continue to adversely affect the classification performance of the model. Yi et al. [12] introduced 2ATT-RNN, a GRU architecture that leverages multiple attention layers. A word-level attention layer extracts sentence representations in combination with a sentence-level attention layer that combines other sentences containing the same drug mentions. However, the inclusion of the negative class in the overall performance metric does not allow for a clear depiction of the effectiveness of the proposed method. Zhang et al. [13] divided the input sentence sequences into three parts according to the position of two drug entities, and applied a hierarchical BiLSTMs to integrate sentence sequences, shortest dependencies paths and attention mechanisms to classify DDIs. The experimental results show improvements over the previous approaches, but continue to underperform in cases where the two drug entities are mentioned over a long distance with each other.

Similarly, Zhou et al. [14] utilized the attention mechanism in a BiLSTM-based architecture. To improve the efficiency of the attention mechanism, the proposed PM-BLSTM system utilizes an additional position embedding to generate the attention weights. The model takes advantage of multi-task learning by predicting whether or not two drugs interact with each other, further distinguishing the types of interactions jointly. The reported results show that the position-wise attention improves the performance but continues to misclassify instances that contain multiple drug mentions. Salman et al. [15] proposed a straightforward LSTM and attention-based architecture and expanded the DDI extraction task to include sentiment-based severity prediction. The sentence-level polarity is extracted using an NLP toolkit and finally classified as either low, moderate or high level of severity for instances that contain at least on DDI. However, the Word2Vec [16] generated word embeddings are context-independent and do not account for the word positions in the sentences. Furthermore, Word2Vec learns word level embeddings, resulting in the same embedding for any learned word, independently of the surrounding context. Therefore, this type of embedding cannot generate representations for words encountered outside the initial vocabulary space, which is a major disadvantage in the DDI corpus.

Recently, Transformer-based Language Models (LM) such as ELMo [17], GPT-2 [18] and BERT [19] achieved state-of-the-art results in general domain NLP. By leveraging the capabilities of the transformers, transfer learning and the self-supervised training approach, biomedical and scientific-domain LMs, such as BioBERT [20] and SciBERT [21], were introduced in the DDI extraction task as well. Mondal [22] incorporated BioBERT as a pre-trained LM and chemical structure representations of drugs, in the form of SMILES, to extract DDIs from text. The proposed approach focused on the encoding and incorporation of the chemical structure information from external sources using a Variational AutoEncoder in an attempt to leverage both entities and sentence-level information. However, the low dimensionality of the final representations used for the Transformer initialization could potentially lead to information loss in longer sentences.

The integration and utilization of knowledge through semantic representations of data aims to mitigate the aforementioned problems [23]. Specifically, in recent years, biomedical knowledge base information represented as Knowledge Graphs (KG) tends to be preferred more and more often. KGs are powerful knowledge representation models which focus on the semantic meaning instead of only on the information structures, modeling the relationships between the graph entities [24]. As a result, KGs provide a homogenized view of data regardless of their origin, allowing for human-interpretable encoding of domain-specific information with the use of node and relation types.

Consequently, Graph Neural Networks (GNN) that take advantage of graph-based structures, in combination with Transformer-based LMs, have seen great success in various general-domain NLP tasks and have been introduced in the DDI extraction task as well. Xiong et al. [25] introduced GCNN-DDI, which utilized dependency graphs in a BiLSTM and GCN architecture to classify the interactions. Shi et al. [26], similar to GCNN-DDI, adopted a GNN and introduced a PageRank based multi-hop relevant words selection strategy for the dependency graph. These approaches rely on the construction of dependency trees (or syntax trees) from the sentences where nodes represent individual words and edges the syntactic dependency paths between words in the sentence's dependency tree. The feature vectors of the nodes are initialized by a pre-trained domain-specific LM, utilizing the POS tag of each word and a BiLSTM to update the initial word embeddings for contextual feature extraction. Both approaches utilize GNNs to improve the representations through the incorporation of dependency relations with the word embeddings. However, while GCNN-DDI uses the raw dependency graph, DREAM additionally enhances it with long-range potential words discovered by PageRank by extending some potential relevant multi-hop neighbors, which have high information transferability.

GNN-based approaches exclusively implement dependency graphs which are complex graph structures where the number of nodes equals the number of tokens in each sentence making the application of GNNs slow and computationally expensive. Additionally, since the benchmark corpus is considered relatively small and imbalanced, proposed approaches try to overcome this limitation by incorporating complicated feature engineering or extending the available information from external sources.

In contrast to the previously reported methods, in this paper, we present a novel KG schema for the DDI classification task that is leveraged by our GNN-based architecture that includes message and non-message passing units. We constructed the KG according to the principles of the Resource Description Framework (RDF) data model where each relation is annotated by an *subject-predicate-object* triplet. The proposed graph structure is built upon the DDI corpus to model the sentence and drug mention relations and is further semantically enhanced by domain-adapting a BERT-based LM pre-trained on large-scale domain-specific corpora that generates the initial state of the graph nodes. Finally, the interactions are classified by utilizing a sub-graph, taking the context of the sentence into consideration.

We evaluated our proposed approach for the classification of DDIs according to the SemEval 2013 shared task [4] on the DDI Extraction 2013 dataset. Experimental results indicate that our KG and GNN-based classification model achieves a state-of-the-art F1-score of 79.14% on the four positive classes, outperforming other methodologies. Additionally, we show that the KG information in combination with negative instance filtering can enhance the performance of our model. Table A1 shows a comparative analysis of the related studies presented in this work and our proposed approach.

The remainder of this paper is organized as follows: in Section 2, we elaborate on the dataset used and describe our proposed approach in detail. In Section 3, we present the experimental setup and results and elaborate on the effectiveness and limitations of our proposed model. Finally, in Section 4, we present our conclusions and directions for future research.

2. Materials and Methods

In this section, we introduce the dataset and the architecture of our proposed graph neural network-based classification model for drug–drug interaction extraction, where, given a sentence containing drug mentions, each drug pair is classified into one of the five possible interaction categories. It consists of three main parts, which are the DDI Knowledge Graph, the BERT-based language model and the GNN-based classification module as shown in Figure 1. Specifically, a knowledge graph based on our proposed DDI task related schema is created where a domain-adapted BERT-based language model is then applied to generate meaningful word representations. This knowledge is then integrated into a selected part of the graph, where a GNN is trained to classify the drug pair relationship.

2.1. Dataset

The DDI–Extraction 2013 corpus [6] is a collection of biomedical texts containing sentences from the DrugBank database and MedLine abstracts. The DrugBank database focuses on providing information on medicinal substances, while MedLine is a more general database of scientific publications from health-related sectors. The corpus has been manually annotated by two expert annotators and is considered the benchmark dataset for the text-based DDI extraction task which includes the recognition of drug named entities and the interaction classification of the drug pairs.

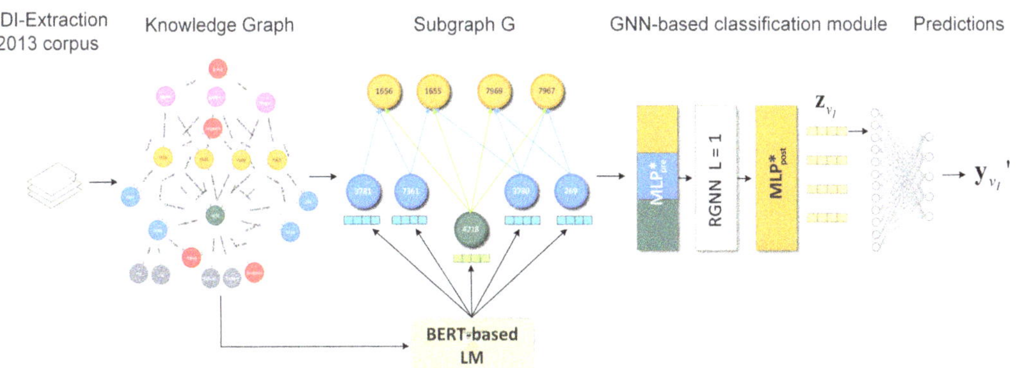

Figure 1. An overview of our proposed architecture for the classification of DDIs.

The dataset contains separate XML files where each one constitutes a document, which is further separated into its individual sentences. For each sentence, the drugs (entity) it mentions are listed and, for each possible pair of them, an interaction pair (relation pair) is defined. Therefore, n drugs define $n(n-1)/2$ pairs of interactions. Highlighted elements are characterized by unique identifiers (id) that reveal their position in the XML tree hierarchy. The corpus is split into a single training set and two separate test sets for both drug recognition and interaction classification tasks. Table 1 provides a summary of the corpus's main features and statistics for the predefined train and test datasets splits.

Drug entities and interactions are classified into categories (types) based on the context of the sentence they are mentioned in. In the majority of them, the named entities concern drugs intended for human use and are classified as "drug", "brand" and "group" types, while other substances are classified as "drug_n". Similarly, the interaction types between two drugs when administered simultaneously are categorized as follows:

- Effect: These are changes in the effect of a substance on the body, such as the appearance of symptoms and clinical findings.; The results of such effects are also referred to as pharmacodynamic properties of drugs.
- Mechanism: Refers to modifications in the absorption, distribution, metabolism and excretion of drugs, characteristics that constitute their pharmacokinetic properties. In other words, it concerns how the concentration of a substance in the body is affected by the presence of the other substances;
- Advice: Refers to descriptions containing recommendations or advice regarding the simultaneous use of two drugs;
- Int: Assigned in the case where the existence of an association between two drugs is mentioned, without any additional information indicating its type;
- Negative: It refers to the absence of interaction between two substances.

2.2. DDI Knowledge Graph

In order to model the DDI-specific Knowledge Graph, we used the RDF standard to create the proposed schema for representing the corpus knowledge. Figure 2 provides an overview of the DDI Knowledge Graph.

According to the RDF principles, the base of a knowledge graph is composed of a set of <Subject, Predicate, Object> statements, with the Subject and Object resources being respectively the initial and terminal nodes of a directed edge. The Predicate resource is considered the label of the edge in question, which is a property that associates the individual resources or serves to assign a value (Object) to some attribute of the Subject. This basic model is extended by defining classes into which the objects of the world belong.

Table 1. The DDI-Extraction 2013 corpus statistics.

		Training Set	Test Set	
			DNER	RC
	Documents	714	112	191
	Sentences	6976	665	1299
Drug Entities	Drug	9425	351	1864
	Group	3399	155	667
	Brand	1437	59	369
	Drug_n	504	120	140
DDIs	Mechanism	1322	-	303
	Effect	1700	-	363
	Advice	827	-	222
	Int	188	-	96
	Negative	23,771	-	4737

In the DDI corpus, all drug entities are annotated by providing the exact drug name and the location in the context of the specific sentence they are found in. Initially, each sentence of a document becomes an instance of the *Sentence* class, with its text preserved intact in the graph as an attribute of the specific node. Additionally, a sentence refers to a set of drugs that are modeled by the *Token* class, which is associated with the *Sentence* through the *contains_token property*. This property has a minimum cardinality of 2, filtering sentences that mention less than two drug entities (i.e., do not contain at least one drug pair). Furthermore, the set of unique drug entities in the collection is described by the *Drug_Class* class and its subclasses are the four types of drugs, which are mutually exclusive.

Finally, the concept of interaction (relationship between two drugs) is also modeled through classes. Typically an RDF statement represents a binary relationship between two resources in the form of a triplet. However, it may be necessary to add additional information regarding the statement resulting in an n–ary relationship. Each possible drug pair of a particular sentence is represented by an *Interaction* helper node. Thus, the information defining an interaction is composed centered on this node, through properties that associate it with other entities (e.g., 1 Sentence instance, 2 Drug_Class, 2 Token). Similar to the *Drug_Class*, its subclasses are based on the five predefined interaction types.

Collectively, a drug entity (Drug_Class), referred to as (found_as) Token in a particular sentence (Sentence), participates (interaction_found) in some pairs of interactions (Interaction). The sentence contains (contains_interaction) the interacting pair, while the Token reference participates (in_interaction) in it.

2.3. Modeling the DDI Relation Classification Task

In order to model the drug–drug interaction classification task and utilize DDI Knowledge Graph representations, a subgraph $G = (V, E, X, K, R)$ of the complete graph is selected. V and K denote the sets of nodes and the classes they belong to, with class instances denoted as k defining a subset of V_k. X constitutes an accompanying matrix of node characteristics (node feature vector), dimension $|V| \times d_{BERT}$, where d_{BERT} denotes the dimension of the BERT-based LM vector representation. Finally, R refers to the types of edges (properties) that associate the nodes, while E is the set of edges of the graph expressed in the form of a coordinate list (coordinate list format—COO).

Each interaction node is associated with a pair of specific drug mentions occurring in a sentence. Figure 3 shows the schema of the subgraph resulting from $K = \{Interaction(I), Token(T), Sentence(S)\}$ and $R = \{in\_interaction, contains\_interaction\}$. Therefore, the target is to classify the Interaction nodes V_I, or otherwise to determine the object of each triplet $< v_I, rdf{:}type, c_{v_I} >$, where c_{v_I} is a subclass of Interaction, from the three elements (i.e., the two tokens and the sentence) that determine its type.

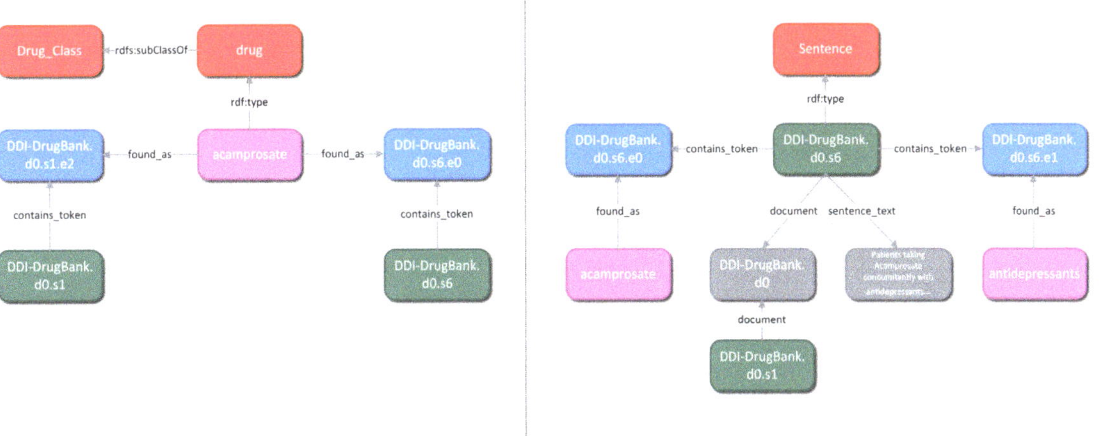

Figure 2. The DDI Knowledge Graph overview based on an example sentence. (**Left**) drug sub-graph; (**Right**) sentence sub-graph.

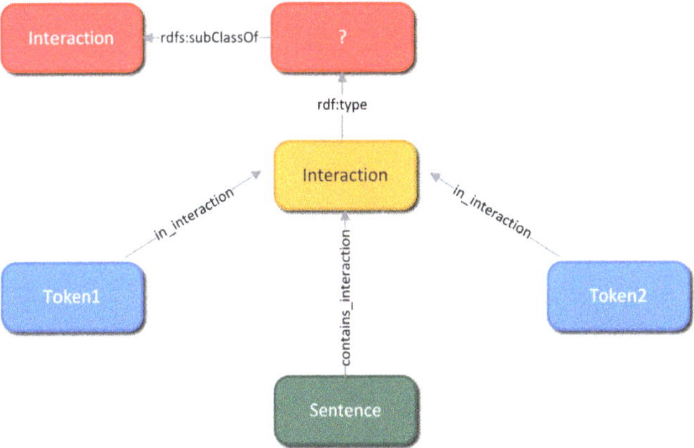

Figure 3. The schema of subgraph G which models the classification of an Interaction node.

2.4. Negative Instance Filtering

The extraction of DDIs from biomedical text is a multi-class classification problem, with *Advice*, *Effect*, *Int* and *Mechanism* being the positive classes and *Negative* being the negative class. The dataset statistics in Table 1 show the highly imbalanced nature of the corpus, in terms of both the positive and negative class distribution and within the four positive classes as well. In particular, the instances of the negative class exceed the positive classes with a ratio of 1:5.9 in the training set. It can be observed that only a part of the pairs labeled as negative explicitly express the knowledge that there is no interaction between the substances (drugs). Conversely, in the vast majority, the negative instances follow the same pattern where a number of drug–pair interactions were labeled in the same sentence, without clarifying the relationship between them. Consequently, the following set of rules was defined to detect them, as such cases can be dismissed.

- Consistent with the assumption that a substance does not interact with itself, pairs involving the same or synonymous drugs are rejected;
- There are sentences in the form of "$drug_1$ [:-]$\ldots drug_1 \ldots drug_2 \ldots$" which provide conflicting knowledge about the type of interaction between two drug entities. Therefore, any pair involving the first occurrence of $drug_1$ will be removed;
- Regular expressions identify patterns, such as quoting a list of drugs or referring to a sub-case, broader category or abbreviation. Additionally, consecutive or overlapping matches are merged to combine regular expressions. Finally, any pairs found within a match are discarded.

The above rule set leads to the rejection of 44.5% and 0.94% of the negative and positive examples in the training set, respectively, with the ratio changing to 1:3.3. Finally, as the corresponding Interaction nodes are not taken into account, they are excluded in the results during the evaluation of our proposed system.

2.5. Language Model-Based Transfer Learning

BERT [19] is an extensively pre-trained, Transformer-based language model, capable of state-of-the-art performance in NLP tasks. The increased ability to understand the conceptual framework that characterizes it is due to the self-attention mechanism, through which the token-level importance is assigned based on the associations between the individual words (tokens) of the sentence. Additionally, due to the training as a masked language model, it is able to perceive the information of the text in a bidirectional manner, allowing the LM to produce vector representations that reflect the syntactic, semantic and grammatical relationships between words.

BERT's architecture is composed of a number of N_{enc} consecutive encoder units with d_{BERT} hidden vector dimension, where $N_{enc} = 12$, $d_{BERT} = 768$ for the base version and $N_{enc} = 24$, $d_{BERT} = 1024$ for the large version. Furthermore, multiple BERT-based variations pre-trained on specific domains have been developed which achieve better performance in domain-specific tasks compared to BERT. As an example, BioBERT [20] and SciBERT [21] are two popular variations of BERT, pre-trained on large text corpora from the biomedical and scientific field, respectively.

2.5.1. Embedding Generation

In the DDI classification task, knowledge about the interaction type of a drug pair is expressed through text. The sentence text t is an associated element of the Sentence node, or otherwise contained in $< v_S, sentence\_text, t >$ triplets. By applying a BERT-based LM to t, it becomes possible to reduce the text to a suitable vector representation, in addition to sharing information among the individual nodes of the subgraph G.

The preparation of t involves the addition of the special tokens [CLS] and [SEP], which mark the beginning and end of the sentence, respectively. Then, each drug in the sentence is replaced by $drugi$, where i is a number. Finally, WordPiece tokenization is applied, through which words outside the BERT vocabulary are broken into individual pieces (subwords) that belong to it.

Furthermore, the LM is used to initially generate word embeddings x_{v_S} (sentence embeddings) for the entire sentence and a set of $x_{v_{T_j}}$ (token embeddings) where each one is a representation of a $Token_j$ contained within the sentence. These vectors are assigned to the respective nodes v_S and v_{T_j}, constituting their feature vectors.

2.5.2. Sentence and Token Nodes Feature Generation

Each sentence contains words that reveal or indicate the type of interaction between two drugs. For example, expressions such as "should (not) be administered", "caution should be used" and "is (not) recommended", are associated with suggestions (advice) when taking more than one drug simultaneously. Therefore, although the expressions show some variety, they are characterized by a high semantic similarity and are expected to correspond to nearby points in the embeddings vector space.

When generating the sentence embeddings x_{v_S}, it is important for the LM to focus on the above type of information and therefore any reference to any drug is replaced by "drug0" (i.e., $i = 0$). Because of the repeated occurrences, the string loses its meaning within the sentence, with the BERT-based LM giving an appropriate weight. Finally, the [CLS] token embedding is chosen for x_{v_S}, which is a centralized representation of the sentence and is often used in classification tasks. However, since a sentence most likely contains more than one drug pair, the interaction classification can not be performed solely on the basis of x_{v_S}, consequently requiring the feature generation for each drug mention in the sentence.

Furthermore, the interaction type does not depend on the specific drug names in the potential drug pair, but on their position within the sentence. In line with previous studies [7], their replacement is expected to aid in noise reduction when generating the *Token* embeddings, increasing the classification performance. To this end, each unique drug entity in the sentence is assigned a sequence number i, according to the order of appearance of its first mention within the sentence. Thus, $Token_j$ references that participate in $< Drug\_Class, found\_as, Token\_j >$ triplets with a common subject are replaced by $drugi$ with a common i. The final feature vector of each node v_{T_j} is obtained by pooling the embeddings of the subwords that compose it. That is, $x_{v_{T_j}} = pooling\{B_{drug\#\#}, B_{\#\#x}\}$, where the pooling method is the average, and B_w is the BERT output for the input w.

2.6. GNN-Based Classification Module

KGs are a complex and dynamic data structure where the notion of fixed ordering does not apply. However, by implementing a message passing framework, GNNs are able to better utilize the KG's underlying graph structure as well as any initially available data for its nodes, compared to other approaches [27]. Specifically, each layer first applies a transformation function on the node feature vectors. The generated messages from each node's neighbors, as well as its own message, are then aggregated to produce a new embedding that encodes additional semantic information provided by the defined relation types. These embeddings can finally be used to perform predictions for the nodes.

Given a selected subgraph G, the classification takes place on one of three types of nodes, namely the set V_I. As a sentence defines a maximum number of pairs according to the contained drug mentions, applying a BERT-based LM for their embedding generation may not be ideal. Instead, word embeddings were generated at the token level, as well as aggregated for the entire sentence that contains them. However, utilizing a GNN allows for the feature generation for each Interaction node.

The vector representation of these nodes is initialized with a null (zero) vector $h_{v_I}^0 = x_{v_I} = 0$, indicating no initial characterization for the Interaction nodes. However, GNNs pay special attention to the current h_v^{l-1} representation of a node when generating h_v^l from layer l. Therefore, when applying a GNN layer, its new embedding results exclusively from the topology of the graph around it, i.e., through transformation and aggregation of x_{v_S} and x_{v_T} from the one neighboring Sentence and the two Tokens nodes, respectively.

As the subgraph G is a heterogeneous graph, the use of a Relational GNN (RGNN) is required. The management of the heterogeneity is based on the logic of parameter distribution according to the type r of the edge that connects to the Interaction node (i.e., relation-specific transformations) [28]. Therefore, the embedding results from the following equation:

$$h_{v_I}^1 = aggr_{r \in R}\{GNN^{l=1;r}(h_{v_I}^0, \{h_u^0, u \in N^r(v_I)\})\}, \tag{1}$$

where $aggr$ is an aggregation function, R the set of edge types and $N^r(v_I)$ the set of neighbors of v_I according to the triplets $< u, r, v_I >$.

The modeling capability of GNNs is determined by the expressive power of the message aggregation functions, making the choice of the appropriate GNN architecture critical to the performance of this shallow network. Therefore, with the utilization of a Graph Isomorphism Network (GIN), the architecture's deep layers are encapsulated within the single layer GNN. Using the sum as the aggregator operation, the above relation is formulated as:

$$h_{v_I}^1 = \sum_{r \in R} MLP^{l=1;r}\left(h_{v_I}^0 + \sum_{u \in N^r(v_I)} h_u^0\right) \tag{2}$$

The architecture's main RGNN element is surrounded by MLP modules, which are defined according to the node classes. Specifically, through the integration of the MLP preprocessing layers, the initial representation of each node is obtained by $h_{v_k}^0 = MLP_{pre}^k(x_{v_k})$, $k \in K$. Similarly, the final vector representation of the Interaction nodes is defined by $z_{v_I} = MLP_{post}^I(h_{v_I}^1)$.

Conclusively, the system is now able to classify each node v_I into one of the $|C| = 5$ interaction classes. First, the probability distribution of the classes is calculated as $y'_{v_I} = softmax(z_{v_I} W^T + b)$, where W and b are the trainable weights and biases parameters, respectively. The dimensions of matrix W are $|C| \times d_{GNN}$, with the hyperparameter $d_{GNN} = \dim(z_{v_I})$ constituting the dimension of the vector space defined by the network. The final classification during the inference is obtained by $c_{v_I} = argmax_{c \in C}\{y'_{v_I}\}$.

3. Results and Discussion

3.1. Experimental Setup

Training is performed in a supervised manner on the labels of the Interaction nodes, resulting from the provided training set by merging the MedLine and DrugBank subsets. This amounts to a total of 17,176 training examples, derived from 3395 sentences and 608 documents after the construction of the knowledge graph. Similarly, the evaluation is performed on the RE task test set, where a separate graph with 3057 Interaction nodes and 604 sentences from 159 documents is created. The domain-adaption of the LMs is trained either on the training set sentences (Sentence Level Domain Adaption—SLDA) or training set paragraphs (Document Level Domain Adaption—DLDA) only, in a self-supervised manner using the Masked Language Modeling task.

Our proposed approach requires the definition of the two main elements of the architecture, the underlying BERT-based LM that will generate the word embeddings and the GNN-based classification module. First, different pre-trained BERT variants were compared, such as the base version of the general domain BERT, the scientific domain SciBERT and the biomedical domain BioBERT. Furthermore, BioBERT, the pre-training of which is in alignment with the DDI domain, was tested on both base and large versions. Additionally, since recent studies show that domain-adapting a LM by pre-training it on the downstream task can potentially offer large gains in the task performance [29], we aligned both SciBERT and BioBERT base to the DDI task corpus and compared their performance.

Having the features of the nodes generated by the BERT-based LMs, it is then necessary to define and train the classification unit. GIN [30] was chosen as the GNN framework, with node embeddings dimensions $d_{GNN} = 256$ for $d_{BERT} = 768$ and $d_{GNN} = 512$ for $d_{BERT} = 1024$, with the internal MLP unit consisting of $l_{GIN} = 3$ consecutive layers. Its performance is also compared to the mean GraphSAGE framework [31]. Additionally, the contribution of a single-level MLP_{pre} of size d_{GNN} and two-level MLP_{post} of sizes $d_{GNN}/2$ and $d_{GNN}/4$ with a drop-rate of 0.4 and ReLU activations are evaluated.

Adam was chosen as the optimizer with a learning rate and weight decay equal to 5×10^{-5} and 5×10^{-4}, respectively. Furthermore, the mini-batch training approach is followed, where the Interaction nodes of the training set are divided into 53 batches of size 324, while the number of epochs is equal to 170. Finally, the cross entropy loss function is used in the context of the multi-class classification problem.

The experiments were conducted on a computer with a single RTX 3090 24 GB graphics card an a 24-core Intel CPU and the LM domain-adaption on a computer with two RTX A6000 48 GB graphics cards and were implemented using the Pytorch library and the Python programming language.

3.2. Evaluation Metrics

Similar to the related studies, the performance of the system was evaluated based on the Precision (P), Recall (R) and micro F1-score ($F1_{mirco}$) metrics on the test set with the four positive classification targets C^+. The ratio of correctly classified instances c to all instances that were classified as c or actually belong to c constitutes the Precision and Recall of class c, respectively. The micro F1-score, which is the harmonic mean of P and R, provides an overall picture of the system without focusing on the individual performance of each class. The metrics are defined by the following formulas, where the number of corresponding cases is denoted by the combination of T (true) or F (false) and P (positive) or N (negative):

$$P_{micro} = \frac{\sum_{c \in C} + TP}{\sum_{c \in C} + (TP_c + FP_c)} \qquad R_{micro} = \frac{\sum_{c \in C} + TP}{\sum_{c \in C} + (TP_c + FN_c)} \qquad F1_{micro} = \frac{2 \times P_{micro} \times R_{micro}}{P_{micro} + R_{micro}}$$

3.3. Overall Comparison

In Table 2, we show the results of our method in comparison to baseline and state-of-the-art DDI classification approaches which reported their overall performance metric based on the four positive classes. These approaches are trained on the same training set and evaluated on the same test set provided by the DDI corpus and follow the same experimental setting without the inclusion of external information. These approaches can effectively be divided into two categories: traditional methods that utilize extensive feature engineering and state-of-the-art neural network-based approaches that aim to learn feature representations automatically based on different architectures. We compare our proposed approach to the traditional method "FBK-irst" presented in [32], which used linear features, path-enclosed tree kernels and linguistic features. For the NN-based approaches, we compared our approach to the following methods:

- "SCNN" [10]—CNN-based architecture with manually designed features;
- "MCCNN" [8]—CNN with multichannel word embeddings;
- "ASDP-LSTM" [13]—Hierarchical RNNs with shortest dependency paths;
- "PM-BLSTM" [14]—Bidirectional LSTM with position-aware attention;
- "GCNN-DDI" [25]—Bidirectional LSTM with GNN that utilized entire dependency graphs;
- "DREAM" [26]—Bidirectional LSTM with GNN that utilized PageRank enhanced dependency graphs.

The experimental results show that our Knowledge Graph-based approach that utilized BioBERT LM achieves the best overall performance for the classification of DDIs. The proposed KG schema with the domain-adapted pre-trained weights and the non-message passing MLPs are the main contributing factors, which will be analyzed in the following subsections. In the four positive classes, our approach achieves the best results in the *Advice*, *Effect* and *Mechanism* classes and a similar score in the *Int* class. In the following sections, we additionally analyze and discuss the various components of our method and their contribution to the overall performance.

Table 2. Overall performance comparison of our proposed method. All values are $F1$ scores (%) and '-' denotes the value was not provided in the published paper. $F1_{micro}$ denotes the overall score on the four positive classes. The highest values are shown in bold.

Method	System	Advice	Effect	Int	Mechanism	$F1_{Micro}$
SVM	FBK-irst	69.20	62.80	54.70	67.90	65.10
CNN	SCNN	-	-	-	-	68.60
CNN	MCCNN	78.20	68.20	51.00	72.20	70.21
LSTM	ASDP-LSTM	80.30	71.80	54.30	74.00	72.90
LSTM	PM-BLSTM	81.60	71.28	48.57	74.42	72.99
GNN	GCNN-DDI	83.50	75.80	51.40	79.40	77.00
GNN	DREAM	84.80	76.10	**55.10**	81.60	78.30
Our method	BioBERT-GIN	**86.45**	**78.46**	54.80	**82.27**	**79.14**

3.4. The Importance of the Pre-Trained Language Model Domain

The initially generated graph node features are inextricably linked to the performance of the GNN [33]. Accordingly, we evaluated the effects of the various BERT-based LMs for the task of relationship classification from biomedical text, by comparing the models that are trained with their respective BERT variant word embeddings, as shown in Table 3.

Based on the overall performance metrics, we define BERT (M9.1), SciBERT (M6.1) and BioBERT (M2.1) as the three baseline approaches with BioBERT (M2.1) achieving the best baseline results. This further validates the fact that domain-specific LMs tend to outperform general-domain LMs on the domain-specific task. However, the general-domain BERT achieves significantly better performance in the underrepresented class Int. As a reminder, sentences that contain the interactions of type Int indicate that there is a relationship between two drugs but no additional information about the relation type. In this context, the performance increase of general-domain BERT can be attributed to the use of non-scientific language when describing these types of interactions.

The best performing model is M3.3, which makes use of the SLDA BioBERT base and the GIN framework surrounded by pre- and post-processing MLPs. Furthermore, it is the only one that achieves R_{micro} and $F1_{micro}$ scores greater than 75 in addition to the maximum value of P_{micro} among all the models. Moreover, it achieves the best F1 scores in most of the classes (4 out of 5), unlike other models that usually excel in just one class (M9.1-3).

At the same time, M8.2 that utilized DLDA SciBERT achieves a comparable score in R_{micro} and $F1_{micro}$ to the best performing models and surpasses 70% of the models in P_{micro} but significantly underperforms on the Int class. We observe an improvement over the SciBERT baseline approach (M6.1-3) proving that domain-adapting SciBERT to the DDI domain leads to a performance gain in the relationship classification task. Similarly, the performance improved significantly when domain-adapting BioBERT (M3.1-3) to the same task with SLDA, indicating that the biomedical-domain pre-trained LM model may benefit from adapting to other tasks in the same domain. Conversely, adapting same LM with DLDA (M4.1-3), a significant performance degradation can be observed, suggesting that the LM benefits mostly from the context of individual sentences and not larger paragraphs.

The overall results show the effectiveness of the biomedical-domain pre-trained BioBERT base LM, especially compared to the general-domain BERT. Furthermore, aligning the BioBERT to the DDI corpus did yield significant improvement and led to performance increase. Similarly, domain-adapting SciBERT with DLDA produced improved task performance. Noticeably, BioBERT large (M5.1-3) performs worse than its base counterparts, especially in the Int class, which warrants further investigation as no interpretable patterns could be found. However, recent findings [34] suggest that fine-tuning noise increases with model size and that instance-level accuracy has momentum leading to larger models having higher variance due to the fine-tuning seed.

3.5. Effectiveness of the GIN Message Aggregation Function Layers

A basic hyperparameter of the GIN module is the number of layers of the internal MLP network (l_{GIN}). Figure 4 shows the system's behavior on the test set as a function of the MLP network depth (number of layers). For $l_{GIN} = 1$, GIN shows comparable performance to GraphSAGE (M1.1—Table 3), which is significantly lower than the best performing model, validating the limiting factor of shallow aggregation functions. However, as l_{GIN} increases beyond four layers, the $F1_{micro}$ score displays a sharp decrease as it detects a higher percentage of existing interactions, combined with an increase in false positives, evidenced by the significant difference in the recall–precision curve slopes.

Therefore, the intermediate values $l_{GIN} = 2$ and 3 are compared. The change in $F1_{micro}$ is negligible, while the recall improvement and precision drop is approximately 4% when increasing the layer depth by one. However, at $l_{GIN} = 3$, a better compromise is made between the two metrics, with a difference of only 1.7%, compared to the corresponding 9.6% for $l_{GIN} = 2$. Moreover, considering the risk of not being able to detect an existing interaction, the behavior of the system with the best recall is considered more appropriate for the current task.

Table 3. Performance comparison of the pre-trained LM. SAGE denotes the models that utilize the GraphSAGE framework. SLDA and DLDA denote the Sentence and Document Level Domain Adaption, respectively, and x denotes the inclusion of the corresponding pre/post-processing MLP unit. The highest values are shown in bold.

	Model	MLP		Metrics per Classification Target				Overall Metrics		
M	**BERT-Based LM**	**Pre**	**Post**	**Advice**	**Effect**	**Int**	**Mech.**	P_{micro}	R_{micro}	$F1_{micro}$
1.1				70.51	67.04	48.21	58.39	71.59	57.37	63.70
1.2	SAGE BioBERT base		x	74.78	72.92	51.61	70.72	72.42	69.47	70.92
1.3		x	x	65.59	70.02	50.00	68.69	75.00	60.39	66.90
2.1				75.00	74.96	52.71	71.56	72.91	71.18	72.04
2.2	BioBERT base		x	76.84	76.09	52.71	71.49	73.16	72.58	72.87
2.3		x	x	82.45	73.46	52.80	74.27	73.42	74.87	74.14
3.1				79.14	76.19	52.71	71.46	72.81	71.28	72.04
3.2	BioBERT SLDA		x	84.45	75.46	53.80	76.27	75.81	74.13	74.96
3.3		x	x	**86.45**	**78.46**	54.80	**82.27**	**84.33**	**75.55**	**79.14**
4.1				72.30	73.14	40.88	67.94	66.67	70.53	68.54
4.2	BioBERT DLDA		x	72.53	73.36	40.88	70.02	66.91	71.84	69.29
4.3		x	x	71.94	70.43	39.37	69.22	63.78	72.76	67.98
5.1				69.23	67.83	15.50	70.66	62.97	66.45	64.66
5.2	BioBERT large		x	70.86	69.54	20.97	71.52	65.48	67.63	66.54
5.3		x	x	66.67	67.19	12.17	71.85	62.58	66.45	64.45
6.1				72.48	71.68	51.24	73.00	70.51	70.79	70.65
6.2	SciBERT		x	73.51	72.82	52.31	73.75	70.03	73.16	71.56
6.3		x	x	73.63	69.86	50.38	70.25	66.67	72.11	69.28
7.1				80.12	69.43	47.06	67.11	67.16	71.32	69.18
7.2	SciBERT SLDA		x	77.35	69.18	43.94	68.60	68.45	69.08	68.76
7.3		x	x	79.06	67.15	40.35	67.10	70.55	65.26	67.81
8.1				78.86	73.24	51.91	70.46	72.52	71.18	71.85
8.2	SciBERT DLDA		x	80.89	74.15	47.93	73.28	72.74	74.08	73.40
8.3		x	x	77.97	71.26	46.55	73.94	72.70	70.79	71.73
9.1				73.24	68.23	58.06	67.24	69.19	67.37	68.27
9.2	BERT base		x	72.24	67.62	**60.00**	70.74	68.25	69.87	69.05
9.3		x	x	68.15	68.42	55.74	68.91	64.25	71.18	67.54

3.6. Effectiveness of Non–Message Passing Units

In addition to adopting the GIN framework to increase the expressiveness of the shallow GNN network, it has been further proposed to incorporate non-message passing units MLP_{pre} and MLP_{post} into the architecture [35]. The contribution of increased model complexity to the performance is confirmed by the significant improvement in the otherwise underperforming GraphSAGE (M1.1). Although not as pronounced, GIN models also appear to benefit.

The performance metrics in Table 3 show the advantage of including the MLP_{post} unit in the M1-9.2 models over the basic M1-9.1 models. Adding the unit yields an average increase of 0.9% in each metric, with eight models achieving better P_{micro}, R_{micro} scores, and seven models achieving better $F1_{micro}$ scores. At the same time, the rest of the models where either precision or recall is affected, the opposite metric ($P_{micro} \iff R_{micro}$) shows an improvement in the order of 2.6% on average, which is always superior to the corresponding drop.

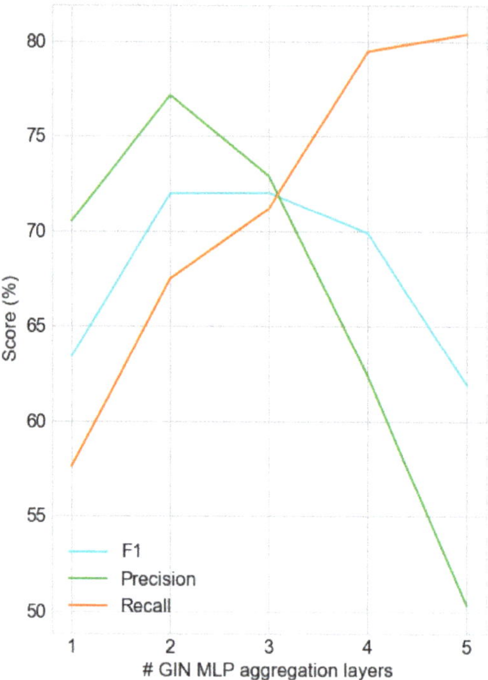

Figure 4. GIN MLP performance compared to the number of aggregation layers using the baseline BioBERT and no additional MLP units.

Conclusively, the three best performing models (M2.3, M3.3 and M8.2) include this unit. Each presents an improvement over its base version (M2.1, M3.1 and M8.1, respectively), while improving on the individual class level as well. Particularly, in at least three of the five classes, there is a clear improvement or at least some stability (decrease ≤ 0.08). This behavior is partially confirmed by the other (underperforming) GIN models, although not universally.

In contrast, the effect of the combination of the two units appears to be negative in the vast majority of models. Through MLP_{pre}, the word embeddings x_v produced by the BERT-based LMs (d_{BERT}) are projected into a smaller dimensional space d_{GNN} before the execution of message passing by GNN. This results in reduced performance in at least two instances. However, obvious exceptions are the models which make use of the BioBERT base architecture. Especially in combination with GIN, the models M2-3.3 outperform their M2-3.1-2 counterparts in every metric, with the exception of the $F1_{Effect}$ metric.

3.7. Effectiveness of Preprocessing

Focusing on the best performing model and its base (M3.3 and M8.2 respectively), the contribution of the preprocessing steps to the data was studied. Particularly, Figure 5 shows the effects of Drug Name Replacement (DNR, Figure 5a,b) and Negative Instance Filtering (NIF, Figure 5c,d) on the $F1$ metric of each positive class and the overall $F1_{micro}$ score.

First, reducing imbalance greatly benefits the *Advice* and *Int* classes in both models. Especially in M3.3 (Table 3), applying NIF improved the Recall by 7.6% and 10.9%, respectively, and the Precision by 3.4% and 6.2% respectively. Simultaneously, a relative robustness is demonstrated in the *Effect* class when restoring the rejected pairs; however, their inclusion appears to favor the *Mechanism* class. Conclusively, the overall $F1_{micro}$ is improved in each case through the rejection of trivial negative instances; however, M3.3 maintains a better balance between Recall and Precision than M8.2, where although no class is perceptibly affected, their removal leads to an increase in false positives.

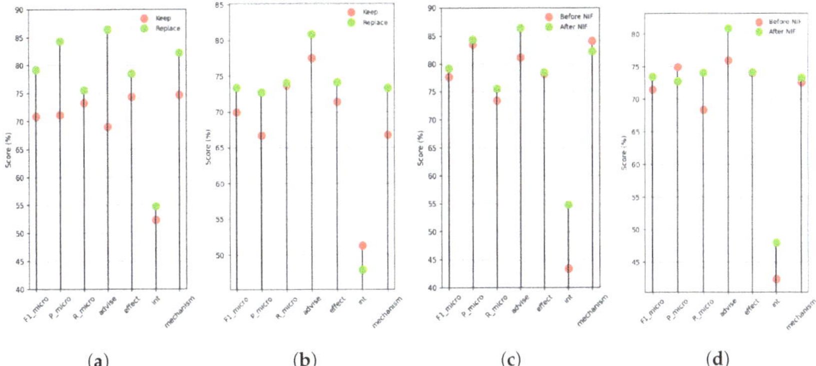

(a)   (b)   (c)   (d)

Figure 5. Performance comparison based on Drug Name Replacement (DNR) and Negative Instance Filtering (NIF), (**a**) M3.3 with DNR, (**b**) M8.2 with DNR, (**c**) M3.3 with NIF, (**d**) M8.2 with NIF.

In contrast, the use of *drugi* over the original drug names provides a clear performance improvement when generating word embeddings with BERT-based LM. This fact confirms the hypothesis that the type of interaction is determined by their syntactic role within the sentence and its content and not by specific substances' names. This method of replacing drug names, also known as entity blinding or entity masking, supports the generalization of the model. Therefore, through these preprocessing steps, noise is reduced in the feature vectors of the Token nodes.

3.8. Error Analysis

To analyze the advantages and limitations of our proposed approach, we compare and analyze the classification results of the best performing model (M3.3) on a few indicative cases (Table 4). Figure 6 shows the confusion matrix, with a total of 330 errors that were made, representing 11% of the 3057 test cases. In addition, 273 (83%) originate from the DrugBank instances and the remaining 57 (17%) from MedLine abstracts, corresponding to 9.8% and 20% of the total interactions of their respective collection.

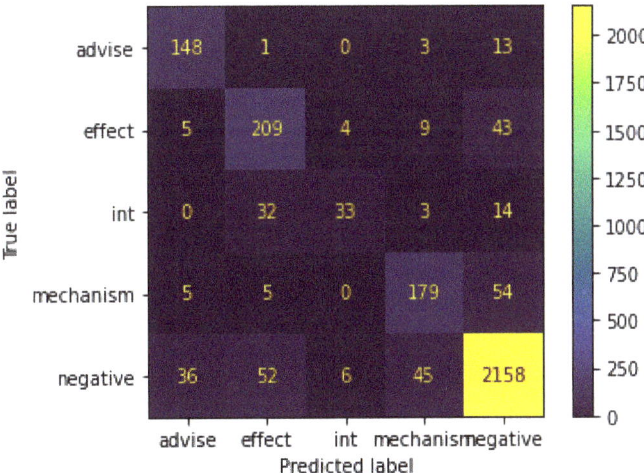

Figure 6. Confusion matrix of our proposed model.

First, misclassifying an existing interaction is the least common type of error, accounting for 20% of the total errors. Furthermore, 48% of these correspond to the case where instances of Int are classified as Effect, which makes this the main source of confusion in this category. The contributing drug pairs were found in only five sentences, with one of them giving 28 of the total 32. This is S1, where although the system can recognize the existence of interactions in the long list of drugs, it fails to predict the types. This is probably due to the presence of "enhance", a word regularly associated with drug pairs with the interaction type Effect. Furthermore, 42% of the errors involve false positive pairs that can occur due to the following cases:

- The possibility of an annotation error is not excluded. Indicatively, in S2, every pair containing e_0 is of type Effect, as predicted by the model. However, any such relationship other than (e_0, e_1) is marked as negative in the data set.
- A substance appears multiple times in the sentence, such as *mebendazole* in S3. The pairs e_0, e_1 and e_0, e_2 are the Mechanism and Negative types, respectively, but both are predicted as Mechanism. An attempt was made to limit these occurrences when rejecting cases, but, as S3 indicates, they do not follow a specific pattern that can be expressed.
- Similar to the case of S1, confusion is caused when the description is made with expressions that refer to another type. In S4, e_0 is the subject of the sentence and declares the simultaneous administration of the mentioned drugs as safe, since they do not interact. However, due to the wording, the system perceives the relationships as Advice.
- Cases where drug mentions are listed and are not covered by regular expressions. e.g., in S5, the existence of "the" excludes the match and makes it impossible to locate the negative pairs (e_0, e_i), $i = 1, \ldots, 5$. However, as a large number of instances has been discarded from the corpus, the model is unable to handle these underrepresented patterns.

However, the most serious form of error concerns the inability to detect existing interactions, or else the existence of false negatives. An obvious source of error is the particularly long sentences, where descriptions are highly complex. The same applies to long sentences that could have been separated into smaller sentences, occurring in 53 related instances. We define sentences as long when they have a length of ≥ 40 tokens (the number of words separated by a space, having replaced drugs with "drug"). S6 is an example of a sentence with a length of 40, where three interactions of the Mechanism type were not detected.

However, there are several instances where misclassification can be attributed to system errors. For example, the interaction in S7 is not found, even though the sentence contains characteristic expressions that suggest that an interaction is being described and does not include any redundant information that might cause any confusion. The existence of such phenomena causes difficulty in the holistic interpretation of the results.

3.9. Data Uncertainty

The main point of uncertainty that may arise in our proposed approach, that is shared with all related works using the DDI corpus, is input-dependent data uncertainty. In this case, the observation noise varies based on the input and is commonly introduced during the data generation process [36]. In order to address this issue, we attempt to deal with the observed inconsistencies during the the pre-processing stage.

Initially, simple entity name transformations are applied by changing the plural form to the singular form when the same substance is referenced and both cases are found in the the Drug_Class set (e.g., "penicillin" and "penicillins"). Consequently, this leads to a total of 122 cases of identified cases.

An additional point of uncertainty concerns the classification of drugs into the four classes as each drug mention generally belongs to a single class. However, cases were observed where the same drug was labeled with different classes throughout the instances found in the dataset. Although the percentage of entities in which this is observed is relatively small, the drug mentions in question participate in a large number of interaction pairs. Specifically, 6023 pairs are identified in which at least one entity with multiple types is found.

Table 4. Indicative cases of misclassified instances. Drug names are denoted in bold and underlined words describe the interaction in the sentence. Subscripts denote the drug name (entity) index in the sentence.

	Sentence
S1	Other drugs which may <u>enhance</u> the neuromuscular blocking action of **nondepolarizing agents** such as **MIVACRON** include certain **antibiotics** e.g., *antibiotics_group*).
S2	**Thalidomide**$_{e0}$ has been reported to enhance the sedative activity of **barbiturates**$_{e1}$, **alcohol**$_{e2}$, **chlorpromazine**$_{e3}$, and **reserpine**$_{e4}$.
S3	Preliminary evidence suggests that **cimetidine**$_{e0}$ inhibits **mebendazole**$_{e1}$ metabolism and may result in an increase in plasma concentrations of **mebendazole**$_{e2}$.
S4	**Pyrimethamine**$_{e0}$ <u>may be used</u> with **sulfonamides**$_{e1}$, **quinine**$_{e2}$ and other **antimalarials**$_{e3}$, and with other **antibiotics**$_{e4}$.
S5	**Dopamine antagonists**$_{e0}$, such as <u>the</u> **neuroleptics**$_{e1}$ (**phenothiazines**$_{e2}$, **butyrophenones**$_{e3}$, **thioxanthines**$_{e4}$) or **metoclopramide**$_{e5}$, ordinarily should not be administered concurrently with **Permax**$_{e6}$ (a **dopamine agonist**$_{e7}$)
S6	The bioavailability of **SKELID**$_{e0}$ is decreased 80% by **calcium**$_{e1}$, when **calcium**$_{e2}$ and **SKELID**$_{e3}$ are administered at the same time, and 60% by some **aluminum**$_{e4}$- or **magnesium**$_{e5}$-containing **antacids**$_{e6}$, when administered 1 hour before **SKELID**$_{e7}$.
S7	**Anticholinergics**$_{e0}$ <u>antagonize the effects</u> of **antiglaucoma agents**$_{e1}$.

Thus, the management of differentiation is sought without rejecting them. For example, in the case of "corticosteroid" where the vast majority of occurrences belong to a specific class, the assumption can be made that all instances should be labeled based on the majority class. In contrast, in the case of "tetracycline", where the class distribution is not clearly in favor of a single class, no such assumption can be made without introducing more uncertainty in the dataset.

The class of a drug entity is defined by a <drug-name, rdf:type, drug-class> triplet, where the object takes its value from the set of the four positive drug classes which should be unique for each instance of the Drug_Class class. However, a small amount of drug names cannot be classified to a single class. Moreover, each individual case is characterized by the name of the drug, as well as the set of the additional classes it was labeled in its various occurrences in the dataset. Therefore, for each one of these classes c, a blank node <_:substance-name_c> and a statement in the form of <_:substance-name_c, rdf:type, c> are included in the KG. Furthermore, in order to emphasize that those individual entities are not independent, the property name is defined which participates in <_:substance-name_c, name, substance-name> triplets. Therefore, entities that share the same value in this attribute are referring to the same substance.

An example of how our proposed approach performs on instances that contain uncertainties can be seen in the example sentence S2 (Table 4) in Section 3.8.

4. Conclusions

In this paper, we propose a Knowledge Graph schema in a Graph Neural Network-based architecture for the classification of Drug–Drug Interactions from biomedical literature, which achieves state-of-the-art performance. Specifically, we presented a Graph Isomorphism Network-based architecture with message passing and non-message passing units that leverage the proposed DDI-specific graph structure that models the knowledge (drug identifiers, names, types and interactions) from the DDI corpus. Token and sentence embeddings are generated for the drug named entities and sentences, respectively, and are passed to the graph, populating the Token and Sentence nodes, taking advantage of the underlying BERT-based LM.

Although our approach achieves state-of-the-art performance in the DDI classification task, the experimental results show that the individual class scores are greatly affected by the underlying LM, indicating that further improvements can be achieved. Based on the results, future work could be directed towards exploring a combination of general and domain-specific corpora to pre-train or domain-adapt a LM to further improve the performance of each positive class. Another direction for future work is to extend our approach with multi-task learning for extracting the entities in combination with interactions that could potentially improve generalization by using the domain information contained in the training signals of the related tasks as an inductive bias.

Author Contributions: Conceptualization, D.Z.; methodology, D.Z. and C.K.; formal analysis, D.Z. and C.K.; investigation, D.Z. and C.K.; software, D.Z. and C.K.; resources, D.Z.; validation, D.Z. and C.K.; writing—original draft preparation, D.Z.; writing—review and editing, D.Z., C.K. and I.V.; visualization, D.Z. and C.K.; supervision, D.Z. and I.V. All authors have read and agreed to the published version of the manuscript.

Funding: This research received no external funding.

Institutional Review Board Statement: Not applicable

Data Availability Statement: The DDI Extraction 2013 corpus is available at https://github.com/isegura/DDICorpus, (accessed on 25 October 2022).

Conflicts of Interest: The authors declare no conflict of interest.

Abbreviations

The following abbreviations are used in this manuscript:

DDI	Drug–Drug Interaction
NN	Neural Networks
GNN	Graph Neural Network
GCN	Graph Convolutional Network
RGNN	Relational Graph Neural Network
KG	Knowledge Graphs
ML	Machine Learning
RDF	Resource Description Framework
RE	Relation(ship) Extraction
PoS	Part-of-Speech
CNN	Convolutional Neural Network
RNN	Recurrent Neural Network
LSTM	Long Short-Term Memory
BiLSTM	Bidirectional Long Short-Term Memory
GRU	Gated Recurrent Unit
DGF	Dependency Graph Features
XML	Extensible Markup Language
LM	Language Model
SLDA	Sentence Level Domain Adapted
DLDA	Document Level Domain Adapted
GIN	Graph Isomorphism Network
MLP	Multi-Layer Perceptron
DNR	Drug Name Replacement
NIF	Negative Instance Filtering
CPU	Central Processing Unit

Appendix A

Table A1. Comparative analysis of related studies and our proposed approach. NIF and DNR denote Negative Instance Filtering and Drug Name Replacement, respectively. Values that are not reported in the published work are denoted with '-'. DGF denotes Dependency Graph Features. Y and N denote Yes and No, respectively.

Reference	Method	Embeddings	Emb. dim.	Features	NIF	DNR	Highlights	Review
Liu et al. [7]	CNN	Order	300	Position	Y	Y	Position embeddings improve performance	Dependent on position distributions of the input
Quan et al. [8]	CNN	CBOW	200	Position	N	Y	Multiple embeddings capture better representations	Errors in long sentences
Liu et al. [9]	CNN	Order	300	Position, DGF	N	Y	Syntactic features for long distance dependencies	Only a small set of large sentences parsed correctly
Zhao et al. [10]	CNN	Word2Vec	-	Position, PoS, DGF	Y	N	Dependency features and position embeddings	Filtering rules lead to limited generalization ability
Wang et al. [11]	LSTM	Word2Vec	100	Distance, DGF	Y	Y	Captures distance and dependency-based features	Low performance on long sentences
Yi et al. [12]	RNN	GloVe	100	Position	N	Y	Multiple attention layers to capture better representations	Semantic ambiguity leads to misclassifications
Zhang et al. [13]	LSTM	Word2Vec	200	PoS, DGF	N	N	Integration of sentence sequences, shortest dependency paths and attention layers	Errors in long sentences
Zhou et al. [14]	LSTM	Word2Vec	300	Position	Y	Y	Additional position embeddings to generate the attention weights	Misclassification of instances containing multiple drug mentions
Salman et al. [15]	LSTM	Word2Vec	100	Position, DGF	N	N	Task expansion to sentiment-based severity prediction	Word positions in the sentences are not taken into account
Mondal [22]	BERT-VAE	BioBERT	300	Chemical Structures	N	N	Utilizes chemical structure representations of drugs	Information loss in longer sentences due to low representation dimensionality
Xiong et al. [25]	LSTM- GCNN	Word2Vec	200	PoS, DGF	Y	N	Dependency features with graph neural network	Complex graph structures impact performance
Shi et al. [26]	LSTM-GCNN	Word2Vec	200	PoS, DGF, PageRank	Y	N	Dependency features with graph neural network and PageRank	Added complexity with feature generation from complex graph structures
Our approach	BERT-GIN	BioBERT	512	KG, DAPT	Y	Y	Novel DDI task-based Knowledge Graph leveraged by a graph neural network without relying on manual feature engineering	Nodes are initialized with domain-adapted representations to better capture sentence context

References

1. Percha, B.; Altman, R.B. Informatics confronts Drug–Drug Interactions. *Trends Pharmacol. Sci.* **2013**, *34*, 178–184. [CrossRef]
2. Hunter, L.; Cohen, K.B. Biomedical Language Processing: What's Beyond PubMed? *Mol. Cell* **2006**, *21*, 589–594. [CrossRef]
3. Wang, Y.; Wang, L.; Rastegar-Mojarad, M.; Moon, S.; Shen, F.; Afzal, N.; Liu, S.; Zeng, Y.; Mehrabi, S.; Sohn, S.; et al. Clinical information extraction applications: A literature review. *J. Biomed. Inform.* **2018**, *77*, 34–49. [CrossRef] [PubMed]
4. Segura-Bedmar, I.; Martínez, P.; Herrero-Zazo, M. SemEval-2013 Task 9: Extraction of Drug–Drug Interactions from Biomedical Texts (DDIExtraction 2013). In *Proceedings of the Second Joint Conference on Lexical and Computational Semantics (*SEM), Volume 2: Proceedings of the Seventh International Workshop on Semantic Evaluation (SemEval 2013)*; Association for Computational Linguistics: Atlanta, GA, USA, 2013; pp. 341–350.
5. Zhang, T.; Leng, J.; Liu, Y. Deep learning for Drug–Drug Interactions extraction from the literature: A review. *Briefings Bioinform.* **2019**, *21*, 1609–1627. [CrossRef] [PubMed]
6. Herrero-Zazo, M.; Segura-Bedmar, I.; Martínez, P.; Declerck, T. The DDI corpus: An annotated corpus with pharmacological substances and Drug–Drug Interactions. *J. Biomed. Inform.* **2013**, *46*, 914–920. [CrossRef] [PubMed]
7. Liu, S.; Tang, B.; Chen, Q.; Wang, X. Drug–Drug Interaction Extraction via Convolutional Neural Networks. *Comput. Math. Methods Med.* **2016**, *2016*, 6918381. [CrossRef]
8. Quan, C.; Hua, L.; Sun, X.; Bai, W. Multichannel Convolutional Neural Network for Biological Relation Extraction. *BioMed Res. Int.* **2016**, *2016*, 1850404. [CrossRef] [PubMed]
9. Liu, S.; Chen, K.; Chen, Q.; Tang, B. Dependency-based convolutional neural network for drug–drug interaction extraction. In Proceedings of the 2016 IEEE international conference on bioinformatics and biomedicine (BIBM), Shenzhen, China, 15–18 December 2016; pp. 1074–1080.
10. Zhao, Z.; Yang, Z.; Luo, L.; Lin, H.; Wang, J. Drug drug interaction extraction from biomedical literature using syntax convolutional neural network. *Bioinformatics* **2016**, *32*, 3444–3453. [CrossRef]
11. Wang, W.; Yang, X.; Yang, C.; Guo, X.; Zhang, X.; Wu, C. Dependency-based long short term memory network for drug–drug interaction extraction. *BMC Bioinform.* **2017**, *18*, 99–109. [CrossRef]
12. Yi, Z.; Li, S.; Yu, J.; Tan, Y.; Wu, Q.; Yuan, H.; Wang, T. Drug-drug interaction extraction via recurrent neural network with multiple attention layers. In Proceedings of the International Conference on Advanced Data Mining and Applications, Foshan, China, 12–15 November 2017; pp. 554–566.
13. Zhang, Y.; Zheng, W.; Lin, H.; Wang, J.; Yang, Z.; Dumontier, M. Drug–drug interaction extraction via hierarchical RNNs on sequence and shortest dependency paths. *Bioinformatics* **2017**, *34*, 828–835. [CrossRef]
14. Zhou, D.; Miao, L.; He, Y. Position-aware deep multi-task learning for drug–drug interaction extraction. *Artif. Intell. Med.* **2018**, *87*, 1–8. [CrossRef] [PubMed]
15. Salman, M.; Munawar, H.S.; Latif, K.; Akram, M.W.; Khan, S.I.; Ullah, F. Big Data Management in Drug–Drug Interaction: A Modern Deep Learning Approach for Smart Healthcare. *Big Data Cogn. Comput.* **2022**, *6*, 30. [CrossRef]
16. Mikolov, T.; Chen, K.; Corrado, G.; Dean, J. Efficient Estimation of Word Representations in Vector Space. *arXiv* **2013**. arXiv:1301.3781.
17. Peters, M.E.; Neumann, M.; Iyyer, M.; Gardner, M.; Clark, C.; Lee, K.; Zettlemoyer, L. Deep Contextualized Word Representations. In Proceedings of the 2018 Conference of the North American Chapter of the Association for Computational Linguistics: Human Language Technologies, New Orleans, AK, USA, 1–6 June 2018; Volume 1 (Long Papers), pp. 2227–2237. [CrossRef]
18. Radford, A.; Wu, J.; Child, R.; Luan, D.; Amodei, D.; Sutskever, I. Language models are unsupervised multitask learners. *OpenAI Blog* **2019**, *1*, 9.
19. Devlin, J.; Chang, M.W.; Lee, K.; Toutanova, K. BERT: Pre-training of deep bidirectional transformers for language understanding. In Proceedings of the Conference of the North American Chapter of the Association for Computational Linguistics: Human Language Technologies, Minneapolis, MN, USA, 2–7 June 2019 . Available online: https://aclanthology.org/N19-1423 (accessed on 20 September 2022).
20. Lee, J.; Yoon, W.; Kim, S.; Kim, D.; Kim, S.; Thus, C.H.; Kang, J. BioBERT: A pre-trained biomedical language representation model for biomedical text mining. *Bioinformatics* **2019**, *36*, 1234–1240. [CrossRef] [PubMed]
21. Beltagy, I.; Lo, K.; Cohan, A. SciBERT: A Pretrained Language Model for Scientific Text. In Proceedings of the 2019 Conference on Empirical Methods in Natural Language Processing and the 9th International Joint Conference on Natural Language Processing (EMNLP-IJCNLP), Hong Kong, China, 3–7 November 2019; pp. 3615–3620. [CrossRef]
22. Mondal, I. BERTChem-DDI : Improved Drug–Drug Interaction Prediction from text using Chemical Structure Information. In Proceedings of the Knowledgeable NLP: The First Workshop on Integrating Structured Knowledge and Neural Networks for NLP, Suzhou, China, 7 December 2020; pp. 27–32.
23. Gu, W.; Yang, X.; Yang, M.; Han, K.; Pan, W.; Zhu, Z. MarkerGenie: An NLP-enabled text-mining system for biomedical entity relation extraction. *Bioinform. Adv.* **2022**, *2*, vbac035. [CrossRef]
24. Ren, Z.H.; You, Z.H.; Yu, C.Q.; Li, L.P.; Guan, Y.J.; Guo, L.X.; Pan, J. A biomedical knowledge graph-based method for Drug–Drug Interactions prediction through combining local and global features with deep neural networks. *Briefings Bioinform.* **2022**, *23*, bbac363. [CrossRef]

25. Xiong, W.; Li, F.; Yu, H.; Ji, D. Extracting Drug–Drug Interactions with a dependency-based graph convolution neural network. In Proceedings of the IEEE International Conference on Bioinformatics and Biomedicine (BIBM), San Diego, CA, USA, 18–21 November 2019; pp. 755–759.
26. Shi, Y.; Quan, P.; Zhang, T.; Niu, L. DREAM: Drug-drug interaction extraction with enhanced dependency graph and attention mechanism. *Methods* **2022**, *203*, 152–159. [CrossRef]
27. Hamilton, W.L.; Ying, R.; Leskovec, J. Representation learning on graphs: Methods and applications. *arXiv* **2017**, arXiv:1709.05584.
28. Schlichtkrull, M.; Kipf, T.N.; Bloem, P.; van den Berg, R.; Titov, I.; Welling, M. Modeling Relational Data with Graph Convolutional Networks. In *The Semantic Web*; Gangemi, A., Navigli, R., Vidal, M.E., Hitzler, P., Troncy, R., Hollink, L., Tordai, A., Alam, M., Eds.; Springer: Berlin/Heidelberg, Germany, 2018; pp. 593–607.
29. Gururangan, S.; Marasović, A.; Swayamdipta, S.; Lo, K.; Beltagy, I.; Downey, D.; Smith, N.A. Don't Stop Pretraining: Adapt Language Models to Domains and Tasks. *arXiv* **2020**, arXiv:2004.10964.
30. Xu, K.; Hu, W.; Leskovec, J.; Jegelka, S. How Powerful are Graph Neural Networks? In Proceedings of the 7th International Conference on Learning Representations, ICLR 2019, New Orleans, LA, USA, 6–9 May 2019.
31. Hamilton, W.; Ying, Z.; Leskovec, J. Inductive representation learning on large graphs. *Adv. Neural Inf. Process. Syst.* **2017**, *30* . Available online: https://proceedings.neurips.cc/paper/2017/file/5dd9db5e033da9c6fb5ba83c7a7ebea9-Paper.pdf (accessed on 20 September 2022).
32. Chowdhury, M.F.M.; Lavelli, A. FBK-irst : A Multi-Phase Kernel Based Approach for Drug–Drug Interaction Detection and Classification that Exploits Linguistic Information. In Proceedings of the Second Joint Conference on Lexical and Computational Semantics (*SEM), Volume 2: Proceedings of the Seventh International Workshop on Semantic Evaluation (SemEval 2013), Atlanta, GA, USA, 14–15 June 2013; pp. 351–355.
33. Duong, C.T.; Hoang, T.D.; Dang, H.T.H.; Nguyen, Q.V.H.; Aberer, K. On node features for graph neural networks. *arXiv* **2019**, arXiv:1911.08795.
34. Zhong, R.; Ghosh, D.; Klein, D.; Steinhardt, J. Are larger pretrained language models uniformly better? comparing performance at the instance level. *arXiv* **2021**, arXiv:2105.06020.
35. You, J.; Ying, R.; Leskovec, J. Design Space for Graph Neural Networks. In Proceedings of the 34th International Conference on Neural Information Processing Systems (NIPS'20), Vancouver, BC, Canada, 6–12 December 2020; Curran Associates Inc.: Red Hook, NY, USA, 2020.
36. Xiao, Y.; Wang, W.Y. Quantifying uncertainties in natural language processing tasks. In Proceedings of the AAAI Conference on Artificial Intelligence, Honolulu, HI, USA, 27 January–1 February 2019; Volume 33, pp. 7322–7329.

applied sciences

Article

Anomaly Detection and Early Warning Model for Latency in Private 5G Networks

Jingyuan Han [1], Tao Liu [1], Jingye Ma [2], Yi Zhou [1], Xin Zeng [2,*] and Ying Xu [3]

[1] China Telecom Research Institute, Shanghai 200120, China
[2] Department of Information and Communication Engineering, Tongji University, Shanghai 200070, China
[3] Jiangxi Telecom, China Telecom, Nanchang 330000, China
* Correspondence: zengxin1@tongji.edu.cn

Abstract: Different from previous generations of communication technology, 5G has tailored several modes especially for industrial applications, such as Ultra-Reliable Low-Latency Communications (URLLC) and Massive Machine Type Communications (mMTC). The industrial private 5G networks require high performance of latency, bandwidth, and reliability, while the deployment environment is usually complicated, causing network problems difficult to identify. This poses a challenge to the operation and maintenance (O&M) of private 5G networks. It is needed to quickly diagnose or predict faults based on high-dimensional data of networks and services to reduce the impact of network faults on services. This paper proposes a ConvAE-Latency model for anomaly detection, which enhances the correlation between target indicators and hidden features by multi-target learning. Meanwhile, transfer learning is applied for anomaly prediction in the proposed LstmAE-TL model to solve the problem of unbalanced samples. Based on the China Telecom data platform, the proposed models are deployed and tested in an Automated Guided Vehicles (AGVs) application scenario. The results have been improved compared to existing research.

Keywords: private 5G networks; anomaly detection; abnormal early warning; autoencoder; long short-term memory; transfer learning

Citation: Han, J.; Liu, T.; Ma, J.; Zhou, Y.; Zeng, X.; Xu, Y. Anomaly Detection and Early Warning Model for Latency in Private 5G Networks. *Appl. Sci.* **2022**, *12*, 12472. https://doi.org/10.3390/app122312472

Academic Editors: George Drosatos, Avi Arampatzis and Pavlos S. Efraimidis

Received: 21 October 2022
Accepted: 18 November 2022
Published: 6 December 2022

Publisher's Note: MDPI stays neutral with regard to jurisdictional claims in published maps and institutional affiliations.

Copyright: © 2022 by the authors. Licensee MDPI, Basel, Switzerland. This article is an open access article distributed under the terms and conditions of the Creative Commons Attribution (CC BY) license (https://creativecommons.org/licenses/by/4.0/).

1. Introduction

With the rapid development of wireless communication technology, the 5G era has arrived. Operator services are gradually transforming from traditional services for the general public to customized services for business customers. Compared with existing traditional networks, private 5G networks have the characteristics of diversification, refinement, and certainty. The business scenarios in 5G involve various industrial firms with more complex and diverse terminal modes, which results in a significant increase in the business requirements for network performance indicators. According to the relevant survey report, the custom network for vehicle networking requires the latency of the vehicle-to-vehicle (V2V) service to be less than 200 ms. A minimum latency of 30 ms is required for automated guided vehicles (AGVs) in a smart factory, whereas the latency of a smart grid application scenario is the most sensitive, having a threshold of less than 15 ms. The impacts of not meeting these thresholds include large-scale production downtime, business interruptions, serious economic losses, and even personnel safety risks.

Also in the actual production of the existing network, such a massive system of customized communication systems faces many challenges. The existing network operations and maintenance (O&M) methods mainly rely on manual experience accumulation, which cannot realize the early warning of faults or proactively defend against fault occurrence. At the same time, the difficulty of O&M through manual experience alone has increased significantly because of the existence of multiple alarms behind the single-alarm phenomenon caused by the virtualization of 5G technology deployment, as well as the explosive growth of monitoring data in the O&M system due to refined management. The long anomaly location cycle and difficulty of root cause tracing seriously affect productivity and customer experience. In the face of these factors, there is an urgent need for efficient, rapid, accurate, and low-cost approaches to meet the growing demand for digital O&M.

The concept of data-driven artificial intelligence for IT operations, intelligent operations, and maintenance (AIOps) [1] was presented in 2016 and has become a trending research direction, aiming to solve the low efficiency of O&M caused by inexperience. It combines traditional O&M with artificial intelligence algorithms, big data and artificial intelligence, and other methods to improve the network's O&M capabilities and reduce human interference in the O&M, with the ultimate goal of achieving unmanned and fully automated O&M. In 2018, the white paper "Enterprise AIOps Implementation Recommendations" was jointly initiated and developed by the Efficient Operations and Maintenance community and the AIOps Standard Working Group [2].

Based on the research work in [3], most existing intelligent O&M frameworks are based on machine learning methods, which mainly include the following three aspects: anomaly detection, anomaly localization, and abnormal early warning. Specific research is presented in the related work in Section 2. Most research in the field of anomaly monitoring determines whether an abnormal state exists at the current moment by calculating the error between the current and normal intervals. Based on association rule mining or decision trees, which are mature algorithms in anomaly localization, data of the top-ranked dimension can be returned by ranking the channel anomalous factors (or errors) within the anomaly duration or by reducing the search space complexity through algorithm design. However, existing anomaly localization studies mainly focus on research from the perspective of supervised classification, which is contrary to the background of most unlabeled data in the actual production environment. Abnormal early warning studies focus on regression prediction for time variable series or by matching the real-time data distribution with failure rules to obtain the probability of failure occurrence. After thorough research, it was unfortunately found that the progress of intelligent O&M research lies only in the field of scientific research, and there are still some difficulties in combining the algorithms of academia with the domain knowledge of the industry and applying them to actual industrial production scenarios.

In the early stage of this study, we completed the construction of the data monitoring platform for private 5G networks and found that the abnormal phenomenon of AGVs driving is strongly related to high latency, but the current O&M is based on manual experience or simple data analysis, but the latency-sensitive scenario requires strong timeliness. Therefore, based on the China Telecom network data platform, this study proposes an anomaly detection and abnormal early warning models for AGV's abnormal driving scenarios in private 5G networks, which efficiently detects high latency using the proposed model (the ConvAE-Latency model). Based on this, the LstmAE-TL model is proposed to realize abnormal early warning with a 15 min interval using the characteristics of long short-term memory (LSTM). Simultaneously, transfer learning was used to solve the problem in which loss cannot converge in the training process of abnormal early warning detection owing to the small sample size. However, during data analysis, it was found that the percentage of abnormal samples was few, owing to private 5G networks running smoothly. Considering that the clustering of few samples cannot prove the correlation in the same class, we only studied anomaly detection and abnormal early warning, and anomaly location was not considered.

The main contributions of this study are as follows:

1. Instead of simulation data [4], the training data of proposed models are from the China Telecom network data platform, and the practicability of the ConvAE-Latency and LstmAE-TL models is verified.
2. Considering latency fluctuation as the important indicator for anomaly detection, compared to other methods [5,6], the ConvAE-Latency model uses the latency fitting module to enhance the correlation between target indicators and hidden features.

3. Transfer learning is applied to solve the problem of fewer abnormal data samples; because of the smooth running of private 5G networks, compared to related works [7,8], the LstmAE-TL model works better.

The remainder of this paper is organized as follows: Section 2 introduces state-of-the-art background information and studies, Section 3 introduces the method used in this study, and Section 4 shows the dataset and experimental results. Finally, Section 5 concludes the paper and discusses future research directions.

2. Related Works

Owing to the growing demand for data-driven operations and maintenance, AIOps [1] has become a prominent research direction; it uses artificial technologies such as big data analysis and machine learning to automate the management of O&M and make intelligent diagnostic decisions through specific policies and algorithms, which will replace the collaborative system of R&D and operations personnel and the automated operations and maintenance system based on expert systems to achieve faster, more accurate, and more efficient intelligent operations and maintenance. However, in the real O&M scenario, the abnormal indicators are not affected by a single dimensional key performance indicator (KPI), but by multi-dimensional abnormal KPIs.These lead to abnormal occurrences. Moreover, monitoring data have the characteristics of large volume, being difficult to clean, and having no labeling of abnormal data; therefore, realizing fast and accurate intelligent O&M is not easy. With the development of complex and refined O&M requirements, abnormaly detection, anomaly location, and abnormal early warning as key technologies in the AIOps field have become prominent research directions.

2.1. Anomaly Detection

Compared with traditional time-series-analysis-based anomaly monitoring models [9,10], most existing studies focus on machine-learning-algorithm-based approaches [6]. Shyu et al. [11] focused on the use of principal component analysis (PCA) for anomaly detection and constructed a prediction model to determine anomalies based on the principal and subprincipal components of normal instances. Liu et al. [12] considered the concept of isolation and proposed the isolation forest (IF) model. They considered that anomalies are sparser than other normal data points and can be isolated and divided into leaf nodes using few splits, making the path between the leaf node where the anomaly is located and the root node shorter.

The autoencoder network has also been applied to detect anomaly metrics using reconstruction errors [6,13,14]. Astha et al. [6] proposed an autoencoder-based anomaly detection and diagnosis method by designing a scoring algorithm to calculate the root-cause occurrence likelihood ranking. Zong et al. [13] proposed a method that combines a deep autoencoder (DAE) and Gaussian mixture model (GMM). The method performs well on several types of public data and outperforms many existing anomaly detection methods; however, it does not consider the temporal correlation and severity of anomalies in real scenarios and is not very robust against noise. Zhang et al. [14] used graph neural networks for the anomaly detection of temporal data using a combination of convolutional layers, convolutional long short-term memory networks (Conv-LSTM), and an autoencoder. This algorithm can eliminate the effect of noise between mutually uncorrelated features; however, the data size increases, leading to a high computation time and high computational load. He et al. [15] trained normal data using a temporal convolutional network (TCN) and used Gaussian distribution fitting for anomaly detection. Munir et al. [16] proposed a method that uses a convolutional neural network (CNN) for time-series prediction and then uses the Euclidean distance between predicted and true values for anomaly detection. In addition, Yahoo, Skyline, Netflix, and others have conducted related studies. Yahoo proposed a single-indicator anomaly detection system that combines three modules: timing modeling, anomaly detection, and alerting [17]. Netman Lab of Tsinghua University and Alibaba and Baidu also conducted related studies, such as the Donut algorithm for periodic temporal data detection based on the variational autoencoder (VAE) model [18] and the apprentice algorithm based on random forests [19].

2.2. Anomaly Location

After using anomaly detection algorithms to detect anomalies in multidimensional monitoring metrics at a certain time, we need to quickly and accurately find the specific dimension that causes these anomalies, that is, anomaly location. Association rule mining [20] and decision trees [21] are two mature anomaly location methods in intelligent O& M. Association rule mining is a rule-based machine learning algorithm that aims to discern strong rules present in a dataset using a number of metrics. When association rules are generated, they can be used to assist with the correlation analysis of events (errors and alerts). Decision trees belong to supervised learning in machine learning and represent mappings between object attributes and object values. Each node in the tree represents a judgment condition for object attributes, with branches representing objects that meet the conditions of the nodes and the leaf nodes of the tree representing the judgment results to which the objects belong. Tsinghua University and Baidu Inc. proposed a root cause analysis algorithm called HotSpot in 2018 [22], which uses Monte Carlo tree search to transform root cause analysis into a large spatial search problem with a heuristic search algorithm to locate the root causes that cause anomalies. Lin et al. proposed a multidimensional monitoring metric for the number of problem reports, called the iDice root cause analysis algorithm [23], which reduces the workload of O&M staff by reducing the combinations of attribute values under each dimension through three pruning strategies.

In addition, Bhagwan et al. proposed the Adtributer algorithm [24] for the KPI of the advertising revenue of a website. This algorithm uses the concepts of explanatory power and surprise to identify the set of the most likely root causes of anomalies after detecting anomalies in the KPI. Pinpoint [25] uses cluster analysis to diagnose the root causes of large dynamic Internet services. Argus [26] uses a hierarchical data structure for aggregation and performance metrics for groups of users with the same attributes and attempts to locate users with performance faults.

2.3. Abnormal Early Warning

Early warning is forward-looking compared to anomaly location; predicting the operational state of network or the faults of industrial equipment, which can improve the efficiency of maintenance and reduce loss. Currently, warning algorithms focus on traditional statistical-based logistic regression [27–29] as well as mathematical prediction techniques in which fuzzy theoretical models [30] , and grayscale models [31] are also widely used.

However, owing to the complexity of prediction scenarios and the explosive increase in data volume, researchers are more inclined toward data-driven intelligent prediction techniques [8,32–35]. Zhiqiang et al. [8] proposed an improved intelligent early warning method based on moving window sparse principal component analysis (MWSPCA) applicable to complex chemical processes. The sparse principal component analysis algorithm was used to build the initial warning model, which was then updated using moving windows to make the warning model more suitable for the characteristics of time-varying data. MALHOTRA et al. [32] proposed a coding and decoding model based on LSTM networks. This method models the time dependence of the time series through LSTM networks and achieves a better generalization capability than traditional methods. Related studies [33,34] proposed a deep belief network fault location (DBN-FL) model based on a deep trust network, which sets a series of fault rule identification templates based on historical fault data, integrated data analysis results, and expert experience in the fault identification process. The probability of fault occurrence is obtained by matching the real-time data distribution with the fault rules. In this study [35], an implementation method for abnormal early warning technology based on the probability distribution and density estimation method was proposed, which calculates the probability value of feature distribution by extracting data features and then distinguishes fault data from normal data according to the probability value. The performance of logistic regression and machine learning methods was evaluated in this study. The results show that these methods effectively improve the prediction accuracy relative to traditional networks [36].

3. Proposed Methods

3.1. Framework

As private 5G networks are the key development direction of operators for business customers, compared with public users, private networks have more demanding requirements for various types of performance. Once faults occur, industrial production will be significantly affected; therefore, timely and efficient troubleshooting is particularly important. This poses a challenge to O&M of private 5G networks. It is needed to quickly diagnose or predict faults based on high-dimensional data of networks and services to reduce the impact of network faults on services.

Based on this research direction, we investigate the limitations of private 5G networks O&M and find that most business scenarios focus on latency-sensitive AGVs and band-width-sensitive video transmission. According to the feedback of O&M staff, AGVs demonstrating abnormal driving behavior occurred several times. Considering that this is highly correlated with the high latency, the O&M staff deployed a probe to private 5G network parks to obtain monitor data. The result shows that, when the AGVs work abnormally, the latency of private 5G network is higher than the specified threshold.

Regarding the latency-sensitive scenario in private 5G networks, this study proposes an intelligent O&M framework, as shown in Figure 1, including anomaly detection and early warning models. First, based on the wireless data of private 5G networks, an anomaly detection model (ConvAE-Latency) based on an autoencoder network is proposed to detect whether the latency of network is high by calculating reconstruction error. Moreover, based on the ConvAE-Latency model, the LstmAE-TL model based on LSTM is proposed, which can predict abnormal behavior after 15 min. In addition, transfer learning was used in the training process of LstmAE-TL to solve the problem of low learning speed and difficult convergence because of a limited number of samples.

3.2. Anomaly Detection

Traditional O&M methods is based on periodic inspection, assign work. Problems exist such as low efficiency and lack of timeliness of resource placement; thus, non-preventive O&M fails to meet the needs of private 5G network. The use of artificial intelligence and big data and other new technologies to achieve active, rapid, and accurate detection is a new trend in the development of O&M.

As the autoencoder is trained as a neural network with unsupervised learning, it can reconstruct the original data by hidden features, and the reconstructed data are made closer to the original data by iterative training. If input data are anomaly, the distribution of reconstruction error is different. Based on this, we propose an autoencoder-based anomaly detection model (ConvAE-Latency) for anomaly detection of latency-sensitive scenarios in private 5G networks. The wireless data of base station are used for training, while probed latency is used for labeling to distinguish normal and abnormal samples. Significantly, only normal samples are used to train, and the results show the ConvAE-Latency model can reconstruct normal data well, while the abnormal samples are detected.

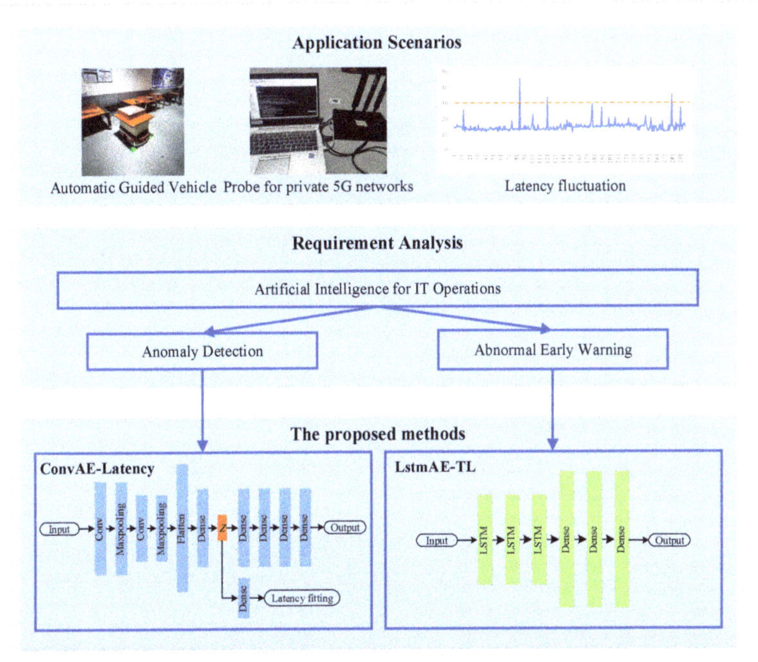

Figure 1. The framework of Intelligent O&M.

The network architecture of the ConvAE-Latency model is shown in Figure 2. It consists of three parts: an encoder, decoder and latency classification network. For the encoder, two convolutional layers are used for dimensionality reduction because of the large dimensionality of the wireless parameters on the base station side, and the training algorithm for the ConvAE-Latency model is shown in Algorithm 1. The encoder maps an input vector x to a hidden representation z by a an affine mapping following a nonlinearity, as shown in Equation (1). Subsequently, the decoding process shown in Equation (2) also maps the potential spatial representation of the hidden layer back to the original input space as the reconstructed output through the nonlinear transformation. The reconstruction goal is given by Equation (3), the difference between the original input vector x and the reconstruction output vector $\hat{x}$. Notably, the reconstruction–error distribution of abnormal samples should be as far away from the normal samples as possible:

$$z = k_\theta^E (W_E x + b_E), \tag{1}$$

$$\hat{x} = k_\sigma^D (W_D z + b_D), \tag{2}$$

$$E_x = \sum_{i=1}^{n} ||x_i - \hat{x}_i||^2, \tag{3}$$

where $x = [x_1, x_2, \cdots, x_n]$, $x \in \mathbb{R}^m$, m represents the wireless parameter dimension, and n represents the number of samples. W and b are the weight matrix and bias vector of the neural network, respectively, and k denotes the nonlinear activation function.

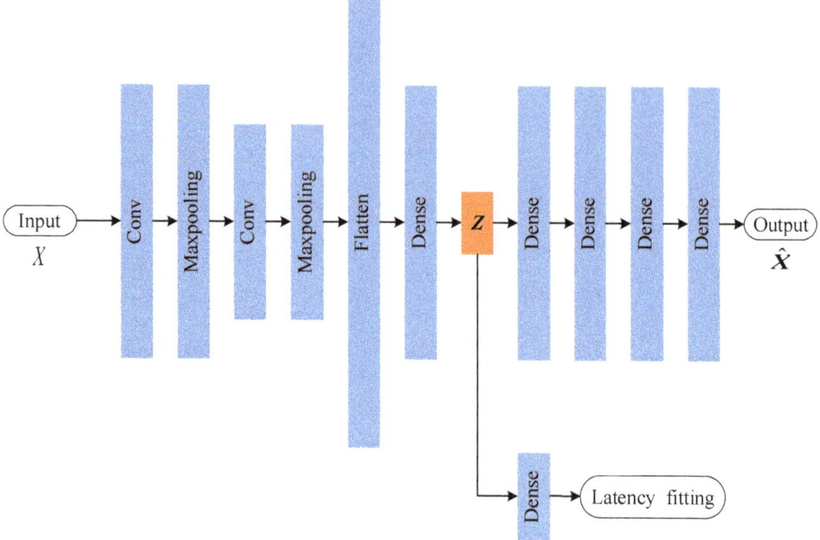

Figure 2. The network architecture of the ConvAE-Latency model.

Considering latency as target indicators in this scenario, the correlation with the hidden features z needs to be enhanced. In addition to reconstructing input, dense networks are used to construct the latency classification network, which ensures that z is strongly correlated with the latency, using the binary cross-entropy loss function with the function set, as shown in Equation (4):

$$L = \frac{1}{N}\sum_i L_i = \frac{1}{N}\sum_i -[y_i \log p_i + (1 - y_i)\log(1 - p_i)], \qquad (4)$$

where y_i denotes the label of sample x_i, that is, the positive class is 1, the negative class is 0, and 30 ms is used as the classification threshold. p_i denotes the probability that sample i is judged to positive class.

Algorithm 1 The ConvAE-Latency model training algorithm

Input: Dataset x, label y
Output: Encoder f_θ, decoder g_σ, classification u_ϕ
1: $\theta, \sigma, \phi \leftarrow$ Initialization parameters;
2: **repeat**
3: $\quad z = f_\theta(x)$;
4: $\quad \hat{x} = g_\sigma(z)$;
5: $\quad E_x = \sum_{i=1}^{n}||x_i - \hat{x}_i||^2$;
6: $\quad L = \frac{1}{N}\sum_i L_i = \frac{1}{N}\sum_i -[y_i \log p_i + (1 - y_i)\log(1 - p_i)]$;
7: $\quad \theta, \sigma, \phi \leftarrow$ Update parameters according to combination of E_x and L;
8: **until** Convergence of parameters θ, σ, ϕ

As shown in Algorithm 2 and Figure 3, the ConvAE-Latency model is implemented as follows:

Algorithm 2 The ConvAE-Latency model based anomaly detection algorithm

---

Input: Dataset x, threshold k
Output: The results of anomaly detection
  1: Pre-training ConvAE-Latency model obtained from Algorithm 1;
  2: $\hat{x} = g_\sigma(f_\theta(x))$ Calculate reconstruct result;
  3: $E_x = \sum_{i=1}^{n}||x_i - \hat{x}_i||^2$ Calculate reconstruction error;
  4: **if** $E_x > $ k **then**
  5:      x is an anomaly;
  6: **else**
  7:      x is not an anomaly;
  8: **end if**

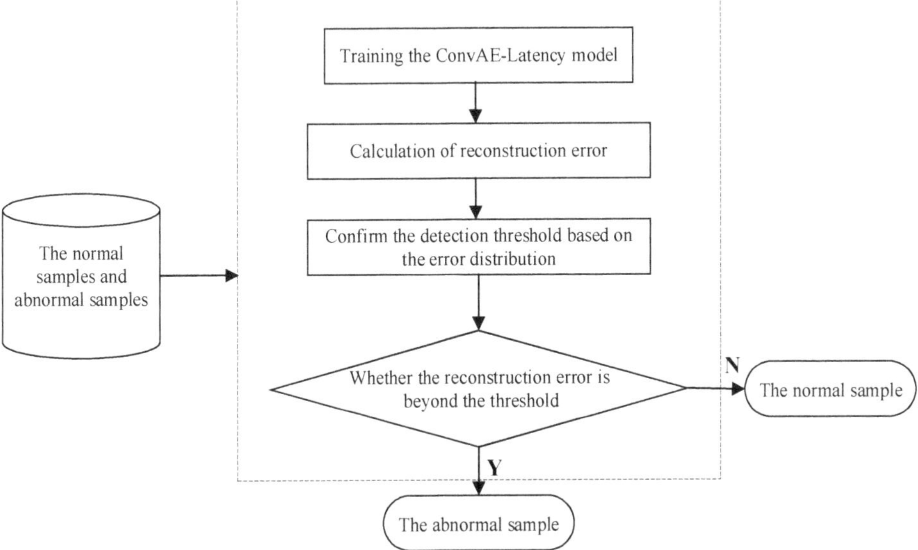

Figure 3. Implementation process of the ConvAE-Latency model.

3.3. Abnormal Early Warning

3.3.1. Network Architecture

In the previous section, we proposed the ConvAE-Latency model, which can effectively shorten the O&M troubleshooting time, compared with manual experience-based O&M methods. However, it only detects the current network status. To further enhance the intelligence of O&M in private 5G networks, abnormal early warning is expected. The possible abnormality is predicted before the network faults occur, so staff can prepare preventive measures as well as maintenance in advance to avoid or reduce the loss caused by the network faults.

LSTM has a memory function in the time dimension, which is often used in prediction schemes based on historical messages. Therefore, based on the ConvAE-Latency model, this study proposes a transfer-learning-based abnormal early warning model (LstmAE-TL), as shown in Figure 4. The model is divided into two parts: the LSTM-based prediction network and ConvAE-Latency model. The specific structure is described as follows.

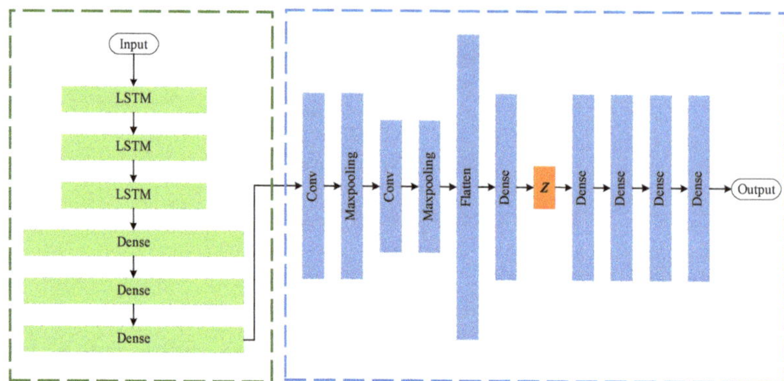

Figure 4. Network architecture of the LstmAE-TL model.

The first part of the proposed model consists of a three-layer LSTM network and dense networks; input data are the same as that for the ConvAE-Latency model. It is noteworthy that the input data need to be obtained by a sliding window algorithm. The output of the first part is the prediction of the wireless data $\hat{w}$ for the next period. The reconstructed function is given by Equation (5). Subsequently, $\hat{w}$ is fed into the previously trained ConvAE-Latency model to predict whether there is high latency, which is the final output. However, the ConvAE-Latency model is frozen and not trained here in order to fit the real situation better because inaccurate prediction will affect its judgment. It is worth noting that the latency fitting module of ConvAE-Latency is not used. The training algorithm for the LstmAE-TL is shown in Algorithm 3:

$$E_w = \sum_{i=1}^{n}||w_i - \hat{w}_i||^2, \tag{5}$$

where $w = (w_{t_1}, w_{t_2}, \cdots, w_{t_5}), w \in \mathbb{R}^{m \times n}$, m represents the wireless parameter dimension, and n represents the number of samples.

Algorithm 3 The LstmAE-TL model training algorithm

Input: Normal dataset w_{normal}, anomaly dataset $w_{anomaly}$
Output: Prediction network h_ξ
  1: Pre-training ConvAE-Latency model obtained from Algorithm 1;
  2: $\xi \leftarrow$ Initialization parameters;
  3: **repeat**
  4:    $E_w = \sum_{i=1}^{n}||w_i - h_\xi(w_i)||^2$ Calculate predict error by only w_{normal};
  5:    $E_x = \sum_{i=1}^{n}||h_\xi(w_i) - g_\sigma(f_\theta(h_\xi(w_i)))||^2$ Calculate reconstruction error of the predict result of w_{normal} by ConvAE-Latency model;
  6:    $\xi \leftarrow$ Update parameters according to combination of E_w and E_x;
  7: **until** Convergence of parameters ξ
  8: Freeze part of parameters in ξ;
  9: **repeat**
10:    $E_w = \sum_{i=1}^{n}||w_i - h_\xi(w_i)||^2$ Calculate predict error by mix $w_{anomaly}$ and part of w_{normal};
11:    $\xi \leftarrow$ Update parameters according to combination of E_w;
12: **until** Convergence of parameters ξ

As shown in Algorithm 4 and Figure 5, the implementation process of the LstmAE-TL model is as follows:

Algorithm 4 The LstmAE-TL model based abnormal early warning algorithm

---

Input: Dataset w, threshold k
Output: The results of abnormal early warning
 1: Pre-training ConvAE-Latency model obtained from Algorithm 1;
 2: Prediction network h_ξ;
 3: $E_x = \sum_{i=1}^{n} || h_\xi(w_i) - g_\sigma(f_\theta(h_\xi(w_i))) ||^2$ Calculate reconstruction error;
 4: **if** $E_x > $ k **then**
 5:    x is an anomaly;
 6: **else**
 7:    x is not an anomaly;
 8: **end if**

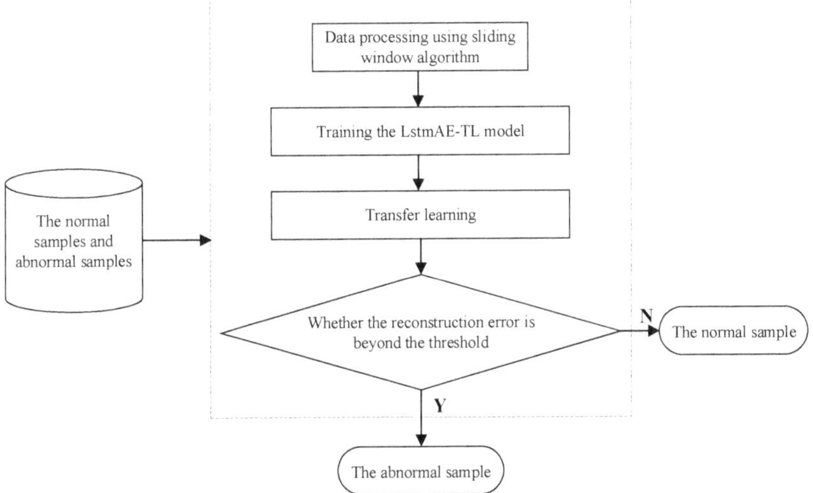

Figure 5. Implementation process of the LstmAE-TL model.

3.3.2. Transfer Learning

For any system, the abnormal state has a small probability compared with normal functionality, and the application scenarios of private 5G networks require high-reliability services, which determines that it is a highly reliable system itself, and its abnormal state is highly improbable. Therefore, the number of anomalous samples is too small in proportion to the samples of normal cases, leading to difficult training of the network and difficulty in fitting the anomalous part of the samples, which is usually referred to as few-shot learning.

For the few-shot learning of anomalous samples, this study used transfer learning to solve this problem. The idea of transfer learning comes from human analogy learning, which aims to apply features learned from a high-quality dataset to other datasets under similar domains. In the environment of this study, although there is a gap between abnormal sample features and normal sample features, both occur in the same network environment and their feature extraction processes are similar; therefore, transfer learning can be used to apply the feature extraction framework learned from normal samples to abnormal samples to improve the accuracy of abnormal sample prediction.

In the previous section, a three-layer dense network was proposed as the temporal feature extraction of wireless base station data and is the shared network part of normal and anomalous samples, which was trained using transfer learning which is not consistent with the training method of anomaly detection. First, the proposed model is trained with normal samples to obtain a prediction network with good performance. Then, the LSTM of LstmAE-TL is frozen and trained again with a small number of normal samples mixed with anomalous samples to achieve accurate prediction of the anomalous part.

3.3.3. Data Sensitivity Analysis

In essence, the proposed abnormal early warning model can be considered a joint composition of two independent networks, and both are trained using the same dataset. Due to the poor quality of the few data present in the dataset, both networks produce large errors, and the errors of both will further superimpose on each other, leading to serious degradation of the model performance.

However, because the two networks are trained separately, the information between them is not interoperable, which leads to the fact that, although the final result error of both networks is small, the locations where the errors are generated are different, which significantly magnifies the small errors in the process of cross-network transfer, thus affecting the performance.

To solve the above two problems, in this study, the loss function of the autoencoder with fixed weights is added as the judgment basis during the training of the LSTM network, that is, it is not only required that the prediction results obtained by the LSTM conform to the true distribution but also that the prediction results can be correctly coded and decoded by the autoencoder.

4. Analysis of Results

4.1. Data Set Description

The dataset was selected from the downlink network status data of a single DU under the exclusive base station of private 5G networks, totaling 16 days, one data point every 15 min, and a total of 1536 data points. The dataset includes PRB, CQI, MCS, etc., the entries of which are listed in Table 1. According to the different pre-processing methods, all the data can be divided into two categories: normalized data, including the total number of PRBs and the total number of CQI reports, which are normalized to the 0–1 range using the maximum and minimum normalization methods, and percentage data, including PRB utilization and retransmission rate. The percentage data were not normalized. The final dataset contained 179 dimensions.

Table 1. Sample wireless data.

Class ID	Meanings
PRB.UsedDl	Total number of downlink PRBs used
PRB.AvailDl	Total number of downlink PRBs available
MAC.TxBytesDl	Number of bits sent at the downlink MAC layer
MAC.TransIniTBDl	Initialize TB count at downlink MAC layer
HARQ.ReTransTBDl	Retransmission TB number
CQI.NumInTable	CQI Report
PRB.NumRbOfTableMCSDl	Table of the number of PRBs used in each level of MCS
PHYCell.MeanTxPower	Transmit power of physical layer
CELL.PowerSaveStateTime	Host standby sleep time

It is worth noting that, to save resources, the base station sleeps for a period of time every day at regular intervals, and the data collected during the sleep period are blank.

4.2. Anomaly Detection Results

For the input of the autoencoder-based anomaly detection model, all the data were randomly divided into training and test sets in a ratio of 2:1, and zeros were added after the existing 179-dimensional KPIs, thus forming a 12×15 matrix, which was fed into the convolutional network and decoded again to 179 dimensions after being compressed and coded to 32 dimensions. Owing to the high dimensionality of the data, MSE was used for the loss function to avoid large local errors. The learning rate was reduced by half every 50 epochs during the training process.

Figures 6 and 7 show the distribution curves and the cumulative distribution function (CDF) of reconstruction error after the autoencoder for abnormal and normal samples, respectively, where red and blue lines represent the abnormal and normal samples, respectively.

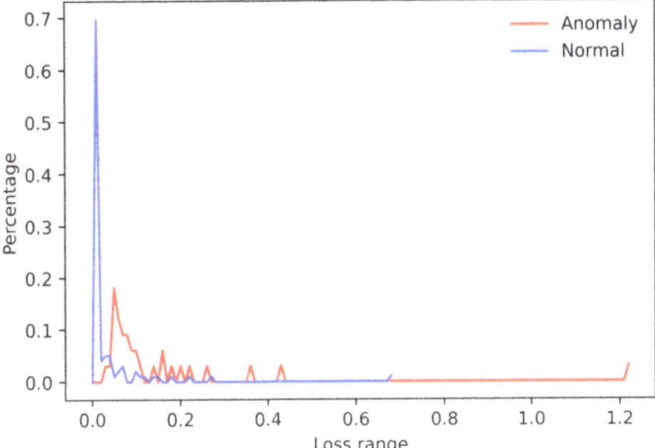

Figure 6. The reconstruction-error distribution for the ConvAE-Latency model.

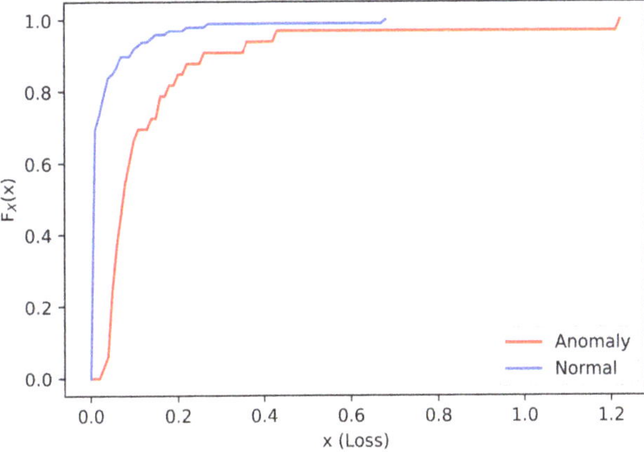

Figure 7. The CDF of reconstruction error for the ConvAE-Latency model.

The results show that there is a significant difference between the error distribution of normal and abnormal samples, and the decoding error of abnormal samples is significantly higher than that of most the normal samples; however, there are also some normal samples with larger errors. This is due to the fact that, when the autoencoder is trained, all the data used in this study are expected to be normal samples, but in fact, only one criterion of latency is used as the screening basis in the data screening process and the actual screened normal samples are obtained, but there may be some abnormal samples that cannot be characterized by latency, and the performance in the training results is attributed to some "fake samples" having a large error.

4.3. Abnormal Early Warning Results

For abnormal early warning, the LstmAE-TL model is proposed, and the training process was divided into two parts: normal training and transfer learning.

Figures 8 and 9 show the distribution curves and the CDF of reconstruction error without the latency fitting module; compared with that, the ConvAE-Latency model works better.

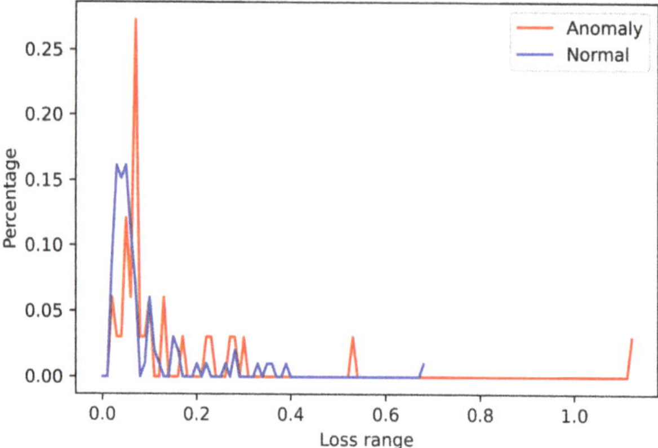

Figure 8. The reconstruction-error distribution without the latency fitting module.

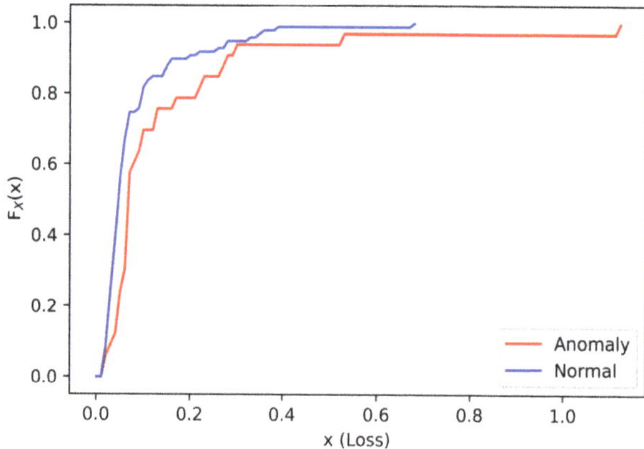

Figure 9. The CDF of reconstruction error without the latency fitting module.

4.3.1. Normal Sample Training

The purpose of normal sample training is to achieve an accurate prediction of the network for normal samples; therefore, the dataset of normal samples is selected and randomly divided into training and test sets according to the ratio of 2:1. The input data were the data of the five time periods before the period to be predicted, and the output data were the 179-dimensional network state data of the current time period. The loss function of the proposed model consists of two parts: LSTM-based abnormal early warning and autoencoder-based abnormal detection. The training first requires the prediction results to match the real situation such that the mean square error (MSE) between the prediction results and the real data are as small as possible. Second, the prediction results must be correctly reconstructed using the autoencoder. Therefore, the prediction results are fed into the abnormal detection network, and the MSE between the reconstructed data and predicted data are calculated again. The learning rate was reduced by half every 100 epochs during training.

Figure 10 shows the MSE of the prediction data based on the normal dataset, and the results indicate that the error of the network appears to increase periodically over time. It is observed that most of the error spikes occur at the beginning of the day, presumably because the base stations are beginning to work during these hours and their performance and service are unstable. Because the base stations are intermittently dormant to save energy, the dormant time data will be set to zero. As mentioned in Section 4.1, these dormant time period data are removed during data pre-processing, which results in discontinuous data timing around the dormant time period and cannot be accurately predicted.

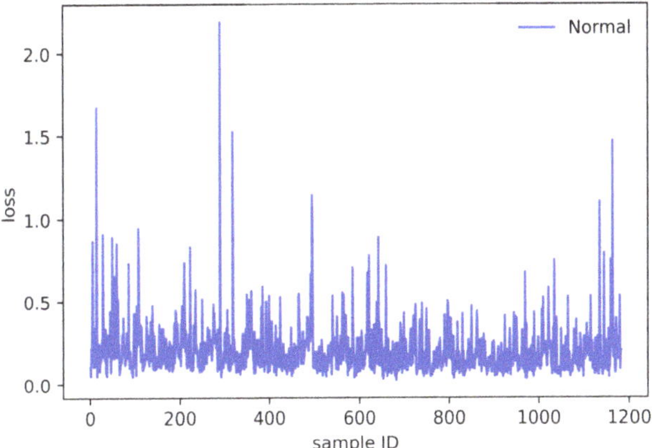

Figure 10. MSE of the prediction data based on the normal dataset.

4.3.2. Effectiveness of Transfer Learning

The purpose of migration learning is to improve the prediction accuracy of the abnormal samples. Based on the early warning network obtained above, this study freezes the three-layer LSTM network and subsequently only trains the three-layer dense network. Similarly, the data were randomly divided into a training and test set at a ratio of 2:1 for all abnormal samples. To enhance the weight of the abnormal samples, each of the samples in the training set was repeated twice and then mixed with an equal number of randomly selected normal samples to form the complete training and test sets. Because the prediction results of anomalous samples are expected to produce large errors when input into the autoencoder (while the opposite is true for normal samples), the process of transfer learning does not use reconstruction error as a loss function and only considers the accuracy of the fitted prediction. The learning rate was chosen to be one-tenth of the initial learning rate in normal learning, and the same learning rate was halved every 100 epochs.

The distribution curves and the CDF of the autoencoder reconstruction errors for abnormal and normal samples are shown in Figures 11 and 12, where red and blue line represent abnormal and normal samples, respectively. Meanwhile, Figures 13 and 14 show the results of abnormal early warning without transfer learning.

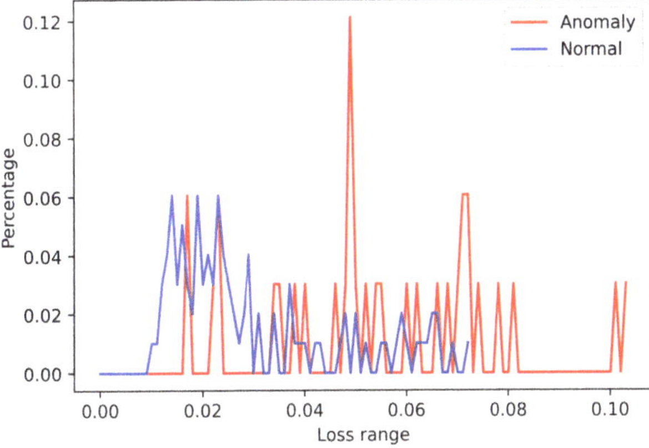

Figure 11. The reconstruction-error distribution for the LstmAE-TL model.

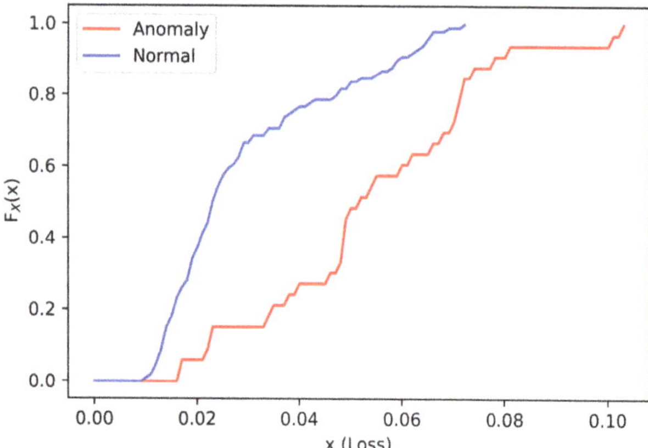

Figure 12. The CDF of reconstruction error for the LstmAE-TL model.

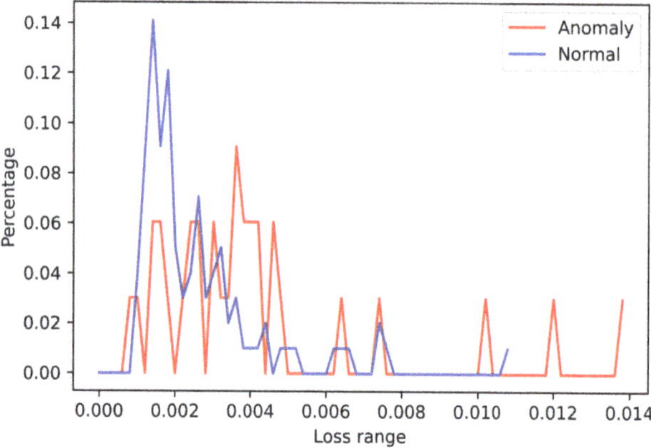

Figure 13. The reconstruction–error distribution without transfer learning.

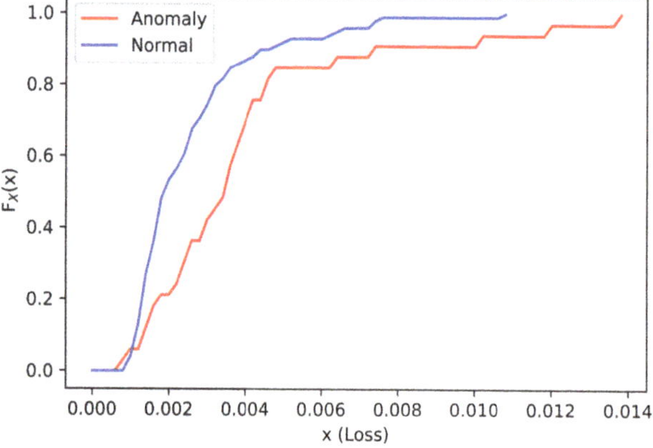

Figure 14. The CDF of reconstruction error without transfer learning.

The results show that there are differences in the distribution of abnormal and normal samples, and the LstmAE-TL works better than training without transfer learning. However, there is an overlap in the distribution of some samples, and the error of normal samples has improved. In transfer learning, to meet the prediction performance of abnormal samples, the prediction of some normal samples is inevitably sacrificed. In addition, the sampling density of the data selected in this study was 15 min, and for some of the abnormal samples, the abnormality may occur within 15 min of the current time period, resulting in it not yet being reflected in the previous time period, which means that these abnormal points do not have temporal characteristics, and the prediction accuracy is poor.

5. Conclusions

To meet the requirements of high-quality operation and maintenance of the private 5G network, this paper proposes the ConvAE-Latency model based on the autoencoder and the LstmAE-TL model based on LSTM. This work is verified at the China Telecom private 5G networks. Transfer learning is introduced in training to solve the problem of few fault samples. The results show that the two models achieve anomaly detection and prediction, and their effects are significantly improved compared with existing research. It is planned to use the hidden features of the proposed model combined with clustering algorithms to achieve anomaly location in the future.

Author Contributions: Methodology, J.H. and J.M.; software, J.M.; validation, J.H. and J.M.; investigation, Y.X.; writing—J.H.; writing—review and editing, Y.Z.; supervision, T.L. and Y.Z.; project administration, X.Z. All authors have read and agreed to the published version of the manuscript.

Funding: This research was funded by the Innovation Network Research Program of China Telecom (Grant No. I202207) and the Research and Development of Atomic Capabilities for 5G Network Services and Operations (Grant No. T202210).

Institutional Review Board Statement: Not applicable.

Informed Consent Statement: Not applicable.

Data Availability Statement: Not applicable.

Conflicts of Interest: The authors declare no conflict of interest.

References

1. Rijal, L.; Colomo-Palacios, R.; Sánchez-Gordón, M. Aiops: A multivocal literature review. In *Artificial Intelligence for Cloud and Edge Computing*; Springer: Cham, Switzerland, 2022; pp. 31–50.
2. Hua, Y. A Systems Approach to Effective AIOps Implementation. Ph.D. Thesis, Massachusetts Institute of Technology, Cambridge, MA, USA, 2021.
3. Notaro, P.; Cardoso, J.; Gerndt, M. A Survey of AIOps Methods for Failure Management. *ACM Trans. Intell. Syst. Technol. (TIST)* **2021**, *12*, 1–45. [CrossRef]
4. Abusitta, A.; Silva de Carvalho, G.H.; Abdel Wahab, O.; Halabi, T.; Fung, B.; Al Mamoori, S. Deep Learning-Enabled Anomaly Detection for IoT Systems. Available online: https://ssrn.com/abstract=4258930 (accessed on 7 November 2022).
5. Chen, Z.; Yeo, C.K.; Lee, B.S.; Lau, C.T. Autoencoder-based network anomaly detection. In Proceedings of the 2018 Wireless Telecommunications Symposium (WTS), Phoenix, AZ, USA, 17–20 April 2018; IEEE: Piscataway, NJ, USA, 2018; pp. 1–5.
6. Garg, A.; Zhang, W.; Samaran, J.; Savitha, R.; Foo, C.S. An evaluation of anomaly detection and diagnosis in multivariate time series. *IEEE Trans. Neural Netw. Learn. Syst.* **2021**, *33*, 2508–2517. [CrossRef] [PubMed]
7. Sun, J.; Wang, J.; Hao, Z.; Zhu, M.; Sun, H.; Wei, M.; Dong, K. AC-LSTM: Anomaly State Perception of Infrared Point Targets Based on CNN+LSTM. *Remote Sens.* **2022**, *14*, 3221. [CrossRef]
8. Geng, Z.; Chen, N.; Han, Y.; Ma, B. An improved intelligent early warning method based on MWSPCA and its application in complex chemical processes. *Can. J. Chem. Eng.* **2020**, *98*, 1307–1318. [CrossRef]
9. Qi, J.; Chu, Y.; He, L. Iterative anomaly detection algorithm based on time series analysis. In Proceedings of the 2018 IEEE 15th International Conference on Mobile Ad Hoc and Sensor Systems (MASS), Chengdu, China, 9–12 October 2018; IEEE: Piscataway, NJ, USA, 2018; pp. 548–552.
10. Pena, E.H.; de Assis, M.V.; Proença, M.L. Anomaly detection using forecasting methods arima and hwds. In Proceedings of the 2013 32nd International Conference of the Chilean Computer Science Society (SCCC), Temuco, Chile, 11–15 November 2013; IEEE: Piscataway, NJ, USA, 2013; pp. 63–66.
11. Shyu, M.L.; Chen, S.C.; Sarinnapakorn, K.; Chang, L. *A Novel Anomaly Detection Scheme Based on Principal Component Classifier*; Technical Report; Department of Electrical and Computer Engineering, University of Miami : Coral Gables, FL, USA, 2003.

12. Liu, F.T.; Ting, K.M.; Zhou, Z.H. Isolation forest. In Proceedings of the 2008 Eighth IEEE International Conference on Data Mining, Pisa, Italy, 15–19 December 2008; IEEE: Piscataway, NJ, USA, 2008; pp. 413–422.

13. Zong, B.; Song, Q.; Min, M.R.; Cheng, W.; Lumezanu, C.; Cho, D.; Chen, H. Deep autoencoding gaussian mixture model for unsupervised anomaly detection. In Proceedings of the International Conference on Learning Representations, Vancouver, BC, Canada, 30 April–3 May 2018.

14. Zhang, C.; Song, D.; Chen, Y.; Feng, X.; Lumezanu, C.; Cheng, W.; Ni, J.; Zong, B.; Chen, H.; Chawla, N.V. A deep neural network for unsupervised anomaly detection and diagnosis in multivariate time series data. In Proceedings of the AAAI Conference on Artificial Intelligence, Honolulu, HI, USA, 27 January–1 February 2019; Volume 33, pp. 1409–1416.

15. He, Y.; Zhao, J. Temporal convolutional networks for anomaly detection in time series. *J. Phys. Conf. Ser.* **2019**, *1213*, 042050. [CrossRef]

16. Munir, M.; Siddiqui, S.A.; Dengel, A.; Ahmed, S. DeepAnT: A deep learning approach for unsupervised anomaly detection in time series. *IEEE Access* **2018**, *7*, 1991–2005. [CrossRef]

17. Laptev, N.; Amizadeh, S.; Flint, I. Generic and scalable framework for automated time-series anomaly detection. In Proceedings of the 21th ACM SIGKDD International Conference on Knowledge Discovery and Data Mining, Sydney, Australia, 10–13 August 2015; pp. 1939–1947.

18. Xu, H.; Chen, W.; Zhao, N.; Li, Z.; Bu, J.; Li, Z.; Liu, Y.; Zhao, Y.; Pei, D.; Feng, Y.; et al. Unsupervised anomaly detection via variational auto-encoder for seasonal kpis in web applications. In Proceedings of the 2018 World Wide Web Conference, Lyon, France, 23–27 April 2018; pp. 187–196.

19. Liu, D.; Zhao, Y.; Xu, H.; Sun, Y.; Pei, D.; Luo, J.; Jing, X.; Feng, M. Opprentice: Towards practical and automatic anomaly detection through machine learning. In Proceedings of the 2015 Internet Measurement Conference, Tokyo, Japan, 28–30 October 2015; pp. 211–224.

20. Ahmed, F.; Erman, J.; Ge, Z.; Liu, A.X.; Wang, J.; Yan, H. Detecting and localizing end-to-end performance degradation for cellular data services. In Proceedings of the 2015 ACM SIGMETRICS International Conference on Measurement and Modeling of Computer Systems, Portland, OR, USA, 15–19 June 2015; pp. 459–460.

21. Nguyen, B.; Ge, Z.; Van der Merwe, J.; Yan, H.; Yates, J. Absence: Usage-based failure detection in mobile networks. In Proceedings of the 21st Annual International Conference on Mobile Computing and Networking, Paris, France, 7–11 September 2015; pp. 464–476.

22. Sun, Y.; Zhao, Y.; Su, Y.; Liu, D.; Nie, X.; Meng, Y.; Cheng, S.; Pei, D.; Zhang, S.; Qu, X.; et al. Hotspot: Anomaly localization for additive kpis with multi-dimensional attributes. *IEEE Access* **2018**, *6*, 10909–10923. [CrossRef]

23. Lin, Q.; Lou, J.G.; Zhang, H.; Zhang, D. iDice: Problem identification for emerging issues. In Proceedings of the 38th International Conference on Software Engineering, Austin, TX, USA, 14–22 May 2016; pp. 214–224.

24. Bhagwan, R.; Kumar, R.; Ramjee, R.; Varghese, G.; Mohapatra, S.; Manoharan, H.; Shah, P. Adtributor: Revenue debugging in advertising systems. In Proceedings of the 11th USENIX Symposium on Networked Systems Design and Implementation (NSDI 14), Seattle, WA, USA, 2–4 April 2014; pp. 43–55.

25. Chen, M.Y.; Kiciman, E.; Fratkin, E.; Fox, A.; Brewer, E. Pinpoint: Problem determination in large, dynamic internet services. In Proceedings of the International Conference on Dependable Systems and Networks, Washington, DC, USA, 23–16 June 2002; IEEE: Piscataway, NJ, USA, 2002; pp. 595–604.

26. Yan, H.; Flavel, A.; Ge, Z.; Gerber, A.; Massey, D.; Papadopoulos, C.; Shah, H.; Yates, J. Argus: End-to-end service anomaly detection and localization from an isp's point of view. In Proceedings of the 2012 Proceedings IEEE INFOCOM, Orlando, FL, USA, 25–30 March 2012; IEEE: Piscataway, NJ, USA, 2012; pp. 2756–2760.

27. Khoshgoftaar, T.M.; Gao, K.; Szabo, R.M. An application of zero-inflated poisson regression for software fault prediction. In Proceedings of the 12th International Symposium on Software Reliability Engineering, Hong Kong, China, 27–30 November 2001; IEEE: Piscataway, NJ, USA, 2001; pp. 66–73.

28. Lessmann, S.; Baesens, B.; Mues, C.; Pietsch, S. Benchmarking classification models for software defect prediction: A proposed framework and novel findings. *IEEE Trans. Softw. Eng.* **2008**, *34*, 485–496. [CrossRef]

29. Nagappan, N.; Ball, T. Static analysis tools as early indicators of pre-release defect density. In Proceedings of the 27th International Conference on Software Engineering, ICSE 2005, St. Louis, MO, USA, 15–21 May 2005; IEEE: Piscataway, NJ, USA, 2005; pp. 580–586.

30. Hu, L.Q.; He, C.F.; Cai, Z.Q.; Wen, L.; Ren, T. Track circuit fault prediction method based on grey theory and expert system. *J. Vis. Commun. Image Represent.* **2019**, *58*, 37–45. [CrossRef]

31. Wende, T.; Minggang, H.; Chuankun, L. Fault prediction based on dynamic model and grey time series model in chemical processes. *Chin. J. Chem. Eng.* **2014**, *22*, 643–650.

32. Malhotra, P.; Vig, L.; Shroff, G.; Agarwal, P. Long short term memory networks for anomaly detection in time series. In Proceedings of the 23rd European Symposium on Artificial Neural Networks, Computational Intelligence and Machine Learning, ESANN 2015, Bruges, Belgium, 22–24 April 2015; Volume 89, pp. 89–94.

33. Yu, A.; Yang, H.; Yao, Q.; Li, Y.; Guo, H.; Peng, T.; Li, H.; Zhang, J. Accurate fault location using deep belief network for optical fronthaul networks in 5G and beyond. *IEEE Access* **2019**, *7*, 77932–77943. [CrossRef]

34. Zhao, X.; Yang, H.; Guo, H.; Peng, T.; Zhang, J. Accurate fault location based on deep neural evolution network in optical networks for 5G and beyond. In Proceedings of the Optical Fiber Communication Conference, San Diego, CA, USA, 3–7 March 2019; p. M3J–5.
35. Mulvey, D.; Foh, C.H.; Imran, M.A.; Tafazolli, R. Cell fault management using machine learning techniques. *IEEE Access* **2019**, *7*, 124514–124539. [CrossRef]
36. Malhotra, R. Comparative analysis of statistical and machine learning methods for predicting faulty modules. *Appl. Soft Comput.* **2014**, *21*, 286–297. [CrossRef]

Article

Blockchain-Based Decentralized Federated Learning Method in Edge Computing Environment

Song Liu, Xiong Wang, Longshuo Hui and Weiguo Wu *

School of Computer Science and Technology, Xi'an Jiaotong University, Xi'an 710049, China;
liusong@xjtu.edu.cn (S.L.); buer_wang@163.com (X.W.); huils20@stu.xjtu.edu.cn (L.H.)
* Correspondence: wgwu@xjtu.edu.cn; Tel.: +86-187-0683-8060

Abstract: In recent years, federated learning has been able to provide an effective solution for data privacy protection, so it has been widely used in financial, medical, and other fields. However, traditional federated learning still suffers from single-point server failure, which is a frequent issue from the centralized server for global model aggregation. Additionally, it also lacks an incentive mechanism, which leads to the insufficient contribution of local devices to global model training. In this paper, we propose a blockchain-based decentralized federated learning method, named BD-FL, to solve these problems. BD-FL combines blockchain and edge computing techniques to build a decentralized federated learning system. An incentive mechanism is introduced to motivate local devices to actively participate in federated learning model training. In order to minimize the cost of model training, BD-FL designs a preference-based stable matching algorithm to bind local devices with appropriate edge servers, which can reduce communication overhead. In addition, we propose a reputation-based practical Byzantine fault tolerance (R-PBFT) algorithm to optimize the consensus process of global model training in the blockchain. Experiment results show that BD-FL effectively reduces the model training time by up to 34.9% compared with several baseline federated learning methods. The R-PBFT algorithm can improve the training efficiency of BD-FL by 12.2%.

Keywords: decentralized federated learning; blockchain; edge computing; stable matching; consensus algorithm

Citation: Liu, S.; Wang, X.; Hui, L.; Wu, W. Blockchain-Based Decentralized Federated Learning Method in Edge Computing Environment. *Appl. Sci.* **2023**, *13*, 1677. https://doi.org/10.3390/app13031677

Academic Editors: George Drosatos, Avi Arampatzis and Pavlos S. Efraimidis

Received: 15 January 2023
Revised: 24 January 2023
Accepted: 26 January 2023
Published: 28 January 2023

Copyright: © 2023 by the authors. Licensee MDPI, Basel, Switzerland. This article is an open access article distributed under the terms and conditions of the Creative Commons Attribution (CC BY) license (https:// creativecommons.org/licenses/by/ 4.0/).

1. Introduction

With the development of new generation information technology such as mobile internet, the number of mobile services and applications is growing exponentially, resulting in the generation of massive amounts of data. It has been shown that there are data security risks in well-known business associations [1]. User data are often stored in the centralized cloud servers of an organization or enterprise and can be accessed without privacy protection, which raises the risk of leakage of user-sensitive data [2]. Google took the lead in proposing a federated learning method [3] to solve the collaborative training problem of privacy protection so that private data can be safely used in a distributed environment. Therefore, federated learning has received extensive attention and research from industry and academia. However, federated learning has its own limitations [4]. It relies on a single centralized server and is vulnerable to a single point of server failure. Additionally, it lacks the incentive mechanism for local devices, which leads to the reduction of the initiative of local devices to participate in the training of federated learning.

Blockchain is a decentralized and auditable ledger technique [5] that has become a solution to replace vulnerable centralized servers in insecure environments. By combining with blockchain, the decentralized federated learning method can be realized. However, in the scenario of combined blockchain and federated learning, distributed servers usually use cloud computing to transmit model data [4]. Since cloud servers are physically far from local devices, which will further increase the delay of network data transmission [6], traditional cloud computing techniques will weaken the role of servers in supporting the training process of federated learning models. Additionally, edge computing has become an efficient computing paradigm that sinks computing and storage resources to the side close to local devices [7,8]. Compared to cloud-based servers, edge servers are closer to local devices and can respond to requests from local devices faster. Applying edge computing to federated learning can further reduce the latency and energy consumption of model training.

However, when combining blockchain with federated learning in the edge computing environment, it also faces the challenge of the high cost of network communication [9]. As the number of local devices is large and keeps growing dramatically, local devices need to effectively offload data and tasks to the appropriate server close to the edge side of the network, optimizing the utilization of edge server resources and maximizing the system efficiency. This requires efficient and stable matching between local devices and edge servers [10]. The matching problem is affected by many factors, such as distance, network bandwidth, and computing power [11]. These factors involve new techniques such as node localization [12], and they need to be considered to reduce the overall system latency and energy consumption in the edge computing environment. Meanwhile, as the core of blockchain technique, the consensus algorithm plays a decisive role in the security and efficiency of blockchain [13–16]. In highly decentralized federated learning with blockchain, all local devices participating in model training will also participate in the consensus process. The communication cost will increase significantly, which is bound to increase the consensus time of the blockchain, thus reducing the efficiency of model training.

To address the problems of centralization and lack of incentives in traditional federated learning, this paper proposes a blockchain-based decentralized federated learning method for edge computing environments, named BD-FL. By designing a stable matching algorithm between local devices and edge servers and an optimized consensus algorithm, BD-FL can effectively reduce the overall system delay and speed up the model training efficiency. This paper makes the following contributions.

- We propose the BD-FL by combining blockchain with federated learning in the edge computing environment. BD-FL uses the distributed characteristics of blockchain and edge computing to solve the problem of a centralized server in that the local device trains the local model and the edge server aggregates the global model. BD-FT also introduces an incentive mechanism to encourage local devices to actively participate in model training, increasing the number of samples and improving the model accuracy.
- We propose a preference-based stable matching algorithm in BD-FT, which binds local devices to appropriate edge servers, improving the utilization of edge server resources and reducing the delay of data transmission. We propose the R-PBFT algorithm, which optimizes the network topology and the consistency protocol and designs a dynamic reputation mechanism, reducing the communication overhead of the blockchain consensus process and improving the model training efficiency.
- We performed extensive simulation experiments to evaluate the proposed BD-FL. Experimental results show that BD-FL effectively reduces the model training time by up to 19.7% and 34.9%, respectively, compared with several federated learning methods with different matching algorithms and a state-of-the-art blockchain-based federated learning method. The R-PBFT algorithm can reduce the communication overhead of the consensus process and improve the training efficiency of BD-FL by 12.2%.

The rest of this paper is organized as follows. Section 2 introduces related work. Section 3 presents the methodology, including the system architecture, BD-FL, and R-PBFT. Section 4 gives the experimental evaluation. Section 5 concludes this paper.

2. Related Work

A centralized network topology has more serious system security issues and higher communication overhead compared with a decentralized distributed network topology. Traditional federated learning adopts a centralized topology with a single centralized server responsible for aggregating the global model. As federated learning takes hold in real-life production, the shortcomings of the centralized topology model have gradually become apparent [17]. In practical applications, the centralized server will put huge pressure on the network bandwidth overhead and also reduce the robustness of the system, which will affect the model training process of federation learning once the server is maliciously compromised. Blockchain is a distributed technique with decentralized characteristics and incentive mechanism [5,18,19], which can effectively solve the above problems of traditional federated learning. In addition, due to being tamper-proof and anonymous, blockchain can guarantee data security. Therefore, a lot of work has been done on decentralized federated learning methods using blockchain.

However, decentralized federated learning with blockchain still faces the challenge of model training efficiency. Kim et al. [20] proposed a blockchain-based federated learning architecture. The local device in this architecture updates the local model based on its available data samples. The architecture uses blockchain to reward updates from local devices, and the reward is proportional to the number of local data samples. This simple reward scheme is not able to accurately reflect the magnitude of the contribution made by the local device to the global model training. Weng et al. [21] designed a federated learning scheme incorporating blockchain incentives. The scheme ensures system reliability by rewarding honest local devices and punishing dishonest ones. They also introduced a consensus protocol based on a committee mechanism, which participates in the consensus process by randomly selecting nodes to form a committee. However, randomly selecting committee nodes is almost negligibly close to a completely random scheme and is probably not optimal. Local devices make significant contributions to the training of the federated learning model, so it is important to reasonably reward local devices. Existing federated learning incentive schemes generally agree that local devices should be fairly rewarded based on the magnitude of their contribution to the model. Jia et al. [22] stated that the most widely used scheme to evaluate the contribution size of local devices is Shapley values (SVs). SVs can fairly distribute rewards for model training, and it is widely used in many fields, such as economics, information theory, and machine learning. However, SV-based reward schemes usually require exponential time to compute, and the computational cost is prohibitive.

In the edge computing environment, the matching problem between local devices and edge servers has an important impact on the data transmission delay and the overall system efficiency. Hu et al. [23] proposed a matching method for mobile edge computing and device-to-device communication environments. This method uses a game model to solve the offloading problem of local devices. Each local device is regarded as a gamer, and the offloading strategy is obtained through a mutual game to make the system reach Nash equilibrium. Wu [24] proposed an intelligent scheduling matching scheme based on a delayed acceptance algorithm. It combines the Hopfield neural network and the decision tree model and quantifies the assignment scheduling problem into a matching optimization problem by defining the cost coefficient between tasks and equipment. Lu et al. [25] proposed an asynchronous greedy matching algorithm, which builds a preference list of both parties based on the utility value between the cooperative node and the requesting node, and uses the greedy strategy for stable matching. The existing related research mainly solves the one-to-one or N-to-N matching problem, and there is less research on stable matching between M devices and N servers in the edge computing environment.

The consensus algorithm, as one of the core ideas of blockchain, can ensure the proper operation of the blockchain, but it has an important impact on communication overhead and model training efficiency. The most commonly used consensus algorithms are proof of stake (PoS) [13], proof of work (PoW) [14], delegated proof of stake (DPoS) [15], and practical Byzantine fault tolerance (PBFT) [16]. The PBFT algorithm is widely used in distributed architectures, but it still suffers from high communication overhead and low reliability of master nodes. Numerous solutions have emerged to address the shortcomings of PBFT. Castro et al. [26] improved the transaction throughput of PBFT by caching blocks, but their method does not perform well on the delay of the consensus process. Zhang et al. [8] made the certificate and other information of blocks clear in time without communication between nodes according to the timestamp in the blockchain, but this method does not consider the optimization of the PBFT consistency protocol. Zheng et al. [27] combined DPoS and PBFT to make the algorithm with dynamic authorization, but the limited bandwidth still reduces the transaction throughput of the algorithm. On the other hand, some studies are dedicated to reducing the time complexity of PBFT. Ma et al. [28] proposed a scheme to verify the consistency of asynchronous Byzantine nodes through a stochastic prediction model. This scheme randomly selects one of the multi-node proposals in each round of consensus and uses the threshold signature algorithm to reduce the communication cost of each round of consensus to $O(n^2)$. However, the random selection method of this scheme may cause security problems. Gao et al. [29] proposed a scalable Byzantine algorithm (FastBFT) that introduces a tree data structure to achieve the optimal time complexity of $O(nlogn)$. However, in the worst case, the time complexity of FastBFT is still $O(n^2)$. Liu et al. [30] improved the consensus efficiency by caching blocks and smart contract techniques, but without reducing the time complexity of PBFT. Wang et al. [31] proposed a PBFT algorithm based on randomly selected collectors. It can reduce the communication cost to a linear level, but if the selected collector is malicious, the communication cost of the algorithm will rise sharply. However, these consensus algorithms are not applicable to real blockchain and edge computing application scenarios, and still have high computational complexity.

In this work, we present BD-FL to solve the single node failure and network communication overhead problems of centralized federated learning. BD-FL introduces an incentive mechanism to increase the contribution of local devices to global model training. We also propose a stable matching algorithm and the R-PBFT algorithm to reduce the number of nodes participating in communication and consensus, which reduces the system delay and improves the model training efficiency.

3. Methodology

3.1. System Architecture

To implement blockchain-based decentralized federated learning, we first designed the system architecture. Figure 1 shows the architecture of BD-FL, which mainly includes the demand release module, aggregation module, training module, and verification module. Since the nodes participating in the consensus in BD-FL are only a limited number of edge servers, the alliance chain is selected as the implementation platform of the architecture.

The demand release module is mainly composed of model demanders, whose main role is to release the demand task and pay the model training fee and the verification fee. After the model demander pays the fee and provides the initial model data, the system will send the relevant information of the initial model to each edge server for download by local devices. The aggregation module is mainly composed of edge servers that are close to local devices or data sources, such as base stations with certain computing and storage capabilities. It will save the gradient parameters of the local model, and other block data uploaded by local devices, and aggregate the global model on edge servers. It will also verify the accuracy of the uploaded gradient parameters and prevent dishonest local devices from maliciously providing wrong information. In the consensus process of blockchain, the local device will not participate, and the edge servers of the aggregation module will participate in the consensus to reduce the system communication delay. The training module is mainly composed of local devices, and its main role is to train local models using local data samples. In the model training, the system will bind local devices with edge servers according to our proposed matching algorithm. The local device will only upload the local model parameters to its bound edge server and only download the global model from its bound edge server. The verification module is also composed of some local devices. In each round of global model aggregation, the edge server sends the received local model gradient parameters uploaded by the local device to the verification device, which uses its own dataset to verify the quality of the local model, and returns the results to its bound edge server.

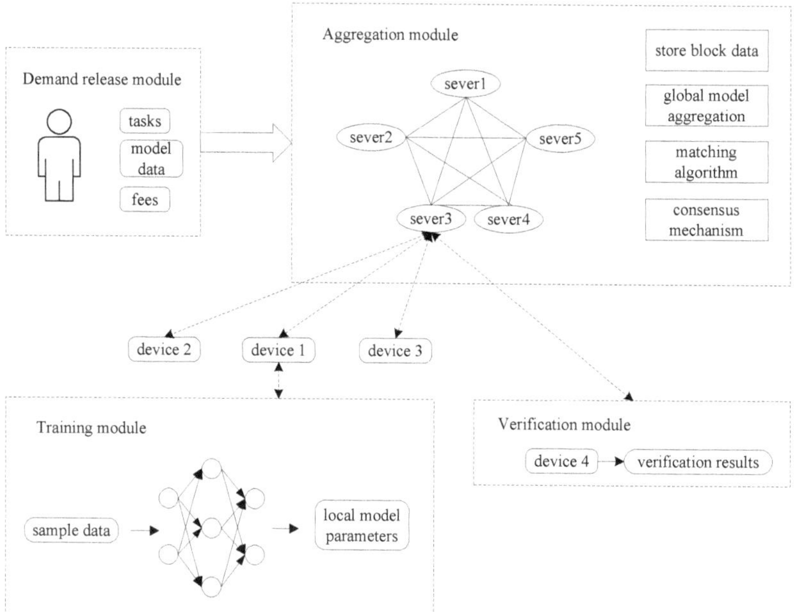

Figure 1. Architecture diagram of the decentralized federated learning.

3.2. Incentive Mechanism

In order to effectively promote local devices to participate in model training and verification, BD-FL has designed an incentive mechanism. During each round of training, the system will store the verification results returned by the verification device, the number of samples uploaded by the training device, and the training time of the device in the block. After a round of global model aggregation, the system will read the data saved on the block, calculate the reward of each training device according to the incentive mechanism, and send it to each local device.

The incentive mechanism gives corresponding rewards or punishments according to the contribution of local devices to model training. During the federated learning process, in order to ensure that the verification devices can give honest reports, their verification results can be re-verified by other devices, and dishonest validation behaviors will be punished. At the same time, in order to improve the fairness of reward distribution, the system will allocate the training fee according to the size of contribution made by each training device. The incentive mechanism introduces two metrics to calculate the final profit of the training device.

The first is the number of data samples owned by the training device. Devices with more data samples contribute more to global training, take longer time to train local models, and cost more. The second one is the accuracy of the model corresponding to the gradient parameters uploaded by the training device. The edge server verifies the model accuracy of the training device using the dataset of the verification device and returns the results to the edge server. An accuracy threshold T is introduced as a standard to measure whether the local model parameters of the training devices are qualified.

The system assigns the training fee S by scoring each training device. The score is related to the training time of the local device (denoted as *trainTime*) and the accuracy of the uploaded local model parameters (denoted as *accValue*). It is calculated by Equation (1), where α and β are the score coefficients, and the sum of them is 1.

$$S = \begin{cases} \alpha * trainTime + \beta * accValue \\ \alpha + \beta = 1 \ (0 < \alpha, \beta < 1) \end{cases}. \tag{1}$$

Equation (2) gives the calculation of the training reward R^k_{train} that an honest training device (device number is denoted as k) should receive in a training round, where m is the number of training devices, and P_{FLM} is the total cost of the model, which is paid to the public account of the system when the demander releases the task.

$$R^k_{train} = \begin{cases} 0 & (accValue < T) \\ P_{FLM} * \left(S_k / \sum_{i=1}^{m} S_i \right) & (accValue \geq T) \end{cases}. \tag{2}$$

3.3. Preference-Based Stable Matching Algorithm

Since local devices need to transmit a large amount of data to edge servers, the network quality, the transmission distance between nodes, the server throughput, and other factors will largely affect the overall delay and energy consumption of the system. We design a preference-based stable matching algorithm in BD-FL, which considers the above factors that affect data transmission, so as to achieve the optimal matching and binding between local devices and edge servers.

In the network environment, local devices and edge servers are abstracted into two sets, which are, respectively, denoted by the set of local devices $K = \{k_1, k_2, \ldots, k_m\}$ and the set of edge servers $S = \{s_1, s_2, \ldots, s_n\}$. For each $k \in K$, it has a matching request, denoted as Equation (3),

$$Q_k = (D_k, Info_k), \tag{3}$$

where D_k represents the data of uplink communication that local device k uploads to the edge server, mainly including information such as local model parameters, version number, local iteration time, etc.; $Info_k$ represents the state information of local device k, such as bandwidth, physical location, etc.

The uplink communication rate v_{ks} determines the data transmission delay, which is expressed as Equation (4) according to [32],

$$v_{ks} = band_k * \log_2 \left(1 + \frac{p_k * g_{ks}}{N_0} \right), \tag{4}$$

where $band_k$ represents the channel bandwidth allocated to the local device k, p_k represents the transmit power of k, g_{ks} represents the channel gain between k and the edge server s, and N_0 represents the noise power of s.

According to [32], the network distance between nodes is related to the channel bandwidth. It reflects the actual distance and network condition between local devices and edge servers. Assuming that $phys_k$ and $phys_s$ are the physical nodes corresponding to the local device and the edge server, respectively, the network distance $distNode_{ks}$ between them is defined as Equation (5),

$$distNode_{ks} = \frac{band_{aver}}{(band_k + band_s)/2} * distPhys_{ks}, \tag{5}$$

where $distPhys_{ks}$ represents the actual distance between $phys_k$ and $phys_s$, which has a certain influence on the reliability of data transmission and the network latency, $band_{aver}$ represents the average bandwidth between them, and $band_k$ and $band_s$ represent the actual bandwidths of $phys_k$ and $phys_s$.

Equations (6) and (7) express the upload time and the upload energy consumption, respectively, after local device k and edge server s are bound.

$$T_{ks} = \frac{D_k}{v_{ks}} \, , \tag{6}$$

$$E_{ks} = p_k * T_{ks} = p_k * \frac{D_k}{v_{ks}} \, . \tag{7}$$

We use the sum of weighted energy consumption and weighted delay, denoted as w_k, as the cost function of k, which is expressed by Equation (8),

$$\omega_k = \mu_k^e * E_k + \mu_k^t * T_k \, , \tag{8}$$

where $\mu_k^e, \mu_k^t \in [0,1], \mu_k^e + \mu_k^t = 1$, μ_k^e and μ_k^t, represent the energy consumption weight and the delay weight of k, respectively. A larger μ_k^e indicates a higher priority of energy consumption, and a larger μ_k^t indicates a higher priority of latency, and vice versa. The priority of energy consumption and delay of the local device can be adjusted by modifying the value of μ.

We define a coefficient matrix W to represent the cost relationship between local devices and edge servers. It is expressed by Equation (9), where w_{ks} denotes the cost when device k is bound to server s, and its value can be calculated by Equation (8).

$$W = (\omega_{ks})_{m*n} = \begin{bmatrix} \omega_{11} & \omega_{12} & \cdots & \omega_{1n} \\ \omega_{21} & \omega_{22} & \cdots & \omega_{2n} \\ \cdots & \cdots & \cdots & \cdots \\ \omega_{m1} & \omega_{m2} & \cdots & \omega_{mn} \end{bmatrix} . \tag{9}$$

We also define an assignment matrix X to represent the matching relationship between local devices and edge servers. It is expressed by Equation (10), where x_{ks} denotes the matching relationship between device k and server s, and $x_{ks} = 1$ means k is bound to s, otherwise, k is not bound to s.

$$X = (x_{ks})_{m*n} = \begin{bmatrix} x_{11} & x_{12} & \cdots & x_{1n} \\ x_{21} & x_{22} & \cdots & x_{2n} \\ \cdots & \cdots & \cdots & \cdots \\ x_{m1} & x_{m2} & \cdots & x_{mn} \end{bmatrix} . \tag{10}$$

The goal of the matching algorithm is to minimize of the total cost of model training, while also satisfying the constraint that the total number of local devices bound to edge server s cannot exceed its maximum number of device bindings. Therefore, we derive the optimization objective function, which is expressed by Equation (11).

$$\min obj = \sum_{k=1}^{m} \sum_{s=1}^{n} \omega_{ks} * x_{ks} \, , \, (x_{ks} = 0 \, or \, 1) \, . \tag{11}$$

We propose a preference-based stable matching algorithm to solve the optimization objective function of cost minimization. Each local device k will establish a preference list of matching degrees for all edge servers. We use a binary (k, s) to represent a match between device k and edge server s. Equation (12) gives the preference function of k for s, and $P_{(k,s)}$ represents the matching degree of (k, s). Equation (13) is the preference list of k, i.e., Γ_k. The matches are sorted in descending order of matching degree, that is, the higher the match in the list, the higher the preference.

$$P_{(k,s)} = \eta E_{ks} + (1 - \eta)T_{ks} \, , \tag{12}$$

$$\Gamma_k = [P_{(k,s)}, \quad \forall s \in S]. \tag{13}$$

Similarly, each edge server s will establish a matching preference list for all local devices, i.e., Γ_s, according to the preference function $P_{(s,k)}$. We only consider the strict partial order, that is, a local device will not have the same matching degree with two edge servers.

The local device tends to match the edge server with the highest matching degree, and the edge server also tends to bind to the local device with the highest matching degree. However, in the matching process, we can only pay more attention to the needs of one side. In order to make the local device have a better experience in the training process, the matching algorithm should meet the needs of the local device as much as possible, that is, the local device has priority to match.

Algorithm 1 describes the preference-based stable matching algorithm. All local devices and edge servers are initialized as unmatched. Then, all devices and servers broadcast their status information to each other, and each device and server establishes its preference lists. Each local device k sends a matching request to the edge server with the highest matching degree according to its Γ_k. If the request is rejected, it sends the matching request to other edge servers again according to their preference lists. Each edge server first places all devices that it receives the matching request into its match list. If the number of devices in the match list is greater than its maximum binding number, the edge server will pre-enroll devices according to its preference list from front to back, and reject the devices with low matching degrees in its Γ_s until the number of devices in the match list equals its maximum binding number. If the number of local devices in the match list is less than the maximum binding number, all devices are reserved. Each edge server repeats updating its match list until all local devices are in the match lists of edge servers. The output matches of the algorithm are Pareto optimal [33]. The time complexity of the algorithm is $O(nm) + O(n + m)$.

Algorithm 1 Preference-based stable matching algorithm.

Input: local device set K, edge server set S, maximum binding number of the server L_s.
Output: match list of edge servers N_s.
 1: All items in K and S are initialized as unmatched;
 2: All devices and servers broadcast their status information to each other;
 3: Each k and s establish the preference lists Γ_k and Γ_s;
 4: $i = 0, N_s = \varnothing, N_k = K$; $//N_k$ is an unmatched device list
 5: **while** $| N_k | > 0$ **do**
 6:    **for** $k \in N_k$ **do**
 7:       $i = i + 1$; $//k$ sends a matching request to the edge server with the highest matching degree in its Γ_k
 8:       k sends Q_k to s_i in its Γ_k;
 9:       $N_s = N_s \cup k$;
10:       $N_k = N_k - k$;
11:    **end for**
12:    **for** $s \in N_s$ **do**
13:       **while** $N_s > L_s$ **do**
14:          $//k'$ is the device with lowest matching degree in its Γ_s
15:          $k' = argmin_{k \in N_s} \Gamma_s$;
16:          $N_s = N_s - k'$;
17:          $N_k = N_k \cup k'$;
18:       **end while**
19:    **end for**
20: **end while**
21: **return** N_s;

3.4. R-PBFT Consensus Algorithm

To improve the model training efficiency of BD-FL, we propose a reputation-based practical Byzantine fault tolerance (R-PBFT) consensus algorithm. In the BD-FL environment, due to the large number of nodes, the traditional PBFT algorithm will lead to a sharp increase in communication cost and network bandwidth consumption in the consensus process, which is easy to cause network congestion and increase the delay of model training. In addition, the master node election in PBFT adopts the modulus calculation, which cannot guarantee the optimal master node of the election, thus affecting the consistency and reducing the reliability and security of the system.

The R-PBFT consensus algorithm combines the characteristics of BD-FL and designs the following improvements to solve the shortcomings of traditional PBFT for BD-FL.

- Remove the client node. In the traditional PBFT algorithm, the request phase and the reply phase occur between the client and the master nodes. However, in the blockchain structure, information is broadcast between nodes in the form of P2P, without the participation of the client. Therefore, we remove the request and reply phases of the client node in the consistency protocol, modify the C/S structure of PBFT to a distributed topology, and divide all nodes into master and slave nodes.
- Optimize the consistency protocol. The five phases of consensus in PBFT are changed to three phases, including the pre-preparation phase, preparation phase, and confirmation phase. In the pre-preparation phase, the master node broadcasts blocks to other slave nodes. In the preparation phase, the slave node broadcasts the block verification results to other slave nodes and master nodes. In the confirmation phase, traditional PBFT requires mutual interaction between nodes. We simplify it as all slave nodes send verification results to the master node, and the master node makes a decision on the consensus results, thus reducing the communication overhead of consensus.
- Introduce reputation mechanism. The main purpose of the reputation mechanism is to make the nodes with high reliability easier to be elected as the master node. Each node will be divided into different reputation levels according to the reputation value, and then each node will be rewarded or punished based on its performance in each round of consensus. According to a preset reputation threshold, nodes can be dynamically transformed in different reputation levels.

When the current round of consensus is completed, the system performs the reputation mechanism to assign each node to different reputation levels according to the reputation value and then selects the node with the highest score among the trusted nodes as the master node for the next round of consensus. Figure 2 shows the execution flow of a round of consensus, where P denotes the master node and S denotes the slave node.

According to the reputation value R of each node, the reputation mechanism divides it into three different reputation levels, namely trusted node, normal node, and unreliable node. R is a real number between 0 and 1, and its size reflects the reliability of the node. For the trusted node, the range of R is (0.8,1], and the node of this level generated valid blocks multiple times. For the normal node, the range of R is (0.3,0.8], and the node of this level generated unqualified blocks, but less often. For the unreliable nodes, the range of R is [0,0.3], and the node of this level generated unqualified blocks many times. The unreliable node will not participate in the election of the master node and the consensus process of the blockchain and only saves block data.

Each node updates its R according to the following rules after a consensus, so as to dynamically transform in different reputation levels. (1) The R of the node will be increased by 0.01 for each successful consensus participation. (2) If the master node successfully generates a valid block, its R will be increased by 0.02. However, if the master node fails or is identified as a malicious node, its R will be deducted by 0.2, and it will be immediately removed from the trusted node. (3) The slave node that correctly overthrows the malicious master node will increase its R by 0.02.

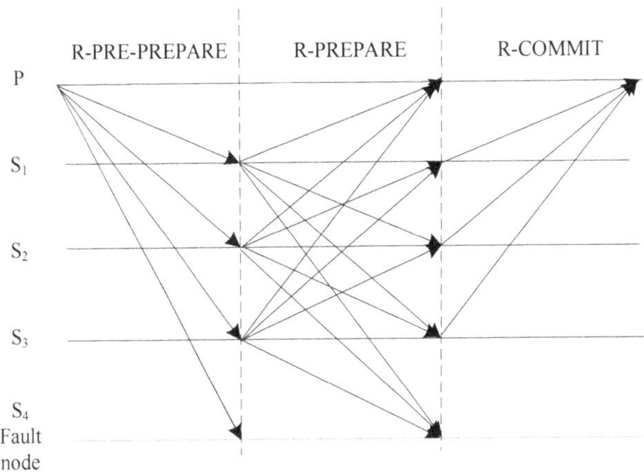

Figure 2. Execution flow of a round of consensus.

The initialization of reputation value is mainly divided into two cases. For the initial nodes of the system, the comprehensive strength of these nodes, such as computing power and network bandwidth, is used as the basis for initializing their reputation values. For the newly added node, other nodes vote to get its initial reputation value according to the comprehensive strength of the new node. Figure 3 shows the dynamic reputation level transformation of nodes.

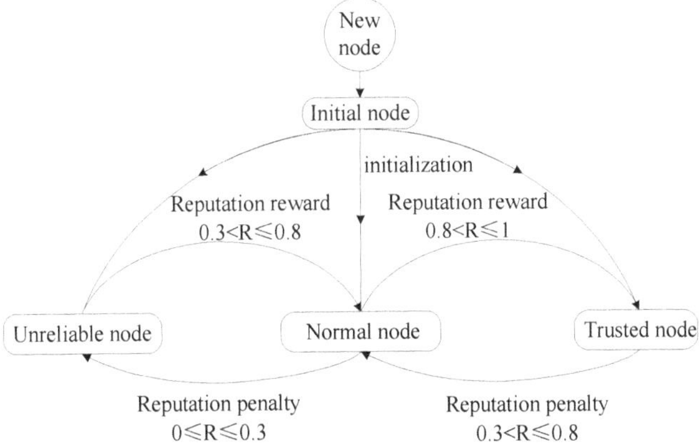

Figure 3. Dynamic reputation level transformation diagram of nodes.

On the one hand, the reputation mechanism can ensure the fairness of R-PBFT. Honest nodes will be rewarded, while malicious nodes will be punished for dishonesty. On the other hand, it can ensure the reliability of R-PBFT and remove malicious nodes in the blockchain network to avoid affecting the security of the system.

3.5. Training of BD-FL

Algorithm 2 describes the training process of the BD-FL. Assuming that the model needs r times of iterative training to reach convergence, the total time complexity of BD-FL is $O(n^2) + O(nm) + O(n + m) + O(1)$.

Algorithm 2 BD-FL training.

Input: local device set K, edge server set S, model demander MQ.
Output: global aggregation model M.
1: Initialization of K and MQ;
2: MQ releases task;
3: **for** $k \in K$ **do**
4:     calculate the matching degree with all edge servers;
5:     bind with a edge server according to the proposed matching algorithm;
6: **end for**
7: **while** the cost of global model is greater than the set threshold **do**
8:     **for** $k \in K$ **do**
9:         download global model M;
10:        train M using local data set;
11:        upload local model parameters, training time, and other information to its bound
            server after training;
12:    **end for**
13:    **for** $s \in S$ **do**
14:        use verification device to verify the authenticity of uploaded data and the accuracy
            of local model parameters;
15:        calculate the scores of local devices according to the incentive mechanism;
16:        **if** it is a master node **then**
17:            participate in the consensus process;
18:            aggregate all local model parameters and update M;
19:            store the global aggregation information, transaction information, scores, and
                etc. into blocks and broadcast in the blockchain;
20:            update the reputation value according to R-PBFT;
21:        **else**
22:            participate in the consensus process or not according to R-PBFT;
23:            receive M from the master node;
24:            notify local devices to download M after consensus;
25:            update the reputation value;
26:        **end if**
27:    **end for**
28: **end while**
29: **return** M;

---

4. Experiments and Results

4.1. Experiment Setting

We built a simulation experiment environment on an Intel(R) Xeon(R) server with two Gold 6248 processors @ 2.50 GHz (Intel Corporation, Santa Clara, CA, USA). We used Java version 1.8 to implement edge servers and the blockchain system, and Python version 3.7 to implement the local device environment. The local device communicates with the edge server through Socket. In the simulation environment, local devices are distributed within the coverage range of a 250 m radius of each edge server. The channel gain between them is modeled as $30.6 + 36.7 log_{10}(distNode_{ks})$ dB using the block fading model. Table 1 shows the simulation parameter settings.

Table 1. Simulation parameter settings.

Simulation Parameters	Value
Network Bandwidth	20 MHz
Shooting Power p_k	200 mW
Power Spectral Density	-95 dbm/Hz
Uplink Data Size D_k	[3000, 4000] kb
μ, η	0.5,0.5

We use the ResNet18 implemented by PyTorch as the federated learning network model, and the dataset is CIFAR-10. In experiments, the CIFAR-10 dataset will be randomly divided into multiple copies with different sizes. The local device will randomly obtain one copy as the local data sample.

4.2. Evaluation of BD-FL

Since the proposed preference-based stable matching algorithm plays an important role in BD-FL to minimize the model cost and reduce the system delay, we first conducted a set of experiments to evaluate the stable matching algorithm. We designed three matching algorithms as the baseline, i.e., the local device first greedy algorithm (DFG), the edge server first greedy algorithm (SFG), and the random matching algorithm (RMA), and applied them to BD-FL for comparison experiments.

For DFG, each local device sends a matching request to the edge server according to its preference list. If the candidate list of the edge server does not reach the maximum number of bindings, the device binds with the server directly, otherwise, it continues to send requests to other servers. For SFG, each edge server sends a matching request to the local device according to its preference list. If the local device is unbound, the server binds with the device directly, otherwise, it continues to send requests to other local devices. For RMA, the local device and the edge server randomly send matching requests to each other. If both parties are unbound, they can bind directly, otherwise, they continue to send requests to other unbound devices or servers.

Figure 4 shows the total system costs of the four matching algorithms with different numbers of nodes. As RMA is a pure random matching algorithm without considering any factors that affect the matching result, it has the highest system cost. The costs of DFG and SFG are also higher than that of the stable matching algorithm. Because they only consider the one-side cost function of the local device or the edge server as the minimization objective, they cannot achieve the overall optimization of the system cost. In contrast, the preference-based stable matching algorithm designs the corresponding preference functions of local devices and edge servers by considering various influencing factors, therefore, both parties are able to match and bind efficiently. The stable matching algorithm achieves the minimum system cost, which reduces the cost by 34.9%, 43.6%, and 69.9% compared to DFG, SFG, and RMA, respectively, on average.

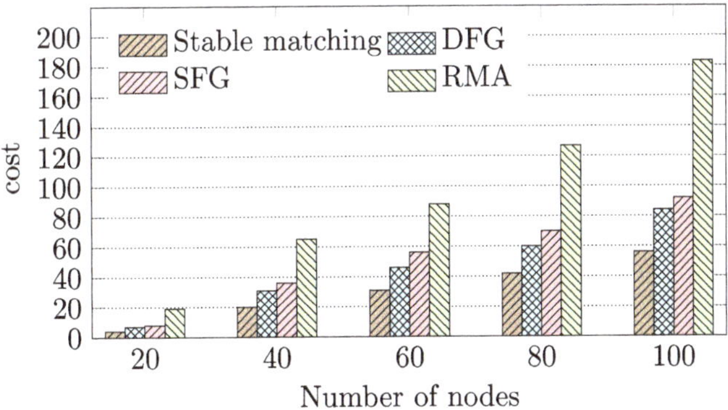

Figure 4. System costs of 4 matching algorithms with different numbers of nodes.

We also counted the number of unstable matches of DFG, SFG, and RMA, which is shown in Figure 5. It can be seen that the number of unstable matches in these algorithms increases with the number of nodes. Due to the randomness of RMA, the number of unstable matches is the largest. DFG and SFG also have multiple unstable matches. The result of unstable matches is consistent with the result of system costs in Figure 4, and the reason is the same. For the stable matching algorithm, the number of unstable matches is always 0, even if the number of nodes in BD-FL increases.

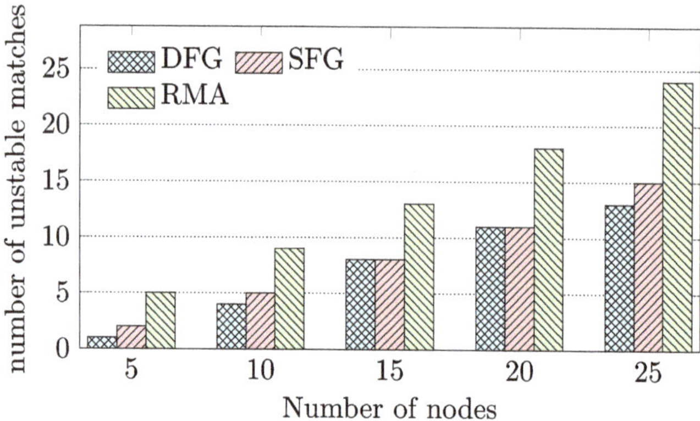

Figure 5. Number of unstable matches of DFG, SFG, and RMA.

To evaluate the model training time of BD-FL with the four matching algorithms, we set the number of edge servers to 4 and the number of local devices to 20 to build the simulation environment. In order to eliminate the influence of the consensus algorithm, all BD-FL models adopt traditional PBFT for blockchain consensus. Figure 6 shows the results of model training time. It can be seen that the training time of BD-FL models with DFG, SFG, and RMA is longer than that of the stable matching algorithm. This is because DFG and SFG only consider the single-side matching on local devices or edge servers, and RMA binds devices to servers in a random way, they cannot achieve the optimal stable matching. According to Figure 5, there are unstable matches in DFG, SFG, and RMA, which will increase the communication delay, and thus lead to a longer total time of model training. On the contrary, BD-FL uses the stable matching algorithm on both sides of the device and the server to utilize the computing resources and minimize the system communication cost. Compared with DFG, SFG, and RMA, the stable matching algorithm reduces the training time of BD-FL by 10.0%, 12.5%, and 19.7%, respectively, for 15 rounds of training. The result of model training is consistent with that of system cost and unstable matches.

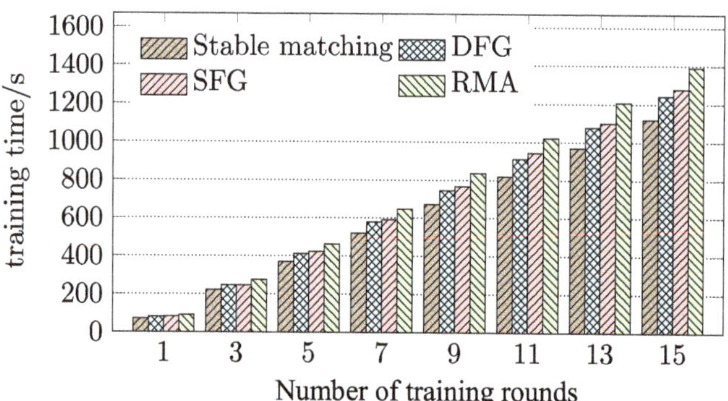

Figure 6. Model training time with different algorithms.

In order to better evaluate the performance of BD-FL, we chose a state-of-the-art blockchain-based federated learning method [17], named FLChain, for comparison. FLChain also applies blockchain techniques to federated learning. However, the nodes in the blockchain are composed of entities registered in FLChain, and all local devices participate in the consensus process of the blockchain.

In the comparison experiments, we set two configurations for BD-FL. In the first configuration, 20 local devices are bound to 4 edge servers in the BD-FL with the stable matching algorithm, denoted as BD-FL1. Additionally, in the second configuration, 20 local devices are bound to 10 edge servers, denoted as BD-FL2. For the network environment of

FLChain, we set 20 local devices and 4 edge servers. The 10 local devices are not bound to edge servers in FlChain for consensus. Figure 7 shows the comparison results of model training time. It can be seen that as the number of global training rounds increases, the training time of all methods grows, but the training time of FLChain grows significantly and is the longest. This is because the more nodes participating in the blockchain consensus, the longer the communication time, and ultimately the longer the model training time. FLChain does not optimize the blockchain consensus process, and all nodes need to participate in consensus, causing a higher communication cost and thus a longer training time. All 20 local devices in FLChain participate in the consensus process, while the numbers of nodes participating in the consensus in BD-FL1 and BD-FL2 are 4 and 10. Therefore, BD-FL1 is the most efficient in the consensus process among these three methods. Experimental results show that BD-FL1 and BD-FL2 reduce the training time by 34.9% and 27.0%, respectively, over FLChain for 150 rounds of global model training.

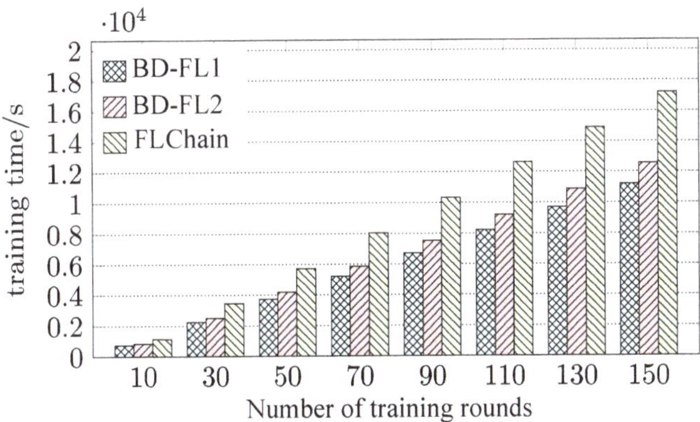

Figure 7. Model training time with different methods.

4.3. Evaluation of R-PBFT

To evaluate the performance of BD-FL with R-PBFT, we compared this method to the BD-FL with PBFT that applies the traditional PBFT algorithm to the consensus process of blockchain. In the experiments, BD-FL uses 20 local devices to bind with 4 edge servers according to the stable matching algorithm for both methods. Figure 8 shows the global model training time of these two methods over different numbers of training rounds. It can be seen that as the number of training rounds increases, the training time of the two methods grows, and the BD-FL with R-PBFT consumes significantly less time than the BD-FL with PBFT. This is due to the fact that R-PBFT streamlines the consistency protocol and eliminates the client nodes to optimize the traditional PBFT for the decentralized federated learning system in the edge computing environment. Thus, R-PBFT can effectively reduce the communication overhead of consensus and improve the global model training efficiency compared to the traditional PBFT. Meanwhile, R-PBFT can better guarantee the security of the system by introducing the reputation mechanism. In the experiments, BD-FL with R-PBFT reduces the training time by 12.2% compared with BD-FL with PBFT for 150 rounds of global model training.

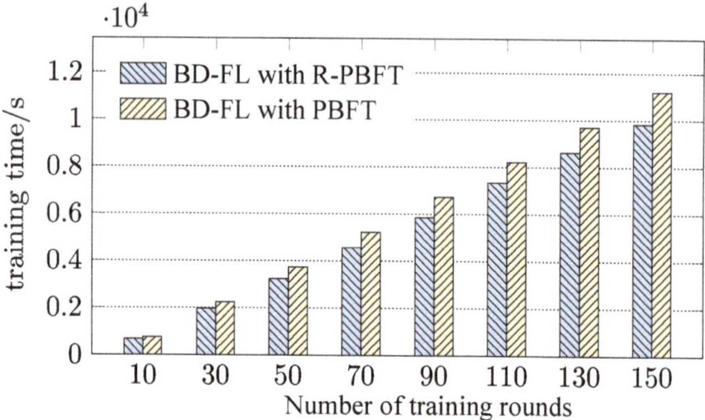

Figure 8. Model training time with different consensus algorithms.

5. Conclusions

In this paper, we mainly proposed a blockchain-based decentralized federated learning method for the edge computing environment. The proposed method joins all edge servers into the blockchain system, and the edge server nodes that obtain bookkeeping rights aggregate the global model to solve the centralization problem of federated learning caused by a single point of failure. In this method, we introduced an incentive mechanism to promote local devices to contribute data samples for model training. To further enhance the system efficiency, we proposed a preference-based stable matching algorithm to bind local devices with appropriate edge servers. For the consensus process of blockchain, we optimized the PBFT algorithm to reduce the communication overhead and enhance the system security, which improves the model training efficiency. Experimental results verified the effectiveness of the proposed method in communication overhead, system delay, and model training efficiency.

In the blockchain consensus process of the proposed method, the information broadcast between edge server nodes is not encrypted, which may lead to the disclosure of local model parameter information of the local models. Avoiding information leakage and improving system security in the consensus process will be one of our future research directions.

Author Contributions: Conceptualization, S.L. and X.W.; formal analysis, S.L.; experiment, S.L. and X.W. and L.H.; writing—original draft preparation, X.W.; writing—review and editing, S.L.; visualization, L.H.; supervision, W.W.; project administration, W.W. All authors have read and agreed to the published version of the manuscript.

Funding: This research was funded by the National Natural Science Foundation of China (No. 62002279 and No. 61972311), Natural Science Basic Research Program of Shaanxi (Program No. 2020JQ-077), and Shandong Provincial Natural Science Foundation (No. ZR2021LZH009).

Institutional Review Board Statement: Not applicable.

Informed Consent Statement: Not applicable.

Data Availability Statement: Not applicable.

Conflicts of Interest: The authors declare no conflict of interest.

References

1. Vinod, D.; Bharathiraja, N.; Anand, M.; Antonidoss, A. An improved security assurance model for collaborating small material business processes. *Mater. Today Proc.* **2021**, *46*, 4077–4081. [CrossRef]
2. Zhang, Q.; Ding, Q.; Zhu, J.; Li, D. Blockchain empowered reliable federated learning by worker selection: A trustworthy reputation evaluation method. In Proceedings of the 2021 IEEE Wireless Communications and Networking Conference Workshops (WCNCW), Nanjing, China, 29 March–1 April 2021; pp. 1–6.

3. Tomovic, S.; Yoshigoe, K.; Maljevic, I.; Radusinovic, I. Software-defined fog network architecture for IoT. *Wirel. Pers. Commun.* **2017**, *92*, 181–196. [CrossRef]
4. Hu, Y.; Niu, D.; Yang, J.; Zhou, S. FDML: A collaborative machine learning framework for distributed features. In Proceedings of the 25th ACM SIGKDD International Conference on Knowledge Discovery & Data Mining, Anchorage, AK, USA, 4–8 August 2019; pp. 2232–2240.
5. Nakamoto, S.; Bitcoin, A. A peer-to-peer electronic cash system. *Bitcoin* **2008**, *4*, 2. Available online: https://bitcoin.org/bitcoin.pdf (accessed on 1 January 2023)
6. Shi, W.; Cao, J.; Zhang, Q.; Li, Y.; Xu, L. Edge computing: Vision and challenges. *IEEE Internet Things J.* **2016**, *3*, 637–646. [CrossRef]
7. Liu, Z.; Cao, Y.; Gao, P.; Hua, X.; Zhang, D.; Jiang, T. Multi-UAV network assisted intelligent edge computing: Challenges and opportunities. *China Commun.* **2022**, *19*, 258–278. [CrossRef]
8. Zhang, X.; Wu, W.; Yang, S.; Wang, X. Falcon: A blockchain-based edge service migration framework in MEC. *Mobile Inf. Syst.* **2020**, *2020*, 8820507. [CrossRef]
9. Guo, F.; Yu, F.R.; Zhang, H.; Ji, H.; Liu, M.; Leung, V.C. Adaptive resource allocation in future wireless networks with blockchain and mobile edge computing. *IEEE Trans. Wirel. Commun.* **2019**, *19*, 1689–1703. [CrossRef]
10. Shahryari, O.K.; Pedram, H.; Khajehvand, V.; TakhtFooladi, M.D. Energy and task completion time trade-off for task offloading in fog-enabled IoT networks. *Pervasive Mob. Comput.* **2021**, *74*, 101395. [CrossRef]
11. Yang, S.; Han, K.; Zheng, Z.; Tang, S.; Wu, F. Towards personalized task matching in mobile crowdsensing via fine-grained user profiling. In Proceedings of the IEEE INFOCOM 2018-IEEE Conference on Computer Communications, Honolulu, HI, USA, 15–19 April 2018; pp. 2411–2419.
12. Bharathiraja, N.; Padmaja, P.; Rajeshwari, S.; Kallimani, J.S.; Buttar, A.M.; Lingaiah, T.B. Elite Oppositional Farmland Fertility Optimization Based Node Localization Technique for Wireless Networks. In Proceedings of the Wireless Communications and Mobile Computing, Dubrovnik, Croatia, 30 May–3 June 2022; Volume 2022.
13. Vasin, P. Blackcoin's Proof-of-Stake Protocol v2. 2014. Volume 71. Available online: https://blackcoin.co/blackcoin-pos-protocol-v2-whitepaper.pdf (accessed on 1 January 2023).
14. Liu, D.; Camp, L.J. Proof of Work can Work. In Proceedings of the 5th Annual Workshop on the Economics of Information Security, Robinson College, University of Cambridge, England, UK, 26–28 June 2006. Available online: https://econinfosec.org/archive/weis2006/docs/50.pdf (accessed on 1 January 2023).
15. Yang, F.; Zhou, W.; Wu, Q.; Long, R.; Xiong, N.N.; Zhou, M. Delegated proof of stake with downgrade: A secure and efficient blockchain consensus algorithm with downgrade mechanism. *IEEE Access* **2019**, *7*, 118541–118555. [CrossRef]
16. Castro, M.; Liskov, B. Practical Byzantine Fault Tolerance. *ACM Trans. Comput. Syst. (TOCS)* **2002**, *20*, 398–461 . [CrossRef]
17. Bao, X.; Su, C.; Xiong, Y.; Huang, W.; Hu, Y. FLChain: A blockchain for auditable federated learning with trust and incentive. In Proceedings of the 2019 5th International Conference on Big Data Computing and Communications (BIGCOM), Qingdao, China, 9–11 August 2019; pp. 151–159.
18. Nofer, M.; Gomber, P.; Hinz, O.; Schiereck, D. Blockchain. *Bus. Inf. Syst. Eng.* **2017**, *59*, 183–187. [CrossRef]
19. Zheng, Z.; Xie, S.; Dai, H.N.; Chen, X.; Wang, H. Blockchain challenges and opportunities: A survey. *Int. J. Web Grid Serv.* **2018**, *14*, 352–375. [CrossRef]
20. Kim, H.; Park, J.; Bennis, M.; Kim, S.L. Blockchained on-device federated learning. *IEEE Commun. Lett.* **2019**, *24*, 1279–1283. [CrossRef]
21. Weng, J.; Weng, J.; Zhang, J.; Li, M.; Zhang, Y.; Luo, W. Deepchain: Auditable and privacy-preserving deep learning with blockchain-based incentive. *IEEE Trans. Dependable Secur. Comput.* **2019**, *18*, 2438–2455. [CrossRef]
22. Jia, R.; Dao, D.; Wang, B.; Hubis, F.A.; Hynes, N.; Gürel, N.M.; Li, B.; Zhang, C.; Song, D.; Spanos, C.J. Towards efficient data valuation based on the shapley value. In Proceedings of the The 22nd International Conference on Artificial Intelligence and Statistics, PMLR, Naha, Okinawa, Japan, 16–18 April 2019; pp. 1167–1176.
23. Hu, G.; Jia, Y.; Chen, Z. Multi-user computation offloading with d2d for mobile edge computing. In Proceedings of the 2018 IEEE Global Communications Conference (GLOBECOM), Abu Dhabi, United Arab Emirates, 9–13 December 2018; pp. 1–6.
24. Wu, D. Research on multi-task multi-device matching algorithm based on machine learning. Master's Thesis, Zhe Jiang University, Hangzhou, China, 2019.
25. Lu, X.; Liao, Y.; Lio, P.; Hui, P. Privacy-preserving asynchronous federated learning mechanism for edge network computing. *IEEE Access* **2020**, *8*, 48970–48981. [CrossRef]
26. Zheng, H.; Guo, W.; Xiong, N. A kernel-based compressive sensing approach for mobile data gathering in wireless sensor network systems. *IEEE Trans. Syst. Man, Cybern. Syst.* **2017**, *48*, 2315–2327. [CrossRef]
27. Ma, S.; Cao, Y.; Xiong, L. Transparent contribution evaluation for secure federated learning on blockchain. In Proceedings of the 2021 IEEE 37th International Conference on Data Engineering Workshops (ICDEW), Chania, Greece, 19–22 April 2021; pp. 88–91.
28. Gao, S.; Yu, T.; Zhu, J.; Cai, W. T-PBFT: An EigenTrust-based practical Byzantine fault tolerance consensus algorithm. *China Commun.* **2019**, *16*, 111–123. [CrossRef]
29. Liu, J.; Li, W.; Karame, G.O.; Asokan, N. Scalable byzantine consensus via hardware-assisted secret sharing. *IEEE Trans. Comput.* **2018**, *68*, 139–151. [CrossRef]

30. Wang, Y.; Song, Z.; Cheng, T. Improvement research of PBFT consensus algorithm based on credit. In *International Conference on Blockchain and Trustworthy Systems*; Springer: Berlin/Heidelberg, Germany, 2020; pp. 47–59.
31. Yu, G.; Wu, B.; Niu, X. Improved blockchain consensus mechanism based on PBFT algorithm. In Proceedings of the 2020 2nd International Conference on Advances in Computer Technology, Information Science and Communications (CTISC), Suzhou, China, 10–12 July 2020; pp. 14–21.
32. Nithya, G.; Engels, R.; Das, H.R.; Jayapratha, G. A Novel-Based Multi-agent Brokering Approach for Job Scheduling in a Cloud Environment. In *Informatics and Communication Technologies for Societal Development*; Springer: Berlin/Heidelberg, Germany, 2015; pp. 71–84.
33. Hochman, H.M.; Rodgers, J.D. Pareto optimal redistribution. *Am. Econ. Rev.* **1969**, *59*, 542–557.

Disclaimer/Publisher's Note: The statements, opinions and data contained in all publications are solely those of the individual author(s) and contributor(s) and not of MDPI and/or the editor(s). MDPI and/or the editor(s) disclaim responsibility for any injury to people or property resulting from any ideas, methods, instructions or products referred to in the content.

Article

Towards Mobile Federated Learning with Unreliable Participants and Selective Aggregation

Leonardo Esteves [1], David Portugal [2,*], Paulo Peixoto [2] and Gabriel Falcao [1]

[1] Instituto de Telecomunicações, Department of Electrical and Computer Engineering, University of Coimbra, 3030-290 Coimbra, Portugal
[2] Instituto de Sistemas e Robótica, Department of Electrical and Computer Engineering, University of Coimbra, 3030-290 Coimbra, Portugal
* Correspondence: davidbsp@isr.uc.pt

Abstract: Recent advances in artificial intelligence algorithms are leveraging massive amounts of data to optimize, refine, and improve existing solutions in critical areas such as healthcare, autonomous vehicles, robotics, social media, or human resources. The significant increase in the quantity of data generated each year makes it urgent to ensure the protection of sensitive information. Federated learning allows machine learning algorithms to be partially trained locally without sharing data, while ensuring the convergence of the model so that privacy and confidentiality are maintained. Federated learning shares similarities with distributed learning in that training is distributed in both paradigms. However, federated learning also decentralizes the data to maintain the confidentiality of the information. In this work, we explore this concept by using a federated architecture for a multi-mobile computing case study and focus our attention on the impact of unreliable participants and selective aggregation in the federated solution. Results with Android client participants are presented and discussed, illustrating the potential of the proposed approach for real-world applications.

Keywords: federated learning (FL); federated averaging (FedAvg); federated SGD (FedSGD); unreliable participants; selective aggregation

Citation: Esteves, L.; Portugal, D.; Peixoto, P.; Falcao, G. Towards Mobile Federated Learning with Unreliable Participants and Selective Aggregation. *Appl. Sci.* **2023**, *13*, 3135. https://doi.org/10.3390/app13053135

Academic Editors: George Drosatos, Pavlos S. Efraimidis and Avi Arampatzis

Received: 2 February 2023
Revised: 24 February 2023
Accepted: 26 February 2023
Published: 28 February 2023

Copyright: © 2023 by the authors. Licensee MDPI, Basel, Switzerland. This article is an open access article distributed under the terms and conditions of the Creative Commons Attribution (CC BY) license (https://creativecommons.org/licenses/by/4.0/).

1. Introduction

Mobile and wearable devices, as well as autonomous vehicles, are just some of the applications that are part of modern distributed networks that generate an abysmal quantity of data every day [1–3]. With increased computing power, energy efficiency, lower latency between communications, greater bandwidth, better data management solutions, and even storage capacity on these devices, as well as concerns about the disclosure of private information, it is becoming increasingly attractive to allocate data locally (e.g., on mobile devices) and push data processing to the edge so that users' private information is not exposed [4].

The concept of edge computing is not new. In fact, computing simple instructions in low-power, distributed devices is an area of research that is several years old. Recently, works have emerged, which consider training machine learning (ML) models centrally but serving and storing models locally in these distributed devices [5].

As the computational and storage capabilities of devices in a distributed network grow, local resources can be used on each device. This has led to an increase in interest in federated learning (FL) [6–8]. Federated learning is a machine learning technique in which an algorithm is trained on multiple decentralized edge devices or servers using local private data samples without sharing them. It has numerous applications, such as healthcare systems, industry, telecommunications, etc., as shown in Figure 1.

This concept shares some similarities with distributed learning. In distributed learning, the training model is distributed across different nodes, while FL assumes a decentralization of both the data and the model and generates a global model that aggregates all the clients' models [9] (see Figure 2).

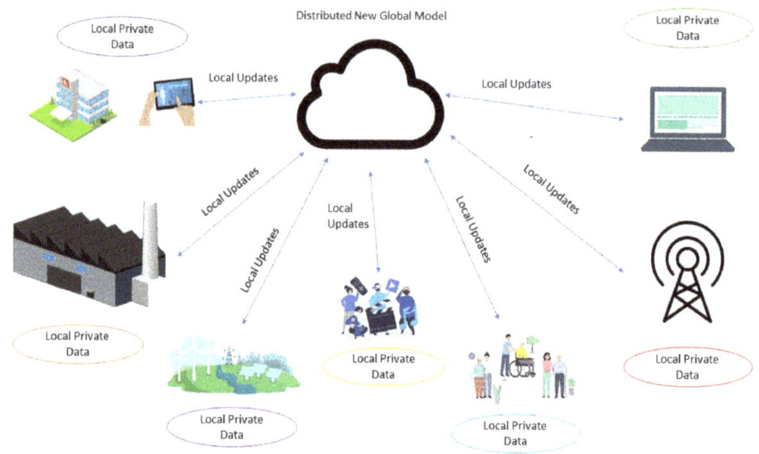

Figure 1. Federated Learning Applications.

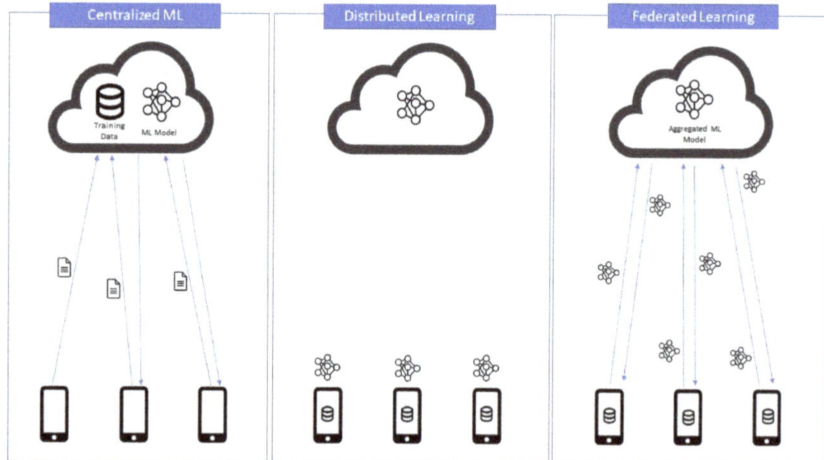

Figure 2. Difference Between Centralized Machine Learning, Distributed Learning and Federated Learning.

With the wide availability of mobile phones, we investigate the feasibility of implementing machine-learning-based techniques in these devices. Specifically, this paper explores federated learning for multimobile computing applications towards real-world deployment scenarios, by focusing on the implementation of a federated solution in Android devices. It also aims to preliminarily investigate different methods for aggregating and updating the global model managed by the central server, using unreliable client participants. In addition, the challenges and drawbacks of the FL process are discussed. Thus, the main contributions are:

- The implementation of a real-world FL application with up to eight Android mobile devices;
- A comparison of results obtained with the developed FL solution against a baseline obtained with a centralized ML approach;

- We demonstrate that a more advanced aggregation method, such as a weighted average instead of an arithmetic average, can significantly improve the overall result of an FL solution;
- Based on existing challenges, we conclude that eligibility criteria can benefit the FL framework in the presence of unreliable participants.

2. Background and Related Work

The federated learning concept aims to build a machine learning model based on data distributed across multiple sites [10]. It offers clients the ability to collaboratively train a common model without sharing personal data stored on their devices [8,11].

FL consists of two processes: model training and model inference. Typically, in an FL framework there are two main players: clients/workers/parties who contribute to the training of the model, and the server where the aggregation and update of the global model is performed. Thus, model training consists of sharing information between the workers and the server. However, the training data can never be shared. On the other hand, during the inference phase, the model is trained based on new data. Lastly, FL provides a mechanism to distribute the benefit of the whole process through all the collaborative parties. As described in [11,12], FL is a framework for building ML models that can be characterized by the following features:

- Two or more parties are interested in jointly creating an ML model. Each party has some data that will be used as input to train the model;
- It is imperative that during model training, the data kept by each party never leave that party;
- The locally trained model can be partially transferred from one party to another under an encryption scheme, so that other parties cannot reproduce the training data of a particular party;
- Theoretically, the performance of the resulting model is a reasonable approximation of an ideal ML model built using data from all parties.

More technically, let $\mathcal{N} = \{1, \ldots, N\}$ be the set of N parties, each of which has a private dataset $D_{i \in \mathcal{N}}$. Each data owner i uses its dataset D_i to train a local model w_i. The local model parameters are then forwarded to the FL server. All collected local model parameters $\mathbf{w} = \cup_{i \in \mathcal{N}} \mathbf{w}_i$ are then aggregated to create a global model W_G [13].

The conventional architecture of an FL system is illustrated in Figure 3. In that system, the parties jointly train an ML model with the assistance of an aggregate server. It is assumed that all parties involved in the process are trustworthy, which means that they use their private data to perform the training and pass the locally obtained model parameters to the FL server in each training round. Evidently, this assumption may not be practical, as discussed further ahead.

Typically, the FL process involves the following three steps [13].

- Step 1 (task initialization): The FL server determines the training assignment, i.e., the application goal and the corresponding data requirements. Moreover, the server determines the hyperparameters of the global model and the training process. Finally, the server transmits the initialized global model $\mathbf{w}_G^0$ to the selected parties.
- Step 2 (local model training and update): Following the global model $\mathbf{W}_G^t$, where t represents the current iteration index, each party individually uses its local data to update the local model parameters $\mathbf{w}_i^t$. The purpose of a party i in iteration t is to find ideal parameters $\mathbf{w}_i^t$ that minimize the loss function $L(\mathbf{w}_i^t)$, for example,

$$\mathbf{w}_i^{t^*} = arg \min_{\mathbf{w}_i^t} L(\mathbf{w}_i^t). \tag{1}$$

Then, the local model parameters are sent to the FL server.

- Step 3 (global model aggregation and update): the FL server aggregates the local models received from all selected parties and sends the updated model parameters $\mathbf{w}_G^{t+1}$ back to the participants.

Steps 2–3 are repeated until the global loss function converges or a desired training accuracy is achieved, i.e.,

$$L(\mathbf{w}_G^t) = \frac{1}{N} \sum_{i=1}^{N} L(\mathbf{w}_i^t). \tag{2}$$

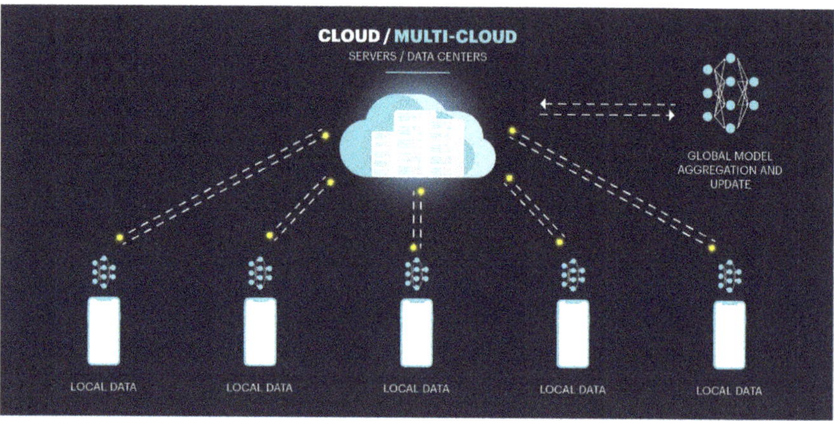

Figure 3. Federated Learning Architecture.

The FL training process can be applied to several ML models using the stochastic gradient descent (SGD) method. Typically, a training dataset contains a set of n data feature vectors $x = \{x_1, \ldots, x_n\}$ and a set of corresponding data labels. In the case of unsupervised learning, there is no data label. $y = \{y_1, \ldots, y_n\}$. Moreover, $\hat{y}_j = f(x_j; \mathbf{w})$ denotes the predicted result from the model $\mathbf{w}$ updated/trained by data vector x_j [13].

Most of the recent successful deep learning [14] applications use variants of SGD for optimization. In fact, many advances can be understood as an adaptation of the structure of the model (and hence the loss function) to facilitate optimization using simple gradient-based methods. Therefore, it is only natural that federated optimization algorithms are built based on SGD.

After training and updating the local model comes an essential part of the FL process: global model aggregation. There are several proposed solutions to this issue.

Experiments by Chen et al. [15] show that large-batch synchronous state-of-the-art SGD optimization outperforms asynchronous approaches, in a data center application. To apply this approach in a federated context, a C-fraction of clients/parties is selected at each round, and the gradient of the loss over all the data held by these parties is computed. Thus, C controls the *global* batch size, with $C = 1$ corresponding to full-batch (nonstochastic) gradient descent. While the batch size selection mechanism is different than selecting a batch by choosing individual examples uniformly at random, the batch gradients g computed by FedSGD still satisfies $\mathbb{E}[g] = \bigtriangledown f(w)$. The baseline algorithm is FederatedSGD (FedSGD) [14].

A typical implementation of FedSGD with $C = 1$ and a fixed learning rate η has each client k compute $g_k = \bigtriangledown F_k(w_t)$, the average gradient on its local data at the current model w_t, and the central server aggregates these gradients and applies the update $w_{t+1} \leftarrow w_t - \eta \sum_{k=1}^{K} \frac{n_k}{n} g_k$, since $\sum_{k=1}^{K} \frac{n_k}{n} g_k = \bigtriangledown f(w_t)$ [16]. An equivalent update is given by $\forall k$, $w_{t+1}^k \leftarrow w_t - \eta g_k$ and then $w_{t+1} \leftarrow \sum_{k+1}^{K} \frac{n_k}{n} w_{t+1}^k$. That is, each client locally performs a gradient descent step on the current model using its local data, and the server takes a weighted average of the resulting models. Within the algorithm, more computation steps can be added to each client by iterating the local update $w^k \leftarrow w^k - \eta \nabla F_k(w^k)$ multiple times before the averaging step [14]. This approach is called `FederatedAveraging` (`FedAvg`).

The number of computational steps is controlled by three key parameters: C, the fraction of clients performing computations in each round; E, the number of training passes each client makes over its local dataset in each round; and B, the local batch size used for client updates. Selecting $B = \infty$ and $E = 1$ is equivalent to FedSGD. In FedAvg, for a client with n_k local examples, the number of local updates per round is given by $u_k = E \frac{n_k}{B}$ [14]. Algorithm 1 is a complete pseudocode of the `FedAvg` procedure.

Algorithm 1: `FederatedAveraging` (`FedAvg`). The K clients are indexed by k; B is the local minibatch size, E is the number of local epochs, and η is the learning rate.

Server executes:
   initialize w_0
   for each round $t = 1, 2, \ldots$ **do**
     $m \leftarrow max(C \cdot K, 1)$
     $S_t \leftarrow$ (random set of m clients)
     for each client $k \in S_t$ **in parallel do**
       $w_{t+1}^k \leftarrow ClientUpdate(k, w_t)$
    $w_{t+1} \leftarrow \sum_{k=1}^{K} \frac{n_k}{n} w_{t+1}^k$

ClientUpdate(k, w): *//Run on client k*
   $B \leftarrow$ split P_k into batches of size B
   for each local epoch i from 1 to E **do**
     for batch $b \in B$ **do**
       $w \leftarrow w - \eta \nabla l(w; b)$
   return w to server

Typical FL solutions prioritize the privacy and security of all the parties involved. In fact, the server does not need to access individual information from each party involved to perform SGD. The server only requires the weighted averages of the update vectors. In [17], the authors proposed the secure aggregation protocol to compute these weighted averages. This method ensured that the server became aware of what types of data were present in the dataset, but without knowing the data that each user contained.

In mobile contexts, devices have limited resources in terms of energy and network connectivity. This introduces some unpredictability in terms of the number of parties that can participate in each round of updates, and the system must be able to respond to disconnections and new participants. Since ML models can be parameterized with millions of different values, model updates can involve a large quantity of data, which imposes a direct cost on users on metered network plans.

Secure aggregation proposes to work with high-dimensional vectors, uses efficient communication strategies, even considering the addition of new parties at each round and being robust to users dropping out, and finally, provides the strongest possible security measures under the constraints of a server-mediated, unauthenticated network model.

Pillutla et al. [18] introduced the use of the robust federated averaging (*RFA*) algorithm. RFA replaces the mean aggregation of *FedAvg* with a geometric median (GM) based robust aggregation oracle. Similar to *FedAvg*, *RFA* trades off communication for local computation by running multiple local steps. The communication efficiency and privacy preservation of *RFA* follow from computing the GM as an iterative secure aggregate.

Challenges

Some challenges distinguish the FL framework from other classical ML problems, such as distributed learning or traditional private data analytics [5]. Below, we describe four of the most important challenges identified for federated learning:

Expensive Communications

Client–client and client–server communication is a critical bottleneck in FL solutions [8]. When paired with privacy concerns over the transmission of raw data, it becomes necessary for the data generated on each side to remain local. It is undeniable that FL networks potentially include a huge number of devices, and network communications can be many orders of magnitude slower than local computations, due to limited resources such as bandwidth, energy, and power [19,20].

To fit a model to the data generated by devices in the federated network, it is important to develop communication-efficient methods that iteratively send small messages or model updates as part of the training process, rather than sending the entire dataset across the network. To reduce communication in such an environment, there are two key aspects to consider: minimizing the total number of communication rounds and the size of messages transmitted in each round [5,20].

Systems Heterogeneity

The storage, computational, and communication capacities of individual devices in federated networks may differ due to differences in hardware (CPU and memory), network connectivity (3G, 4G, 5G, and Wi-Fi), and power (battery level) [19]. In addition, the network size and systems-related constraints on each device typically result in only a small fraction of the devices being active at once. Moreover, it is not uncommon for an active device to lose connection with the server (dropout) in a given iteration due to connectivity or power constraints [8].

These system-level characteristics dramatically increase the number of challenges, such as straggler mitigation and fault tolerance [5]. Proposed FL approaches must therefore anticipate a low amount of participation, tolerate heterogeneous hardware, and be robust enough to dropped devices in the communication network [5,8].

Statistical Heterogeneity

Devices often generate and collect data that are not identically distributed across the network, e.g., smartphone users have a varied use of language in the context of a next-word prediction task [5]. Furthermore, the number of data points may vary significantly between devices, and there can be an underlying statistical structure present that captures the relationship among devices and their associated distributions [21]. This data-generation paradigm frequently violates the independent and identical distribution (i.i.d.) assumptions commonly used in distributed optimization and can increase the complexity of problem modeling, theoretical analysis, and empirical evaluation of solutions [5].

Although the canonical FL problem aims to learn a single global model, there are other alternatives such as learning different local models simultaneously via multitask learning frameworks [21]. In this respect, there is also a close connection between leading approaches for FL. This multitasking perspective allows personalized or device-specific modeling, which is often a more natural approach to handling statistical heterogeneity in the data for better adaptation [5,8].

Privacy Concerns

FL makes a step toward protecting data generated on individual devices by sharing model updates, e.g., gradient information, instead of raw data. However, sending model updates throughout the network during the training process can expose sensitive information, either to third parties or to the central server [22].

Although recent methods aim to improve the privacy of FL using tools such as secure multiparty computation (SMC) or differential privacy, these approaches often provide privacy at the cost of reduced model performance or system efficiency [17,22]. Understanding these tradeoffs, both theoretically and empirically, is a significant challenge in implementing private FL systems [5].

FL is already a case study for many applications. Simsek et al. [23] presented a study on AI-driven autonomous vehicles, and in [24], an energy efficient distributed analytics solution for IoT was proposed. In addition, Malekzadeh et al. [25] used FL for the transmission of sensor-data transformations.

The next section details the implementation of a mobile FL framework using Android devices as a case study, focusing on the impact of unreliable participants and selective aggregation methods in a multimobile computing application. Since this is the focus of the work, we do not explicitly address some of the challenges mentioned above, such as client heterogeneity or privacy concerns.

3. Proposed Federated Learning Infrastructure

In what follows, we describe an implementation which aims to serve as a guide to design a real-world multimobile FL solution.

3.1. Training Parameters

For our FL implementation, there were 4 controllable variables that allowed the testing of various training scenarios.

The first parameter (**num_clients**) represented the total number of clients participating in the federated solution. The dataset was divided among these clients so that each client received a predetermined number of dataset elements.

The second parameter represented the number of rounds that the training should take (**num_rounds**), and the third parameter represented the total number of local training rounds on each selected client's device (**epochs**). The fourth and last parameter was the batch size (**batch_size**). Data were loaded per batch, so it was important to choose a value that did not allow too little data to be loaded at once, as this would degrade training performance by not utilizing the full processing power of the client devices. However, it was also important not to select a value that was too high, as this could lead to data overload on the clients.

After the initial definition of **num_clients** and **num_rounds**, in each communication round (a communication round in federated learning represents the server–client and client–server data exchange associated to the training performed at each client), training on the client's devices with the local dataset took place, according to the number of **epochs** and the **batch_size**. Then, the server aggregated the individual models by weighting the corresponding parameters into one global model.

Below, we present the training parameters chosen in this work. The values for **num_rounds** were set empirically, since for the chosen dataset (see Section 5), we achieved an accuracy after 100 rounds that was approximately equal to the maximum achieved by Zhu et al. [26]. The **num_clients** was limited by the resources available for Android deployment in our work. The values of **epochs** and **batch_size** were fixed to eliminate their influence on the results and to allow a comparison of different configurations under the same test conditions.

```
num_rounds = {10,30,500,100}
num_clients = {2, 4, 8}
epochs = 5
batch_size = 32
```

3.2. Model Architecture

The chosen architecture of the neural network was **VGG19**. This network can be seen as a successor to **AlexNet** [27], which builds on its predecessors and improves them by using deep convolutional neural layers to achieve a higher accuracy [28].

The architecture of the VGG19 network provides 19 weight layers (16 CNN layers, 3 fully connected layers, and a final layer for a softmax function) as seen in Table 1. It assumes 224 RGB channels as input, leading to a 3D matrix of size $224 \times 224 \times 3$. In a first preprocessing stage, the average RGB value calculated for the whole training set is subtracted from each pixel. A 3×3 kernel with a stride size of 1 pixel is used, allowing the entire image to be covered. Spatial padding is used to preserve the spatial resolution of the image. A max-pooling is performed over a 2×2 pixel windows with a stride of 2. This is followed by a rectified linear unit (**ReLu**) to introduce a nonlinearity that improves the classification performance of the model and the computation time compared to previous models that used **tanh** or **sigmoid** functions. Finally, there are three fully connected layers. The first two have a size of 4096 and the last one has 1000 channels. The final layer is a **softmax** function. This network is widely used for image classification in various fields, such as face recognition.

Table 1. Configuration of the VGG-19 Network.

Input (224×224 RGB Image)
conv3-64 + ReLu
conv3-64 + ReLu
maxpool
conv3-128 + ReLu
conv3-128 + ReLu
maxpool
conv3-256 + ReLu
conv3-256 + ReLu
conv3-256 + ReLu
conv3-256 + ReLu
maxpool
conv3-512 + ReLu
conv3-512 + ReLu
conv3-512 + ReLu
conv3-512 + ReLu
maxpool
conv3-512 + ReLu
conv3-512 + ReLu
conv3-512 + ReLu
conv3-512 + ReLu
maxpool
FC-4096 + ReLu
FC-4096 + ReLu
FC-1000 + ReLu
softmax

3.3. Training the FL Model

At this stage, a global model and the individual client models were initialized. In this solution, SGD was used as an optimizer for all client models. The training process took place within a cycle, i.e., training was performed by the clients before each round of communication. After receiving the updates from the local modules, the aggregation of the weights was performed on the server according to a specific aggregation method, which updated the global model that was then used to validate the training. This process continued for **num_rounds**, i.e., 100 communication rounds for all the tests performed. Accordingly, Algorithm 2 summarizes the FL training process in our implementation.

Algorithm 2: Training steps of our FL implementation.

---

for *i in num_rounds* **do**
    for *j in num_clients* **do**
       | Perform local training
    end
    Send the global model to the server
    Aggregate the global model
    Validate the global model
end

3.4. System Design

The server–client and client–server communication was done through a REST server. Since we focused exclusively on the impact of unreliable participants and alternative aggregation methods, a REST server presented a simple way to communicate through POST messages with a JSON formatted body. This allowed us to perform a proof-of-concept study while ignoring aspects that were outside the scope of this work, such as protecting clients and the server from external attacks.

Another major point of our solution was the implementation of all the training logic on several clients using a mobile Android application.

The logic implemented on the server included a function that aggregated the weights coming from the clients and a function to test the updated global model. Figure 4 presents a high-level diagram of the interaction between server and clients in the proposed solution.

Finally, it only remains to define the last point of our solution: the deployment of clients on Android devices. Acquiring multiple devices to test this solution would represent a significant investment. Therefore, we decided to use Android mobile device emulation. There are numerous emulation software applications, such as BlueStacks (https://www.bluestacks.com/, accessed on 28 February 2023), Nox Player (https://www.bignox.com/, accessed on 28 February 2023), MEmu Play (https://www.memuplay.com/, accessed on 28 February 2023), etc., as well as Android Studio (https://developer.android.com/studio, accessed on 28 February 2023) itself where the Android application (`.apk`) development is done.

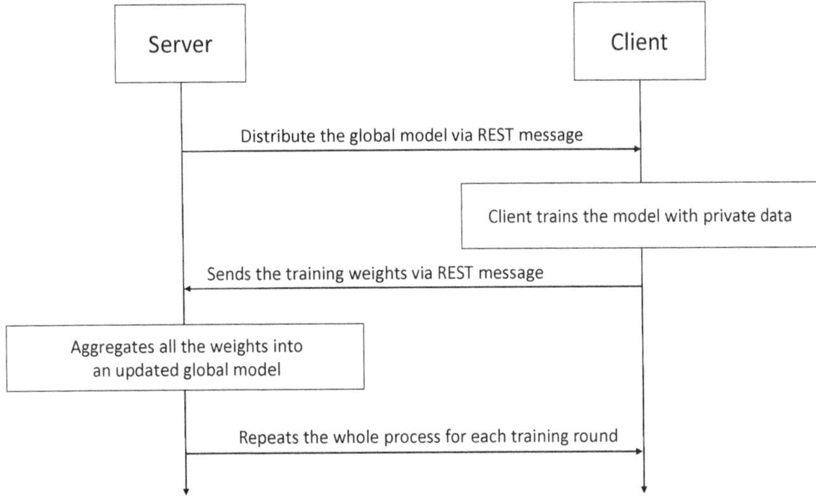

Figure 4. Federated Learning Server–Client Interaction in the Proposed Solution.

3.5. Development and Implementation

In this section, we first describe the implementation of the FL server and then we provide details about the implementation of the mobile clients.

REST Server (Server Implementation)

As seen before, the REST server started the communication rounds and monitored whether all updated weights were sent to the server by the clients. This server also sent the updated model to all clients.

The services supported by the REST API were as follows:

- A GET method to retrieve the latest shared model on the server;
- A POST method to upload the updates from the clients;
- A GET method to retrieve the information about the current round.

The first training round started immediately after the server was initialized. Once the local training was finished, the server asynchronously received updates from all clients. When a client sent the updated weights to be aggregated to the global model, the server stored them in a file in JSON format. This file contained all the information about each client's training rounds known so far.

After the server received information from all clients, it proceeded to aggregate the weights to create the global model. In a production environment, the server should have protection mechanisms to prevent it from receiving new information during the model aggregation process, as this can lead to system degradation. It was assumed that all available clients participated in each training round. Therefore, this paper did not consider the elimination or addition of clients during the training process.

Once the model aggregation was complete, the server sent it back to all participating clients in response to the previously received update message.

Android Application (Client Implementation)

An Android client application was developed using Android Studio based on the above-mentioned strategy. A functional .apk file was created that allowed any user to install the application for local training on an Android mobile device.

This application had to be able to communicate with the server. To do this, each client sent a REST message to an endpoint created for the server described above and with a known IP address. The clients sent weights from the local training and received in a response the updated global model to be used in the next round.

The client side implementation consisted of four phases:

- Setting up the connection to the server and loading the client dataset into memory;
- Receiving the global model through a REST message;
- Performing training on the client dataset;
- Sending the resulting weights to the server through a REST message.

The training function was developed using TensorFlow Federated for Android (https://www.tensorflow.org/federated, accessed on 28 February 2023). For our implementation, tests were initially performed with Android devices emulated in Android Studio. These tests showed that the software application would consume large amounts of computational resources and would therefore affect scalability with a larger number of clients. After further testing with alternative solutions, BlueStacks software was selected as it proved to scale better.

Like any other emulator, BlueStacks creates a virtual version of an Android device that runs locally on any computer. On this virtual device, the application was installed to run the developed FL client.

The multi-instance manager lets a user create multiple instances of BlueStacks. Using this feature, one can emulate several Android devices to test scenarios with multiple clients.

Ideally, one should consider the computational capacity and energy efficiency of each client, since in most real-world cases, client devices are heterogeneous. This has a significant impact on training, as there are clients that perform their local training faster than others. This not only affects the total training time but also the ability of each client to train with a considerably large dataset. However, in our work, all devices were considered homogeneous. The effect of using training devices with different capabilities is therefore reserved for future work.

4. Evaluation Design

The final goal of this paper was to implement a multimobile Android-based solution for federated learning. To reach this goal, we outlined a succession of tests to perform.

First, the dataset to be used for experimental evaluation was selected, and a baseline method was run to serve as a reference for the remaining results. Then, tests were run with 2, 4, and 8 clients, with the selected dataset evenly distributed

In a real-world implementation, clients do not have the same quantity of data available for training. Therefore, we also tested an uneven distribution of the dataset across all clients to get a clearer picture of how a federated solution would perform in such scenarios.

In all the tests described previously, the arithmetic mean was used as the aggregation method. Alternatively, we repeated these tests using an aggregation method that used a weighted average. The weighted average benefited clients with a larger dataset over ones that had less data, so conclusions about the effects of selective aggregation could be drawn by directly comparing the results.

We then performed cross-validation tests that allowed us to test the generalization capability of the implemented solution, i.e., we evaluated the results obtained with different parts of the dataset.

Finally, we experimented with another, more complex dataset to validate the implemented solution and the proposed alternative aggregation method. Figure 5 shows a diagram of the experimental methodology for the results discussed in Section 5.

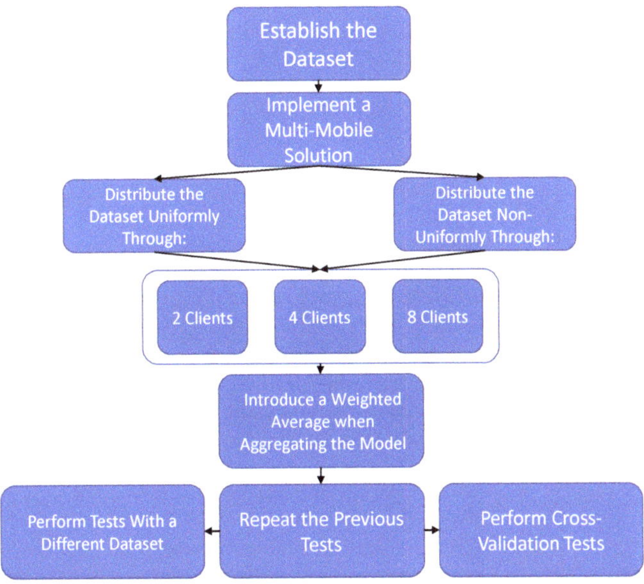

Figure 5. Experimental Methodology.

In all experiments, the performance measure used was the average accuracy after multiple rounds. This was calculated by the server, which progressively recorded the arithmetic mean of the accuracy achieved in each round. Moreover, each test configuration was repeated for five trials.

5. Results and Discussion

After a thorough survey of existing datasets, the CIFAR10 dataset was selected due to its appropriate balance between complexity and size. This allowed a clear analysis of the accuracy curve per round of an FL solution, while ensuring that the tests performed did not take too long and thus a reasonable number of configurations could be tested.

The CIFAR10 dataset consists of 60,000 color images of 32×32 pixels, divided into 10 different classes. There are 50,000 images available for training and 10,000 images for testing [29].

For all tests performed, the parameters mentioned in Section 3.1 were used.

5.1. Machine Learning Baseline Test

To make accurate comparisons between the models, a baseline ML method was run locally with the same parameters, i.e., the same dataset (CIFAR10), the same neural network (VGG 19), the same training rounds (100), epochs (5), and batch size (32).

According to [26], the VGG19 network can achieve up to 94.71% accuracy on CIFAR10 with sufficient training rounds. Based on this, our preliminary test was performed on a local machine to establish a baseline. Note that we used 100 rounds of training, as this was an adequate balance between training time and achieved accuracy.

After running these tests with the above parameters, an average accuracy of 92.87% was achieved. That is, we assumed 93% as the baseline value for the locally achieved accuracy.

5.2. Uniform Distribution of the Dataset

Here, we performed tests while splitting the dataset evenly across the Android devices (see Table 2).

Table 2. Average accuracy results obtained with 2 devices for a uniformly distributed dataset.

Elapsed Rounds	Average Accuracy
10	79%
30	88%
50	90%
100	91%

Although the training time was not analyzed, it is important to note that the average training time for each test performed in this scenario was about 24 h. This long training time was quite expected since the server and the Android devices were emulated on the same machine. In an ideal scenario, each node computes only its own training, and this is done entirely in parallel.

This first experiment was conducted to verify whether an FL solution could achieve a similar maximum accuracy as the same solution using centralized ML. From Table 2, it can be seen that the maximum accuracy achieved was close to the maximum accuracy mentioned in the previous subsection.

5.3. Nonuniform Distribution of the Dataset

Next, we examined how a client with an incomplete dataset affected the accuracy across rounds. In this work, only the set of images available for local training at each client determined whether the dataset was incomplete. In general, a dataset may be incomplete under various conditions, such as the quality of the images used, e.g., blurred images, pixelated images, etc., which is reserved for future work on this topic.

With this in mind, three test cases were conducted (see Table 3). In the first case, two clients performed the training, one using half of the CIFAR10 dataset (25,000 images) and the second using only 10 images (one image for each class in CIFAR10).

Table 3. Average accuracy for nonuniformly distributed datasets with 2 devices (one with 50% of CIFAR10 images and another with 0.0002%), 4 devices (15%, 25%, 60%, and 0.0002% of CIFAR10 images) and 8 devices (two clients with 7.5%, two with 12.5%, two with 30%, and two with 0.0002% of CIFAR10 images).

	# Devices		
Elapsed Rounds	**2**	**4**	**8**
10	27%	22%	12%
30	34%	43%	26%
50	42%	78%	44%
100	48%	90%	89%

In the second case, the number of clients was increased to four. The dataset was divided unevenly among them, but one of the clients saw its dataset allocated with only 0.0002% of the full training batch of CIFAR10 (one image for each class in CIFAR10). The other three clients had 15%, 25%, and 60% of the dataset, respectively.

In the last scenario, the number of clients was increased to eight. Two clients had 7.5% of the dataset, two had 12.5%, and two had 30%. The remaining two clients had 10 images each, i.e., one image for each class in CIFAR10.

The first case led to very poor accuracy results. This was to be expected since the training was performed on only half of the entire dataset. The second and third cases showed that increasing the number of clients and consequently each client having a smaller local dataset, together with introducing clients that hardly contributed to the overall training, still allowed an accuracy close to the maximum achieved in previous tests. However, it could also be observed that the accuracy in the first training rounds was significantly worse than in the previously presented case.

From this point on, no more tests were performed with just two clients. In order to test the implemented solution for as many clients as possible, we proceeded only with tests with four and eight clients.

5.4. Proposing a Weighted Average for Model Aggregation

The preceding results demonstrated the need to use an aggregation method that benefited those clients that contributed the most to the global model. This could be achieved by giving more weight to the contributions of clients with a more complete dataset, i.e., clients that predictably contributed more to a higher accuracy of the model.

Therefore, a new aggregation model was tested to meet this premise. Instead of an arithmetic average with equal weights for each client, we introduced a weighted average where the weights were computed as a fraction of the number of images each client had locally compared to the total number of images used for training.

Equation (3) represents the weighted average used hereafter.

$$w \leftarrow \sum_{k=1}^{K} \frac{n_k}{n} w^k, \tag{3}$$

where n represents the total number of images considered for training in all clients in a round and n_k represents the number of images in each client k. With this aggregation method, little importance is given to clients with a small number of images.

Once again, we performed tests with a nonuniform distribution of the dataset with four clients: one client with 15%, one with 25%, one with 60% of the CIFAR10 images, and one client with only 10 images; and with eight clients: two with 7.5%, two with 12.5%, two with 30% of the dataset, and two clients with 10 images for each class of CIFAR10 (see Table 4).

Figure 6 shows that using a weighted average in the aggregation of the global model improved the accuracy of the training in the first rounds compared to an aggregation method using an arithmetic average with both four and eight clients. However, given a sufficient number of rounds of training, nearly identical accuracy was achieved at the end of training in both cases. As expected, training accuracy converged faster in both cases when a weighted average was used, as clients with large assigned datasets were favored over clients with only a few images that had almost no effect on model aggregation.

To further evaluate the performance of our solution for as many clients as possible, only the tests with eight clients were performed in the following sections.

Table 4. Average accuracy for a weighted average aggregation method using nonuniformly distributed datasets with 4 devices (15%, 25%, 60%, and 0.0002% of CIFAR10 images) and 8 devices (two clients with 7.5%, two with 12.5%, two with 30%, and two with 0.0002% of CIFAR10 images).

	# Devices	
Elapsed Rounds	**4**	**8**
10	73%	36%
30	84%	55%
50	88%	79%
100	91%	90%

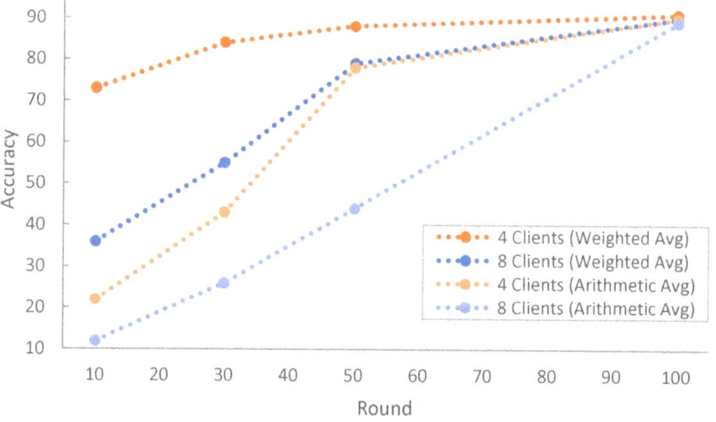

Figure 6. Training 100 Rounds with 4 and 8 Devices (Weighted Average vs Arithmetic Average).

5.5. Cross-Validation

A cross-validation test was performed to obtain a more comprehensive evaluation of the proposed approach. In this case, the 50,000 training images were divided into five batches of 10,000 images. These five batches were again divided into two batches of 12.5%, two batches of 7.5%, and two batches of 30% images (see Figure 7). A training run was performed for each batch of 10,000 images. The aim of that test was to validate the presented solution for different image samples used for training, demonstrating the generalization capacity of the proposed model.

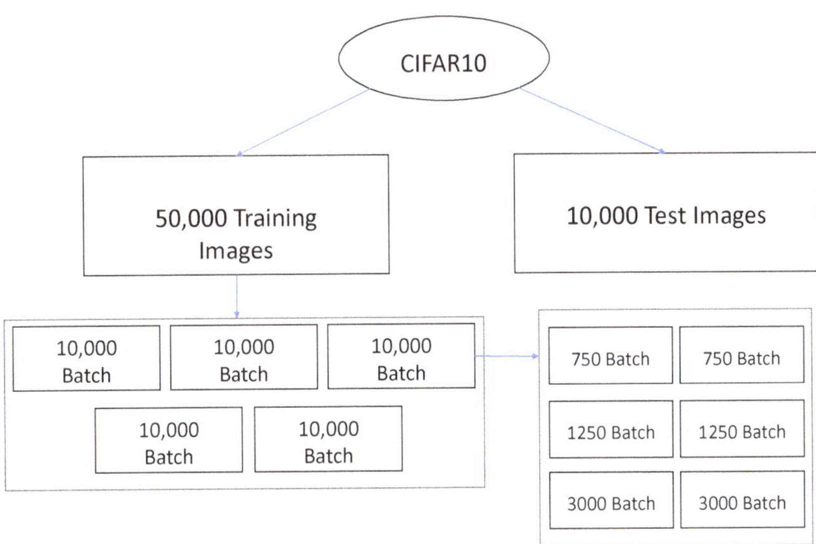

Figure 7. Division of the CIFAR10 dataset for the cross-validation tests.

Table 5 shows the average results obtained for the two aggregation models discussed. The results show that the presented solution was valid for a wide range of cases and not only for the cases presented in this study, since the maximum accuracy obtained after 100 rounds of training in previous tests (91%) and the maximum accuracy obtained after 100 rounds of training with the cross-validation tests (90%) were only slightly different.

Table 5. Results obtained with cross-validation for 8 devices.

Elapsed Rounds	Weighted Average Accuracy	Arithmetic Average Accuracy
10	28%	8%
30	42%	17%
50	74%	42%
100	90%	89%

5.6. Increasing the Complexity of the Dataset

The use of the weighted average aggregation led to an overall improvement in training. However, for CIFAR10, both the weighted average and the arithmetic average achieved a similar maximum accuracy. Therefore, we performed a final test on a larger and more complex dataset—CIFAR 100 [30]—to further evaluate the performance of our FL solution.

This dataset has the same number of training and test images as CIFAR10. The main difference is in the number of image classes. While CIFAR10 considers 10 image classes, CIFAR100 considers 100 image classes.

In Table 6, we present the results obtained by repeating the tests with eight clients with an unbalanced dataset for both the arithmetic average and the weighted average aggregation methods.

Table 6. Results obtained from 8 clients (two clients with 7.5%, two with 12.5%, two with 30%, and two with 0.0002%) with the more complex CIFAR100 dataset.

Elapsed Rounds	Weighted Average Accuracy	Arithmetic Average Accuracy
10	6%	2%
30	21%	12%
50	32%	24%
100	63%	51%

Note that for the same rounds of training with this dataset, only a maximum accuracy of 63% was achieved. This was to be expected since the dataset had a much higher complexity. Sterneck et al. [31] showed that with the CIFAR100 dataset, the **VGG19** CNN achieved a maximum accuracy of 70.3% after 100 training rounds, which was obtained with a balanced dataset and a centralized machine learning method with access to all training data. In our case, since we performed distributed training with unreliable clients (due to the lack of dataset balance), we would need a much larger number of rounds to converge closer to that maximum accuracy.

These results support the use of aggregation methods that benefit those clients who contribute more to training.

6. Lessons Learned

Although the advantages of using FL are obvious, it is also important to note that a solution based on an FL architecture has some potential weaknesses that need to be addressed, such as:

- The need for communication between clients and server also brings the need to deal with asynchronous client connection or disconnection (client dropout), which will have an impact on the system. For these cases, both the server and the FL architecture need to be extremely resilient. If the system is not prepared to handle connection dropouts or a large influx of users during a communication round, the end result may be significantly degraded or the system may suddenly fail.
- The need to protect the privacy of every client also brings with it the need to protect against cyberattacks or malicious clients. Therefore, all communications must be secured, as well as the access to the dataset at each client.
- In a federated solution, there is no prior knowledge of the data that each client contains locally. This means that there may be clients that do little to improve the accuracy of the global model. Moreover, in most cases, the clients are heterogeneous, meaning that not all clients have the same computational power or energy efficiency. Accordingly, some clients may take significantly longer to train locally than others. In these cases, the training time in each round depends on the slowest client training and transmitting its local updates.

However, FL has very important strengths compared to centralized ML. Above all, client data protection is the focus of the concept presented. Moreover, two main advantages arise: (*i*) distributed training, i.e., it is no longer necessary to have a superpowerful machine on the server to run computationally intensive algorithms, and training becomes easier while being performed in parallel; (*ii*) the need for distributed training implies the decentralized use of data. A federated solution has the potential to use a dataset that grows exponentially with the number of clients connected to the system, i.e., the model becomes more complete and robust and responds more accurately to a wider variety of situations.

Based on the analysis performed, this work suggests the inclusion of admission criteria in the implementation of an FL solution. This would allow the introduction of several requirements in the system, for example:

- Ensure that all clients have sufficient computing capacity to make a positive contribution to the training;
- Ensure that all clients have sufficient data to ensure model convergence;
- Ensure that all clients are trustworthy, i.e., do not pose a risk to the privacy of others;
- Ensure that all clients have a stable and secure connection to the server.

We strongly believe that the inclusion of admission criteria should be further explored as it would improve the overall quality of an FL architecture.

We also believe that FL could benefit from other aggregation methods. A clear example of this is reported in [32], where a protocol with user selection was proposed to reduce the negative effect of correlation within the training data of nearby participants, which impaired the optimization of the FL process. In our work, we demonstrated that using a slightly more complex method improved the overall accuracy of the solution. In addition to a weighted averaging of each client's contribution, other approaches could be explored, such as:

- Performing a validation of the local model in each client and giving preference to the clients that achieve a higher accuracy locally;
- Maintaining and updating a historical score on the server side to weight each client's contribution according to training performance over time;
- Sending updates to the server along with the relevant client status (e.g., connection status, local dataset size, etc.) to continuously adjust the aggregation method.

7. Conclusions

This paper presented a multimobile Android-based implementation of a federated learning solution and studied the impact of unreliable participants and selective aggregation in the federated solution towards the deployment of mobile FL solutions in real-world applications.

After establishing a baseline using a traditional ML approach, we showed that we were able to achieve similar performance using an FL solution with two clients and a uniformly distributed dataset. When the dataset was not evenly distributed among participants, a lower accuracy was observed in the early stages of training, and the network typically required a higher number of rounds before performance converged. After the introduction of a weighted average selective aggregation, we found a significant improvement in convergence when using a dataset unevenly distributed among clients. However, the accuracy achieved for four and eight clients after the 100 rounds of training was not different from the previous tests. This was due to the complexity and dimension of the dataset, which led to a training accuracy near the accuracy that was close to the known maximum accuracy when enough training rounds were performed.

Afterwards, we ran cross-validation tests to demonstrate the generalizable nature of our results. Finally, we conducted experiments with a more complex dataset and found that using the weighted average aggregation method not only improved the accuracy across rounds but also improved the maximum accuracy achieved after 100 rounds of training.

By comparing an arithmetic average aggregation method and a weighted average aggregation method, we concluded that there was room for improvement in the aggregation methods used for FL. In the context of the existing challenges of FL, we also concluded that FL would benefit from the introduction of admission criteria for clients.

For future work, it would be important to test the implemented solution for a larger number of clients (i.e., higher than eight) and further analyze the impact of the number of clients in the training process and communications. In addition, our experience has shown that the proper handling of disconnections and newly added clients during training is critical for real-world deployment. Other opportunities to be explored include using heterogeneous clients with different computational capacities, investigating the impact of clients with unfavorable data (e.g., blurred or pixelated images), improving the aggregation method to further optimize model convergence, testing the federated solution with even larger and more complex datasets to validate the applicability of the model to other use cases, and exploring cybersecurity techniques and secure connections for privacy preservation to enable sustainability for all participants.

Author Contributions: Conceptualization, D.P., P.P. and G.F.; methodology, L.E., D.P., P.P. and G.F.; software, L.E.; validation, L.E.; formal analysis, L.E., D.P., P.P. and G.F.; investigation, L.E.; resources, D.P., P.P. and G.F.; data curation, L.E.; writing—original draft preparation, L.E. and D.P.; writing—review and editing, D.P., P.P. and G.F.; visualization, L.E. and D.P.; supervision, D.P., P.P. and G.F.; project administration, G.F.; funding acquisition, D.P., P.P. and G.F. All authors have read and agreed to the published version of the manuscript.

Funding: This work was supported by Instituto de Telecomunicações and Fundação para a Ciência e a Tecnologia, Portugal, under grants UIDB/50008/2020 and EXPL/EEI-HAC/1511/2021. This work was also supported by Instituto de Sistemas e Robótica (Project UIDB/00048/2020), funded by the Portuguese Science Agency "Fundação para a Ciência e a Tecnologia" (FCT).

Institutional Review Board Statement: Not applicable.

Informed Consent Statement: Not applicable.

Data Availability Statement: The data presented in this study are available on request from the first author.

Conflicts of Interest: The authors declare no conflict of interest.

Abbreviations

The following abbreviations are used in this manuscript:

AI	Artificial intelligence
CPU	Central processing unit
CNN	Convolutional neural network
FedAvg	Federated averaging
FL	Federated learning
FedSGD	Federated stochastic gradient descent
GM	Geometric median
IP	Internet Protocol
JSON	JavaScript Object Notation
ML	Machine learning
ReLu	Rectified linear unit
RGB	Red green blue
REST	Representational state transfer
RFA	Robust federated averaging
SMC	Secure multiparty computation
SGD	Stochastic gradient descent

References

1. Duranton, M.; De Bosschere, K.; Coppens, B.; Gamrat, C.; Hoberg, T.; Munk, H.; Roderick, C.; Vardanega, T.; Zendra, O. HiPEAC vision 2021: high performance embedded architecture and compilation. *Eur. Netw. High-Perform. Embed. Archit. Compil.* **2021**, *1*, 5–29.
2. Coughlin, T. 175 Zettabytes By 2025. 2018. Available online: https://www.forbes.com/sites/tomcoughlin/2018/11/27/175-zettabytes-by-2025/?sh=59d7f05f5459 (accessed on 28 February 2023).
3. Data Creation and Replication Will Grow at a Faster Rate than Installed Storage Capacity, According to the IDC Global DataSphere and StorageSphere Forecasts. Available online: https://www.businesswire.com/news/home/20210324005175/en/Data-Creation-and-Replication-Will-Grow-at-a-Faster-Rate-Than-Installed-Storage-Capacity-According-to-the-IDC-Global-DataSphere-and-StorageSphere-Forecasts (accessed on 28 February 2023).
4. Satyanarayanan, M. Mahadev (Satya) Satyanarayanan—Edge Computing: a New Disruptive Force, Keynote at SYSTOR 2020. Available online: https://www.youtube.com/watch?v=7D2ZrMQWt7A (accessed on 28 February 2023).
5. Li, T.; Sahu, A.K.; Talwalkar, A.; Smith, V. Federated learning: Challenges, methods, and future directions. *IEEE Signal Process. Mag.* **2020**, *37*, 50–60. [CrossRef]
6. Mcmahan, H.B.; Ramage, D. Communication-Efficient Learning of Deep Networks from Decentralized Data. *Artif. Intell. Stat.* **2017**, *54*, 10.

7.	Shokri, R.; Shmatikov, V. Privacy-preserving deep learning. In Proceedings of the Proceedings of the 22nd ACM SIGSAC Conference on Computer and Communications Security, Denver, CO, USA, 12–16 October 2015; pp. 1310–1321.
8.	Bonawitz, K.; Eichner, H.; Grieskamp, W.; Huba, D.; Ingerman, A.; Ivanov, V.; Kiddon, C.; Konečný, J.; Mazzocchi, S.; McMahan, H.B.; et al. Towards federated learning at scale: System design. *arXiv* **2019**, arXiv:1902.01046.
9.	Srinivasan, A. Difference between distributed learning versus Federated Learning Algorithms. 2022. Available online: https://www.kdnuggets.com/2021/11/difference-distributed-learning-federated-learning-algorithms.html (accessed on 28 February 2023).
10.	Greengard, S. AI on Edge. *Commun. ACM* **2020**, *63*, 18–20. [CrossRef]
11.	Yang, Q.; Liu, Y.; Chen, T.; Tong, Y. Federated Machine Learning: Concept and Applications. *ACM Trans. Intell. Syst. Technol.* **2019**, *10*, 1–19. [CrossRef]
12.	Yang, Q.; Liu, Y.; Cheng, Y.; Kang, Y.; Chen, T.; Yu, H. Federated learning. *Synth. Lect. Artif. Intell. Mach. Learn.* **2019**, *13*, 1–207.
13.	Lim, W.Y.B.; Luong, N.C.; Hoang, D.T.; Jiao, Y.; Liang, Y.C.; Yang, Q.; Niyato, D.; Miao, C. Federated Learning in Mobile Edge Networks: A Comprehensive Survey. *IEEE Commun. Surv. Tutorials* **2020**, *22*, 2031–2063. [CrossRef]
14.	Courville, I.G.; Bengio, Y.; Aaron. Deep learning by Ian Goodfellow,Yoshua Bengio, Aaron Courville. *Nature* **2016**, *29*, 1–73.
15.	Pan, X.; Chen, J.; Monga, R.; Bengio, S.; Jozefowicz, R. Revisiting Distributed Synchronous SGD. *arXiv* **2017**, arXiv:1604.00981.
16.	Hard, A.; Rao, K.; Mathews, R.; Ramaswamy, S.; Beaufays, F.; Augenstein, S.; Eichner, H.; Kiddon, C.; Ramage, D. Federated learning for mobile keyboard prediction. *arXiv* **2018**, arXiv:1811.03604.
17.	Bonawitz, K.; Ivanov, V.; Kreuter, B.; Marcedone, A.; McMahan, H.B.; Patel, S.; Ramage, D.; Segal, A.; Seth, K. Practical secure aggregation for privacy-preserving machine learning. In Proceedings of the 2017 ACM SIGSAC Conference on Computer and Communications Security, Dallas, TX, USA, 30 October–3 November 2017; pp. 1175–1191. [CrossRef]
18.	Pillutla, K.; Kakade, S.M.; Harchaoui, Z. Robust Aggregation for Federated Learning. *IEEE Trans. Signal Process.* **2022**, *70*, 1142–1154. [CrossRef]
19.	van Berkel, C.H.K. Multi-Core for Mobile Phones. In Proceedings of the 2009 Design, Automation & Test in Europe Conference & Exhibition, Nice, France, 20–24 April 2009; pp. 1260–1265.
20.	Konečný, J.; McMahan, H.B.; Yu, F.X.; Richtárik, P.; Suresh, A.T.; Bacon, D. Federated learning: Strategies for improving communication efficiency. *arXiv* **2016**, arXiv:1610.05492.
21.	Smith, V.; Chiang, C.K.; Sanjabi, M.; Talwalkar, A.S. Federated Multi-Task Learning. In *Advances in Neural Information Processing Systems*; Guyon, I., Luxburg, U.V., Bengio, S., Wallach, H., Fergus, R., Vishwanathan, S., Garnett, R., Eds.; Curran Associates, Inc.: Dutchess County, NY, USA, 2017; Volume 30.
22.	McMahan, H.B.; Ramage, D.; Talwar, K.; Zhang, L. Learning differentially private recurrent language models. In Proceedings of the 6th International Conference on Learning Representations, Vancouver, BC, Canada, 30 April–3 May 2018; pp. 1–14.
23.	Simsek, M.; Boukerche, A.; Kantarci, B.; Khan, S. AI-driven autonomous vehicles as COVID-19 assessment centers: A novel crowdsensing-enabled strategy. *Pervasive Mob. Comput.* **2021**, *75*, 101426. [CrossRef]
24.	Valerio, L.; Conti, M.; Passarella, A. Energy efficient distributed analytics at the edge of the network for IoT environments. *Pervasive Mob. Comput.* **2018**, *51*, 27–42. [CrossRef]
25.	Malekzadeh, M.; Clegg, R.G.; Cavallaro, A.; Haddadi, H. Privacy and utility preserving sensor-data transformations. *Pervasive Mob. Comput.* **2020**, *63*, 101132. [CrossRef]
26.	Zhu, C.; Ni, R.; Xu, Z.; Kong, K.; Huang, W.R.; Goldstein, T. GradInit: Learning to Initialize Neural Networks for Stable and Efficient Training. *Adv. Neural Inf. Process. Syst.* **2021**, *34*, 16410–16422.
27.	Krizhevsky, A.; Sutskever, I.; Hinton, G.E. Imagenet classification with deep convolutional neural networks. *Adv. Neural Inf. Process. Syst.* **2012**, *25*, 84–90. [CrossRef]
28.	Kaushik, A. Understanding the VGG19 Architecture. 2020. Available online: https://iq.opengenus.org/vgg19-architecture/ (accessed on 28 February 2023).
29.	Yadav, H. Preserving Data Privacy in Deep Learning: Part 1. 2020. Available online: https://towardsdatascience.com/preserving-data-privacy-in-deep-learning-part-1-a04894f78029 (accessed on 28 February 2023).
30.	Krizhevsky, A.; Hinton, G. Learning multiple layers of features from tiny images. Master's Thesis, Department of Computer Science, University of Toronto, Toronto, ON, Canada, 2009.
31.	Sterneck, R.; Moitra, A.; Panda, P. Noise Sensitivity-Based Energy Efficient and Robust Adversary Detection in Neural Networks. *IEEE Trans. Comput.-Aided Des. Integr. Circuits Syst.* **2021**, *41*, 1423–1435. [CrossRef]
32.	Sun, C.; Wu, S.; Cui, T. User Selection for Federated Learning in a Wireless Environment: A Process to Minimize the Negative Effect of Training Data Correlation and Improve Performance. *IEEE Veh. Technol. Mag.* **2022**, *17*, 26–33. [CrossRef]

Disclaimer/Publisher's Note: The statements, opinions and data contained in all publications are solely those of the individual author(s) and contributor(s) and not of MDPI and/or the editor(s). MDPI and/or the editor(s) disclaim responsibility for any injury to people or property resulting from any ideas, methods, instructions or products referred to in the content.

Article

Wireless Traffic Prediction Based on a Gradient Similarity Federated Aggregation Algorithm

Luzhi Li [1], Yuhong Zhao [1,*], Jingyu Wang [1,*] and Chuanting Zhang [2]

[1] School of Information Engineering, Inner Mongolia University of Science and Technology, Baotou 014010, China; liluzhilm@163.com
[2] Department of Electrical and Electronic Engineering, University of Bristol, Bristol BS8 1UB, UK
* Correspondence: zhaoyuhong35@163.com (Y.Z.); 13734728816@126.com (J.W.)

Abstract: Wireless traffic prediction is critical to the intelligent operation of cellular networks, such as load balancing, congestion control, value-added service promotion, etc. However, the BTS data in each region has certain differences and privacy, and centralized prediction needs to transmit a large amount of traffic data, which will not only cause bandwidth consumption, but may also cause privacy leakage. Federated learning is a kind of distributed learning method with multi-client joint training and no sharing between clients. Based on existing related research, this paper proposes a gradient similarity-based federated aggregation algorithm for wireless traffic prediction (Gradient Similarity-based Federated Aggregation for Wireless Traffic Prediction) (FedGSA). First of all, this method uses a global sharing enhanced data strategy to overcome the data heterogeneity challenge of multi-client collaborative training in federated learning. Secondly, the sliding window scheme is used to construct the dual channel training data to improve the feature learning ability of the model; In addition, to improve the generalization ability of the final global model, a two-layer aggregation scheme based on gradient similarity is proposed. The personalized model is generated by comparing the gradient similarity of each client model, and the central server aggregates the personalized model to finally generate the global model. Finally, the FedGSA algorithm is applied to wireless network traffic prediction. Experiments are conducted on two real traffic datasets. Compared with the mainstream Federated Averaging (FedAvg) algorithm, FedGSA performs better on both datasets and obtains better prediction results on the premise of ensuring the privacy of client traffic data.

Keywords: wireless traffic prediction; federal learning; FedAvg; deep learning; gradient similarity

Citation: Li, L.; Zhao, Y.; Wang, J.; Zhang, C. Wireless Traffic Prediction Based on a Gradient Similarity Federated Aggregation Algorithm. *Appl. Sci.* **2023**, *13*, 4036. https://doi.org/10.3390/app13064036

Academic Editor: Christos Bouras

Received: 22 February 2023
Revised: 14 March 2023
Accepted: 17 March 2023
Published: 22 March 2023

Copyright: © 2023 by the authors. Licensee MDPI, Basel, Switzerland. This article is an open access article distributed under the terms and conditions of the Creative Commons Attribution (CC BY) license (https://creativecommons.org/licenses/by/4.0/).

1. Introduction

According to the State Ministry of Industry and Information Technology, as of the end of August 2022, 1,854,000 5G base stations have been completed and opened nationwide. Indeed, 5G networks play an essential role in realizing application scenarios such as AR, Telematics, and 4K TV, but 5G base station construction also faces problems such as difficulty and long investment cycles, etc. If we can accurately understand the different demands and growth trends of network traffic in each region, we can allocate network resources and reasonably plan the construction of 5G base stations. Meanwhile, the development and application of big data and artificial intelligence technologies are very effective in improving the quality of service (QoS) of access and core networks [1]. The application of artificial intelligence in the convergence of communication networks is significant for the accurate prediction of wireless traffic. Wireless traffic prediction estimates future traffic data volumes based on historical data and provides a decision basis for communication network management and optimization [2], and, based on the predicted traffic data, proactive measures can be taken to alleviate network congestion and improve network operation performance. In addition, common heterogeneous service requirements can be well met in future 6G communication networks by wireless traffic prediction at a lower cost [3].

Existing centralized traffic prediction methods require large and frequent interactions to share data from various regions for learning prediction. In real-world applications, it is difficult to achieve sufficient data sharing across enterprises due to multi-level privacy factors, i.e., the existence of data silos. A large number of data interactions also poses a huge communication overhead and risk of privacy leakage, and the centralized training data model poses a huge challenge to the computational and storage capacity of the central server. The implementation and application of Federal Learning (FL) provide a new way of thinking for traffic prediction models. Specifically, FL provides a distributed training architecture that can be jointly applied with many machine learning algorithms, especially deep neural networks, based on which local data can be effectively learned and global models can be obtained by iteratively aggregating local training models, which can also share the data information of the clients while protecting the privacy of the training client data, thus obtaining more accurate prediction results.

FL has been studied in the field of wireless traffic prediction, but there are still many challenges and problems. First, the client data of collaborative learning in FL has certain heterogeneity, i.e., Non-Independent Identically Distribution (Non-IID) characteristics, and the effective solution of the data heterogeneity problem is a prerequisite for the effective execution of the federal learning algorithm. In addition, the generalization performance of the global model generated by the final aggregation in FL traffic prediction largely determines the model prediction capability. In 2017, the federal average [4] algorithm proposed by Google uses the average aggregation approach for the integration of model parameters across edge nodes, but the strategy does not consider the differences between edge computing nodes, and the average global aggregation weights will undoubtedly reduce the generalization effect of the global model. Based on the above problems, this paper proposes a new wireless network traffic prediction method, called Federated Gradient Similarity-based Aggregation Algorithm for Wireless Traffic Prediction (FedGSA), which can collaboratively train multiple base stations and provide them with high-quality prediction models, including an enhanced data strategy based on global incremental integration, a two-channel training data scheme using sliding window construction, and a gradient aggregation mechanism to cope with data heterogeneity and global model generalization in FL.

Paper Organization and Contribution:

In this paper, we study the application of federation learning in wireless network traffic prediction. To achieve this goal, this paper addresses research-related issues in the following article sections. In Section 2, we discuss recent developments and applications of federation learning and wireless network traffic prediction. Then, in Section 3, we discuss the specific implementation details of the methodological techniques used in our proposed framework. Subsequently, in Section 4, we show the details of the experiments and the conclusions of the comparative analysis with existing methods.

Based on the proposed FedGSA framework and the descriptions in previous articles, the contributions of this paper can be summarized as follows:

- To balance the individuality of clients and the correlation characteristics among multiple clients to obtain a global model with better generalization capability, we propose a two-layer global aggregation scheme based on gradient similarity, which quantifies the client similarity relationship by calculating the Pearson correlation coefficient of each client's gradient to guide the weighted aggregation on the server side;
- To address the problem of statistical heterogeneity between traffic patterns collected by different clients, which can lead to difficulties in generalizing the global model, we introduce a quasi-global model and use it as an auxiliary tool in the model aggregation process;
- Considering the time-dependent characteristics of base station network traffic, we use a sliding window strategy here to represent the traffic of each time slot as a two-channel Tensor matrix, and divide the historical traffic data into adjacent time traffic data and periodic daily historical traffic data;

- We conducted validation experiments on two publicly available real datasets and compared and analyzed the experimental results with existing experimental methods.

2. Related Work

As the present work is closely related to wireless traffic prediction and FL, we re-view the most related achievements and milestones of these two research topics in this section.

2.1. Federated Learning

Federated Learning (FL), first proposed by Google in 2016, provides a collaborative training architecture based on deep learning. FL is a distributed learning framework in which raw data is collected and stored on multiple edge clients, and model training is locally performed on the clients, and then the models are progressively optimized to learn the models through client interaction with a central server. Its classical architecture diagram is shown in Figure 1.

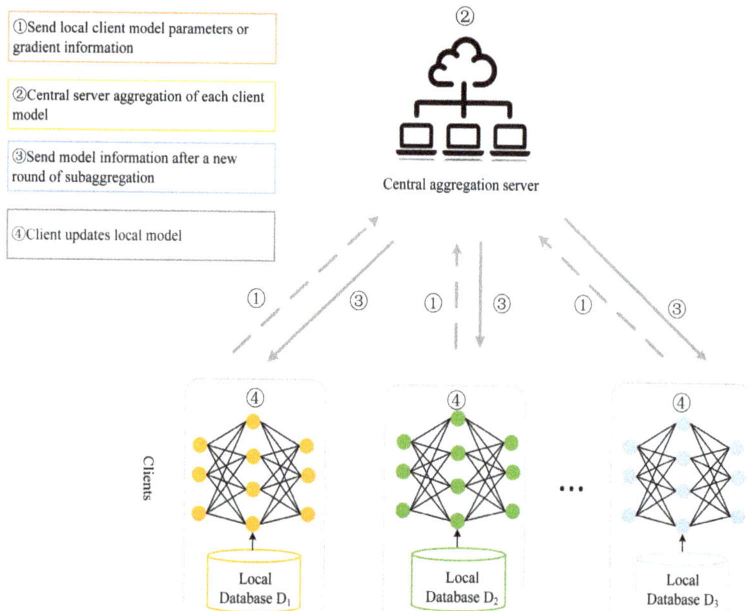

Figure 1. Classic architecture of federated learning.

As shown in Figure 1, the global model is initialized by the client application and trained based on local data, and the local model is obtained after the local training is completed, and its parameters or gradient information is uploaded to the central server, which aggregates the local client model based on the aggregation algorithm to generate a new global model, and then sends the new round of global models to each edge client again to iterate the above process until the final global model is obtained.

It is shown that the performance of federated learning is similar to that of centralized learning when the client data have Independent Identically Distributed (IID) characteristics. However, when the multi-client data are Non-IID, the performance of the federation learning algorithm is significantly reduced. Therefore, solving the problem of statistical heterogeneity of data is an urgent prerequisite to be addressed before deploying federated learning algorithms. To address this issue, a data-sharing strategy is proposed in the literature [5] for creating a globally shared data subset to integrate the local data features of the participating training clients to overcome the data heterogeneity challenge faced by FL. In addition, how to aggregate each client model to the global model while ensuring its generalization capability is a critical issue in federation learning. FedAvg is currently the mainstream federation aggregation scheme, the core idea of which is to weight the average of each local model participating in the aggregation according to the ratio of the amount of data each client has to the total training data, and its process can be described as:

Assuming K clients participate in federation training, each client has multiple training data volumes n_k and local model weights w_k^{t+1} in the $(t + 1)$st global iteration, and the FedAvg aggregation approach can be expressed as Equation (1):

$$w_G^{t+1} = \sum_{k=1}^{K} \frac{n_k}{n} w_k^{t+1} \tag{1}$$

where w_G^{t+1} is the global model parameters after the $t + 1$ st communication aggregation, n denotes the total amount of data for K clients and $\sum_{k=1}^{K} n_k = n$, and w denotes the parameters of the local model of the kth client at $t + 1$ communication rounds.

However, in the federation learning aggregation process, FedAvg's aggregation approach by averaging the model parameters or gradients across clients is difficult to guarantee the generalization ability of the global model generated by the final aggregation, in addition to the fact that federation learning cannot observe the amount of data from the clients of edge computing nodes. Therefore, the aggregation weight assignment using the actual data volume is difficult to achieve, and the quantity of data does not represent the quality of data. To address this problem, this paper proposes a federated aggregation scheme based on gradient similarity, which considers the similarity of gradient information among individual clients and performs two-level aggregation of client models based on similarity knowledge. The simulation experiments show that the scheme can achieve better results.

2.2. Wireless Traffic Prediction

Accurate traffic modeling and forecasting capabilities play an important role in wireless services, and research related to wireless traffic forecasting has received significant attention. Wireless traffic prediction is essentially a time series prediction problem. The solution methods can be broadly classified into three categories, namely, simplex methods, parametric methods, and non-parametric methods.

The historical average method and the simplex method are the representatives of the first type of method. The two methods use the average value of historical data and the last observation as the prediction value, respectively. This type of prediction method does not require complex calculations and is easy to implement, but it cannot capture the hidden patterns of the wireless traffic and the prediction performance is relatively poor.

For the second category, i.e., parametric methods, tools based on statistics and probability theory are used to model and forecast wireless services, among which the most classical method is the Auto-Regressive Integrated Moving Average (ARIMA). In order to characterize the self-similarity and burstiness of wireless traffic, the authors explored ARIMA and its variants in the literature [6,7]. In addition to ARIMA models, literature [8–10] explored alpha-stable models, entropy theory, and covariance functions to perform wireless traffic forecasting, respectively.

With the rapid development of machine learning and artificial intelligence techniques, nonparametric methods have become a strong contender among wireless traffic prediction methods. In particular, research on wireless traffic prediction based on deep neural networks has attracted great attention. In the literature [11], the authors designed a traffic prediction model based on a multi-channel sparse long-term short-term memory network to capture multi-source network traffic information and improve the ability of deep neural network models to capture important features. In [12], the authors designed a Generative Adversarial Network (GAN) traffic prediction method and separately captured traffic spatio-temporal features and base-station-type features, input the spliced features into the composite residual module to generate predicted traffic, judge the generated traffic by the discriminative network, and then generate highly accurate predicted traffic by the generative network after the game confrontation between the generative network and the discriminative network.

To effectively extract spatial and temporal features, a joint spatio-temporal prediction model based on neural networks has been proposed in the literature [13], which uses graph convolutional networks to extract complex topological spatial features in a targeted manner, while using gated cyclic units to extract temporal features of the traffic. City-scale wireless traffic forecasting is also studied in the literature [14], where the authors introduce a new forecasting framework by modeling the spatio-temporal correlation of cross-domain datasets.

The above work mainly uses centralized wireless traffic prediction, and to address the problems of communication overhead, privacy leakage, and data silos in centralized prediction schemes, this paper implements wireless traffic prediction through distributed architecture and federated learning.

3. Proposed Framework and Methods

This section may be divided into subheadings. It should provide a concise and precise description of the experimental results, their interpretation, as well as the experimental conclusions that can be drawn.

3.1. Overview

In this section, we describe the proposed FedGSA framework in detail. Figure 2 shows the overall model framework of FedGSA; specifically, FedGSA has the following steps.

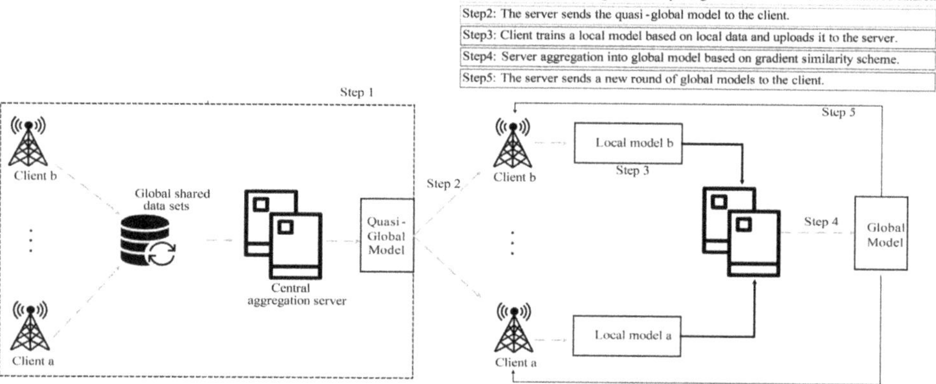

Figure 2. FedGSA Overall Architecture.

1. First, clients share local augmented data to form an augmented dataset based on global incremental integration, and a central aggregation server is trained to generate a quasi-global model based on this dataset and apply it to each client.

2. After each client applies the quasi-global model locally, a sliding window scheme is used to generate local two-channel network traffic data for each client, and then the client executes a local training procedure and passes the local model parameter information to the central aggregation server after the local training is completed.

3. Finally, the central server performs a two-level weighted aggregation of each client's network model based on the gradient similarity of each client, and finally generates a global model.

3.2. Enhanced Data Strategy Based on Global Incremental Integrations

The variability of base station traffic patterns and the mobile characteristics and communication behaviors of users within the base station range further expand the model diversity of wireless services, and the wireless traffic data from different base stations are highly heterogeneous and non-IID in nature. It is shown that Non-IID client-side data leads to a degradation of the performance of the federation learning algorithm, since the weight differences of the client-side model parameters should be considered when performing model aggregation at the server side. Therefore, this paper uses an augmented data strategy based on global incremental integration to overcome the traffic data heterogeneity challenge by creating a small augmented dataset using the original wireless traffic dataset and generating a global shared dataset.

The augmentation strategy in this paper is as follows. The dataset is first partitioned into weekly slices based on temporal indexes. For weekly traffic, statistical averages are calculated for each time point and the obtained results are considered as augmented data, and finally, the augmented data are normalized as shown in Figure 3.

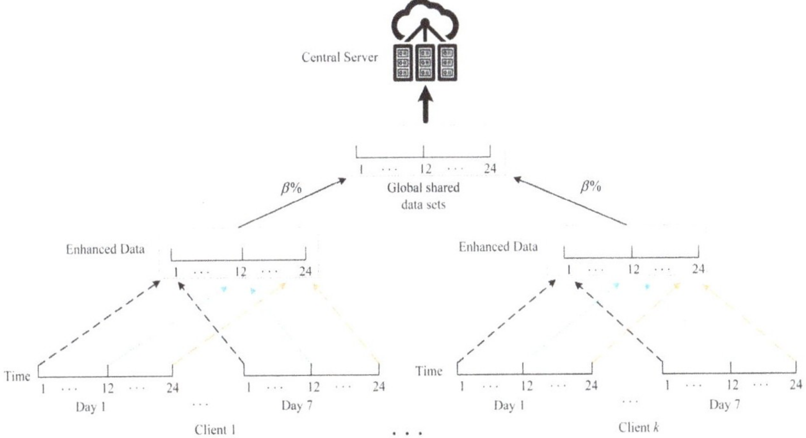

Figure 3. Enhanced data strategy.

As can be seen from Figure 2, the employed enhancement strategy is easier to implement and generate enhanced data than the traditional time-domain or frequency-domain time-series data enhancement strategies [15]. It has been experimentally proven to provide an effective solution to the problem of data heterogeneity.

During the training process, each base station sends a small fraction of its augmented dataset, say $\beta\%$, to the central server to eventually generate a global dataset that obeys the original client data distribution. The size of the augmented data is much smaller compared to the size of the original data. Based on this augmented dataset, a quasi-global model can be trained and used as prior knowledge for all clients, and the model is trained using the augmented data for all clients rather than the original data. Even so, due to the high similarity between the augmented data and the original data, the model can still be used as prior knowledge for all clients.

3.3. Constructing Two-Channel Sliding Window Training Data

The wireless traffic prediction service in general is: given K base stations, the local wireless traffic data of each base station can be represented as a dataset, as in Equation (2):

$$d^k = \left\{ d_1^k, d_2^k, d_3^k, \cdots, d_z^k \right\} \tag{2}$$

where the total time interval is Z; and assuming that $\hat{d}_z^k$ is the target service volume to be predicted, then the wireless service prediction problem can be described as Equation (3).

$$\hat{d}_z^k = f(d_{z-1}^k, d_{z-2}^k, \cdots, d_1^k; w) \tag{3}$$

where $f(\cdot)$ denotes the chosen prediction model and w denotes the corresponding parameter. The prediction model $f(\cdot)$ can be in linear form (e.g., linear regression) or in nonlinear form (e.g., deep neural network).

For wireless network traffic prediction techniques, to reduce data complexity, partial historical traffic data are usually used as input features, and considering that the base station network traffic is time-dependent, the traffic of each time slot can be represented as a two-channel Tensor matrix [16,17].

Therefore, based on the wireless traffic dataset d^k, a set of input-output pairs $\left\{ x_i^k, y_i^k \right\}_{i=1}^n$, where x_i^k denotes the historical traffic data associated with y_i^k, can be obtained using a sliding window scheme, and x_i^k is partitioned into two time channels, i.e., adjacent time and cycle time channels, which represent the predicted target time adjacent time traffic and cycle time traffic for the corresponding time points, respectively, as in Figure 4.

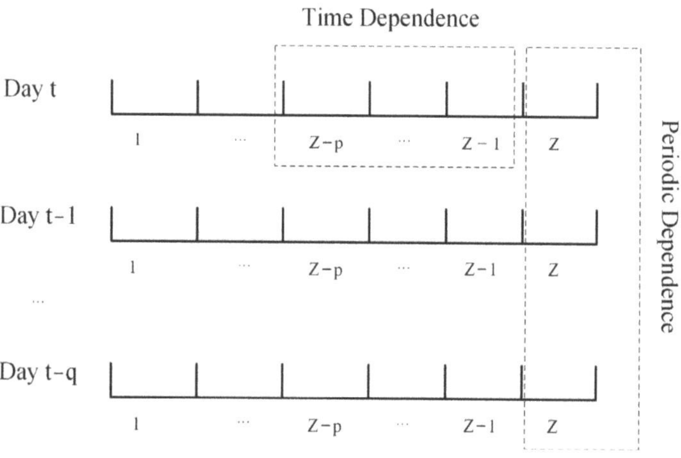

Figure 4. Two-channel training data.

Defining p as the adjacent time point sequence dependence length, the flow of the adjacent time series can be expressed as $\left\{ d_{z-1}^k, \cdots, d_{z-p}^k \right\}$, q as the periodic time dependence length, and the periodic historical sampling flow can be expressed as b, μ is periodic. The flow prediction target of this paper is the flow value at the next time point, so Equation (3) can be described as Equation (4):

$$\hat{y}_i^k = f\left(x_i^k; w \right) \tag{4}$$

The objective of the experiment is to minimize the prediction error on K clients, so the objective of the traffic prediction can be described as solving for the parameter w under the optimal solution in Equation (5):

$$\min_{w} \mathcal{I}(w) = \frac{1}{Kn} \sum_{k=1}^{K} \sum_{i=1}^{n} \mathcal{I}\left(f\left(x_i^k; w \right), y_i^k \right) \tag{5}$$

where $\mathcal{I}$ is the loss function, which can be expressed as $\left| f\left(x_i^k; w \right) - y_i^k \right|$.

Long Short-Term Memory (LSTM)) has the powerful ability to model time series datasets, so this paper selects a LSTM Long Short-Term Memory network as the network model, sets two LSTM network layers, corresponding to the input adjacent time point dependent sequence traffic data and periodic time series dependent data in turn, after which the output data features of each channel are spliced, and finally, the features are mapped to the prediction by a linear layer.

3.4. Global Aggregation Based on Gradient Similarity

The aggregation process in FL is a key part of model training, and the quality of aggregation directly affects the strength of the generalization ability of the final generated global model. The goal of central server-side model aggregation is to obtain a global model with strong generalization capability across all clients, which should balance client personalization and correlation characteristics across multiple clients. To achieve this, the global model should find a balance between capturing the personalized traffic patterns of the clients and the public shared traffic patterns.

In literature research, it is found that similarity-based weighted fusion schemes have a wide range of applications in machine learning, such as natural language processing and transformers in image vision [18], where similarity knowledge can tap potential correlations among different clients, and FedGSA quantifies client similarity relationships by calculating Pearson correlation coefficients for each client gradient to guide the server-side weighting of client models for aggregation.

The Pearson correlation coefficient is used to describe the degree of linear correlation between two variables, i.e., the larger the absolute value of the correlation, the stronger the correlation, and the value is $[-1, 1]$. The Pearson correlation coefficient is the ratio of the covariance to the standard deviation, and the Pearson correlation coefficient for a set of data (x, y) is calculated as:

$$\rho_{x,y} = \frac{\Sigma xy - \frac{\Sigma x \Sigma y}{N}}{\sqrt{(\Sigma x^2 - \frac{(\Sigma x)^2}{N})(\Sigma y^2 - \frac{(\Sigma y)^2}{N})}} \tag{6}$$

where N denotes the number of values of the variable.

Here, we use gradient information to measure the similarity between individual client models, rather than based on the original traffic data of each client itself. The central server uses the similarity relationship between the gradients of the individual client models to guide the clients in generating personalized models, thus helping to moderate their impact on global aggregation (i.e., reduce variance), and the aggregation principle of FedGSA can be described as follows:

Assuming K clients involved in training, after t rounds of global iterations, then in $t + 1$ rounds, each client is trained based on the quasi-global model obtained under T rounds using local data to obtain its local model parameters $\{w_k^{t+1}\}_{k=1}^{K}$, and the central server in the aggregation phase has two layers of aggregation for the client models:

The first layer of aggregation aims to capture the similarity relationship between each client and quantify the impact of each client on the global model by assigning weights to its Pearson correlation coefficients among the clients, and for each client, a personalized model is formed based on its gradient similarity relationship with other clients using Equation (7):

$$\widetilde{w}_k^{t+1} = \sum_{r=1}^{K} \rho_{k,r} w_r^{t+1} \tag{7}$$

The second layer performs aggregation among the personalized models: the central server generates a new round of quasi-global models based on the final aggregation of Equation (8):

$$w_G^{t+1} = \frac{1}{K} \sum_{k=1}^{K} \widetilde{w}_k^{t+1} \tag{8}$$

where $\rho_{k,r}$ denotes the Pearson correlation coefficient between two models of K clients, $\widetilde{w}_k^{t+1}$ denotes the new personalized model parameters of each client obtained in $t + 1$ round after the weighting operation of comparing the gradient similarity of each client in the round, and w_G^{t+1} denotes the quasi-global model parameters finally generated in the round.

Algorithm 1 describes the execution process of FedGSA:

Algorithm 1: FedGSA Implementation Process

> **Input:** The wireless traffic data $\{x^k, y^k\}_{k=1}^{K}$; the clients share the enhanced data and form the globally shared dataset D_s; the percentage of selected clients is δ, and the client learning rate is η.
>
> **Output:** Global Model w_G.
>
> **Obtain** w_Q by using D_s//Get the quasi-global model w_Q and send it down to the client
>
> 1 **for** each round $t = 1, 2, ...,$ do
> 2 $m \leftarrow \max(K \cdot \delta, 1)$
> 3 $S_t \leftarrow$ Randomly selected m-base stations
> 4   **for** each client $k \in S_t$ do
> 5 $w_k^{t+1} \leftarrow w_k^t - \eta w^t$//Client local update model
> 6   **end for**
> 7   **Obtain** w_k^{t+1}//Get the client local model
> 8   **Obtain** $\widetilde{w}_k^{t+1}$ by using Equation (5)//First aggregation to obtain personalized models for each client
> 9 **end for**
> 10 **Obtain** w_G^{t+1} byusing Equation (6)//Second aggregation, get the global model
> 11 **Return** w_G

---

4. Experiments and Conclusions

In this paper, two real datasets were selected to learn and train clients by combining federal learning mechanisms with LSTM long and short-term memory networks. To verify the feasibility and effectiveness of the method in this paper, some traditional network models based on LSTM, Lasso, Support Vector Regression (SVR), and FedAvg traffic prediction methods were selected for comparative analysis. Except for the shallow learning algorithm, the FedAvg algorithm and the structure of this experimental network remained consistent.

4.1. Dataset and Evaluation Metrics

This paper used the Trento and Milano telecommunication activity datasets provided by Telecom Italia in the European "Big data challenge" [19,20], and used the network traffic records of these two regions as the raw data for traffic prediction. The cellular networks for cellular user activity recorded traffic every ten minutes for two months from 11 January 2013 to 1 January 2014. For the experiments in the following subsections, the network traffic was resampled to hourly to avoid data sparsity issues.

To evaluate prediction performance, three widely used regression metrics were adopted in this paper, i.e., mean squared error (MSE), mean absolute error (MAE), and R-squared score:

1. Mean Absolute Error (MAE): Is the average of the absolute error, which can better reflect the actual situation of the prediction value error. The range is $[0, +\infty)$, as in Equation (9):

$$MAE = \frac{1}{n}\sum_{i=1}^{n}|y_i - \hat{y}_i| \tag{9}$$

2. Mean Square Error (MSE): Is the square of the difference between the true value and the predicted value, and then the average of the summation is used to detect the deviation between the predicted and true values of the model, and its range is as in Equation (10):

$$MSE = \frac{1}{n}\sum_{i=1}^{n}(y_i - \hat{y}_i)^2 \tag{10}$$

where, in MAE and MSE, $\hat{y}_i$ denotes the predicted value of wireless traffic at the time i and y_i denotes the true value at the corresponding time.

3. R-squared score: the R-squared score is applied to regression problems with values between 0 and 1. The closer to 1 indicates a better fit and is generally expressed as R^2, as in Equation (11):

$$R^2 = 1 - \frac{\left(\sum\limits_{i=1}^{m} (\hat{y}^{(i)} - y^{(i)})^2\right)\Big/ m}{\left(\sum\limits_{i=1}^{m} (y^{(i)} - \overline{y})^2\right)\Big/ m} \tag{11}$$

The numerator represents the Residual Sum Of Squares (RSS) and the denominator represents the Total Sum of Squares (TSS).

4.2. Experimental Settings and Overall Results

Experiments were conducted with 100 randomly selected cells from each dataset, and eight weeks of traffic data were randomly selected for the experiments, where the traffic of the first seven weeks was used for training the prediction model and the traffic of the last week was used for testing. In constructing the two-channel training samples using the sliding window scheme, both the temporal channel dependency length and the periodic channel dependency length were set to 3. A total of 100 rounds of communication were conducted between the local client and the central server, and the initial learning rate was set to 0.001, the local training batch size was set to 20, and 10% of the total samples were randomly selected in each round for the client samples to locally participate in the training and report the results of the last round, and had different results according to the different data sharing ratios in the shared data strategy; see Table 1. It can be seen from Table 1 that even if only 1% of the augmented data were shared, the performance of FedGSA, the method proposed in this paper, still outperformed other baseline methods in both datasets.

Table 1. Comparison of MSE and MAE prediction performance of different methods on two datasets.

Methods	Trento		Milano	
	MAE	**MSE**	**MAE**	**MSE**
Lasso	1.5391	5.9121	0.5475	0.4380
SVR	1.0470	5.9080	0.2220	0.1036
LSTM	1.1193	4.6976	0.2936	0.1697
FedAvg	1.0668	4.7988	0.2319	0.1096
FedGS A ($\beta = 1\%$)	1.0455	4.5269	0.2322	0.1089
FedGS A ($\beta = 50\%$)	0.9723	4.2330	0.2285	0.1078
FedGS A ($\beta = 100\%$)	0.9572	4.0257	0.2260	0.1054
Improve	↑10%	↑16%	↑3%	↑4%

Specifically, for the results on the Trento dataset, the present experimental algorithm (FedGSA) provides MAE and MSE gains of 10% and 16%, respectively, compared to the best performing method in the baseline (i.e., FedAvg). Similarly, for the Milano dataset, the FedGSA performance gains (MAE, MSE) are 3% and 4%, respectively. Furthermore, observing Table 1, it can be found that the prediction performance of FedGSA keeps improving with the increase in the shared enhanced data size, i.e., as shown in Figure 5a,b. This is because the initialized quasi-global model can better capture the traffic patterns when more data samples are available. Unless otherwise stated, the following experimental results in the article default to the results at $\beta = 100\%$.

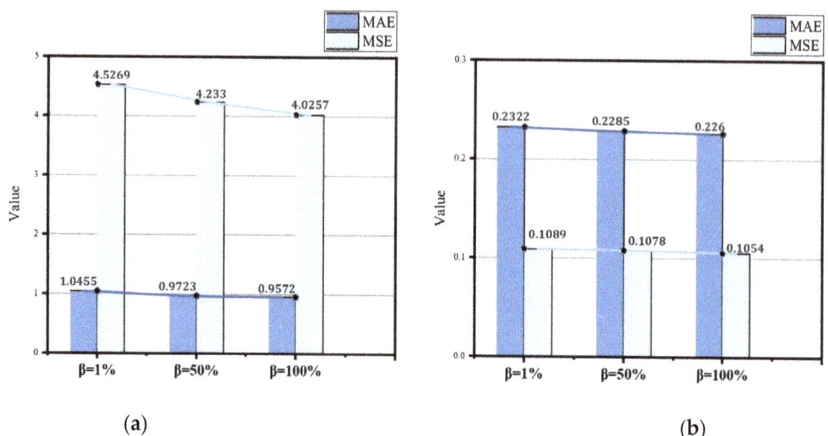

Figure 5. (**a**) describes the performance indicators on the Trento dataset; (**b**) describes the performance indicators on the Milano dataset.

To further evaluate the prediction capabilities of different algorithms, comparisons between the predicted and real network traffic values derived using different prediction algorithms for randomly selected base stations on the Trento and Milano datasets are presented in Figures 6 and 7, respectively, which include the Cumulative Distribution Function (CDF) results of the absolute prediction errors, and this experiment chooses FedAvg as the benchmark for performance comparisons because it achieves the best performance among all the baseline methods in Table 1. As can be seen in Figures 6 and 7, the FedGSA prediction capability outperforms the popular FedAvg algorithm.

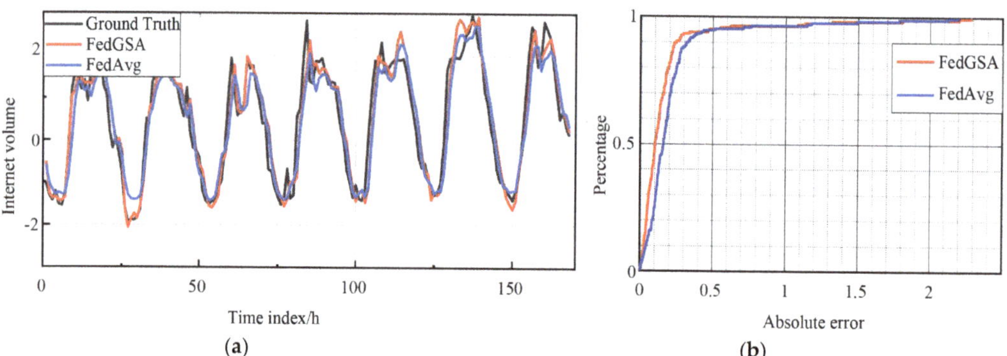

Figure 6. (**a**) describes the traffic comparison on the Trento dataset; (**b**) describes the results of the on cumulative distribution function (CDF) on the Trento dataset.

For the prediction error, in the Trento dataset, for example, FedGSA has about 95% errors less than 0.3, while FedAvg has about 89%, and FedGSA outperforms FedAvg in predicting the peak fluctuation of flow values. Based on the above evaluation, it can be concluded that the algorithm of this experiment can obtain more accurate prediction results than the baseline method.

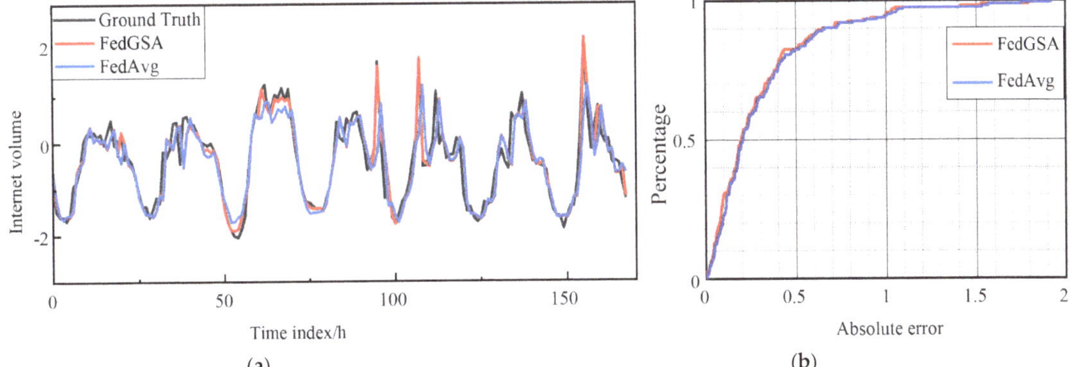

Figure 7. (**a**) describes the traffic comparison on the Milano dataset; (**b**) describes the results of the cumulative distribution function (CDF) on the Milano dataset.

4.3. Communication Rounds versus Prediction Accuracy

In FL or any other distributed learning framework, communication resources are often more valuable than computational resources, and fewer communications are preferred. Therefore, in this subsection, we report the prediction accuracy along with each communication cycle (epoch) and use the R-squared fraction to indicate the accuracy as it reflects how well the model predicts the true value of the network traffic [21]. As shown in Figure 8, we can observe that FedGSA achieves a higher prediction accuracy on both datasets, in addition to the fact that FedGSA requires fewer communication rounds to reach a certain accuracy; for example, for the Milano dataset, after 30 communication rounds, FedGSA achieves an accuracy of about 82% for wireless network traffic, and as for FedAVG, the prediction accuracy is about 75% Therefore, we consider that our proposed method has higher communication efficiency, which is also one of the important metrics for evaluating federation learning methods.

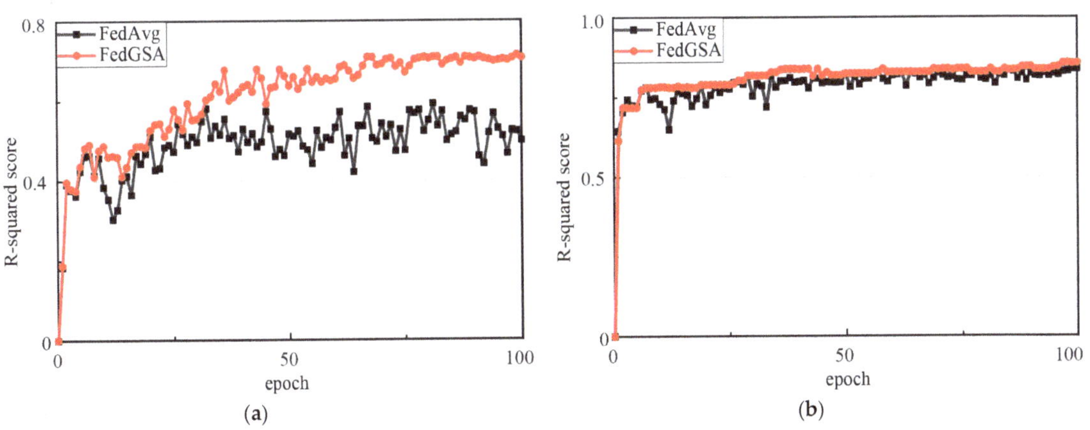

Figure 8. (**a**) Prediction accuracy and communication rounds on the Trento dataset. (**b**) Prediction accuracy and communication rounds on the Milano dataset.

5. Conclusions

In this paper, we propose a model gradient-based similarity aggregation scheme for federation learning and wireless network traffic prediction and name the framework

FedGSA, which uses similarity knowledge to construct individualized models for each client and realize model aggregation for each client to improve the generalization ability of the final global model. Experiments are conducted to predict the base station network traffic based on LSTM long short-term memory network on two real network traffic datasets, and the enhanced data scheme and sliding window strategy are combined to further overcome the problem of high data heterogeneity during FL training and improve the prediction capability. Compared with the current mainstream federal average algorithm, the method proposed in this paper achieves good results in simulation experiments.

Author Contributions: L.L.: methodology, formal analysis, data curation and validation and writing—original draft. Y.Z.: methodology, writing—review and editing and project administration. J.W.: Funding acquisition and Supervision. C.Z.: conceptualization, methodology and formal analysis. All authors have read and agreed to the published version of the manuscript.

Funding: This work was supported by Natural Science Foundation of Inner Mongolia Province of China (Grant No. 2022MS06006, 2020MS06009) and supported by Basic research funds for universities of Inner Mongolia Province of China (Grant Name. Application of Federated Learning for Privacy protection in wireless traffic prediction).

Institutional Review Board Statement: Not applicable.

Informed Consent Statement: Not applicable.

Data Availability Statement: The data in the study are from the literature, which is publicly accessible.

Conflicts of Interest: The authors declare no conflict of interest.

References

1. Yao, M.; Sohul, M.; Marojevic, V.; Reed, J.H. Artificial intelligence defined 5G radio access networks. *IEEE Commun. Mag.* **2019**, *57*, 14–20. [CrossRef]
2. Xu, Y.; Yin, F.; Xu, W.; Lin, J.; Cui, S. Wireless traffic prediction with scalable Gaussian process: Framework, algorithms, and verification. *IEEE J. Sel. Areas Commun.* **2019**, *37*, 1291–1306. [CrossRef]
3. Kato, N.; Mao, B.; Tang, F.; Kawamoto, Y.; Liu, J. Ten challenges in advancing machine learning technologies toward 6G. *IEEE Wirel. Commun.* **2020**, *27*, 96–103. [CrossRef]
4. McMahan, B.; Moore, E.; Ramage, D.; Hampson, S.; y Arcas, B.A. Communication-efficient learning of deep networks from decentralized data. In Proceedings of the 20th International Conference Artificial Intelligence and Statistics, Ft. Lauderdale, FL, USA, 20–22 April 2017; pp. 1273–1282.
5. Zhang, C.; Dang, S.; Shihada, B.; Alouini, M.-S. Dual attention-based federated learning for wireless traffic prediction. In Proceedings of the IEEE INFOCOM 2021-IEEE Conference on Computer Communications, Virtual, 10–13 May 2021; pp. 1–10.
6. Yantai, S.; Minfang, Y.; Oliver, Y.; Jiakun, L.; Huifang, F. Wireless traffic modeling and prediction using seasonal arima models. *IEICE Trans. Commun.* **2005**, *88*, 3992–3999.
7. Zhou, B.; He, D.; Sun, Z. Traffic predictability based on ARIMA/GARCH model. In Proceedings of the 2006 2nd Conference on Next Generation Internet Design and Engineering, Valencia, Spain, 3–5 April 2006; pp. 8–207.
8. Li, R.; Zhao, Z.; Zheng, J.; Mei, C.; Cai, Y.; Zhang, H. The learning and prediction of application-level traffic data in cellular networks. *IEEE Trans. Wirel.* **2017**, *16*, 3899G–3912. [CrossRef]
9. Li, R.; Zhao, Z.; Zhou, X.; Palicot, J.; Zhang, H. The prediction analysis of cellular radio access network traffic: From entropy theory to networking practice. *IEEE Commun. Mag.* **2014**, *52*, 234–240. [CrossRef]
10. Chen, X.; Jin, Y.; Qiang, S.; Hu, W.; Jiang, K. Analyzing and modeling spatio-temporal dependence of cellular traffic at city scale. In Proceedings of the 2015 IEEE International Conference on Communications (ICC), London, UK, 8–12 June 2015; pp. 3585–3591.
11. Zhang, Z.; Wu, D.; Zhang, C. Study of Cellular Traffic Prediction Based on Multi-channel Sparse LSTM. *Comput. Sci.* **2021**, *48*, 296–300.
12. Gao, Z.; Wang, T.; Wang, Y.; Shen, H.; Bai, G. Traffic Prediction Method for 5G Network Based on Generative Adversarial Network. *Comput. Sci.* **2022**, *49*, 321–328.
13. Song, Y.; Lyu, G.; Wang, G.; Jia, W. SDN Traffic Prediction Based on Graph Convolutional Network. *Comput. Sci.* **2021**, *48*, 392–397.
14. Zhang, C.; Zhang, H.; Qiao, J.; Yuan, D.; Zhang, M. Deep transfer learning for intelligent cellular traffic prediction based on cross-domain big data. *IEEE J. Sel. Areas Commun.* **2019**, *37*, 1389–1401. [CrossRef]
15. Wen, Q.; Sun, L.; Song, X.; Gao, J.; Wang, X.; Xu, H. Time series data augmentation for deep learning: A survey. *arXiv* **2020**, arXiv:2002.12478.
16. Zhang, C.; Zhang, H.; Yuan, D.; Zhang, M. Citywide cellular traffic prediction based on densely connected convolutional neural networks. *IEEE Commun. Lett.* **2018**, *22*, 1656–1659. [CrossRef]

17. Zhang, J.; Zheng, Y.; Qi, D. Deep spatio-temporal residual networks for citywide crowd flows prediction. In Proceedings of the AAAI Conference on Artificial Intelligence, San Francisco, CA, USA, 4–9 February 2017.
18. Vaswani, A.; Shazeer, N.; Parmar, N.; Uszkoreit, J.; Jones, L.; Gomez, A.N.; Kaiser, Ł.; Polosukhin, I. Attention is all you need. *Adv. Neural Inf. Process. Syst.* **2017**, *30*, 1843–1846.
19. Telecom Italia. *Telecommunications-SMS, Call, Internet-TN*; TIM: Rome, Italy, 2015. [CrossRef]
20. Telecom Italia. *Telecommunications-SMS, Call, Internet-MI*; TIM: Rome, Italy, 2015. [CrossRef]
21. Cameron, A.; Windmeijer, F. R-squared measures for count data regression models with applications to health-care utilization. *J. Bus. Econ. Stat.* **1996**, *14*, 209–220.

Disclaimer/Publisher's Note: The statements, opinions and data contained in all publications are solely those of the individual author(s) and contributor(s) and not of MDPI and/or the editor(s). MDPI and/or the editor(s) disclaim responsibility for any injury to people or property resulting from any ideas, methods, instructions or products referred to in the content.

Article

Leveraging Dialogue State Tracking for Zero-Shot Chat-Based Social Engineering Attack Recognition

Nikolaos Tsinganos, Panagiotis Fouliras and Ioannis Mavridis *

Department of Applied Informatics, University of Macedonia, 156 Egnatia Str., 54636 Thessaloniki, Greece
* Correspondence: mavridis@uom.edu.gr

Featured Application: Methods and techniques demonstrated in this work can be used to increase the effectiveness of chat-based social engineering attack detection systems.

Abstract: Human-to-human dialogues constitute an essential research area for linguists, serving as a conduit for knowledge transfer in the study of dialogue systems featuring human-to-machine interaction. Dialogue systems have garnered significant acclaim and rapid growth owing to their deployment in applications such as virtual assistants (e.g., Alexa, Siri, etc.) and chatbots. Novel modeling techniques are being developed to enhance natural language understanding, natural language generation, and dialogue-state tracking. In this study, we leverage the terminology and techniques of dialogue systems to model human-to-human dialogues within the context of chat-based social engineering (CSE) attacks. The ability to discern an interlocutor's true intent is crucial for providing an effective real-time defense mechanism against CSE attacks. We introduce in-context dialogue acts that expose an interlocutor's intent, as well as the requested information that she sought to convey, thereby facilitating real-time recognition of CSE attacks. Our work proposes CSE domain-specific dialogue acts, utilizing a carefully crafted ontology, and creates an annotated corpus using dialogue acts as classification labels. Furthermore, we propose SG-CSE BERT, a BERT-based model following the schema-guided paradigm, for zero-shot CSE attack dialogue-state tracking. Our evaluation results demonstrate satisfactory performance.

Keywords: social engineering; dialogue-state tracking; chatbot

Citation: Tsinganos, N.; Fouliras, P.; Mavridis, I. Leveraging Dialogue State Tracking for Zero-Shot Chat-Based Social Engineering Attack Recognition. *Appl. Sci.* **2023**, *13*, 5110. https://doi.org/10.3390/app13085110

Academic Editors: George Drosatos, Avi Arampatzis and Pavlos S. Efraimidis

Received: 25 February 2023
Revised: 14 April 2023
Accepted: 15 April 2023
Published: 19 April 2023

Copyright: © 2023 by the authors. Licensee MDPI, Basel, Switzerland. This article is an open access article distributed under the terms and conditions of the Creative Commons Attribution (CC BY) license (https://creativecommons.org/licenses/by/4.0/).

1. Introduction

Language stands as one of humanity's most remarkable accomplishments and has been a driving force in the evolution of human society. Today, it is an indispensable component of both our professional and social lives. Dialogue, as a term [1], refers to interactive communication between two or more individuals or groups in the context of human language. It is a two-way intentional communication that can take on spoken or written forms. Language assumes a prominent role, especially in the context of cybersecurity and social engineering, as malicious users employ it to deceive and manipulate unsuspecting individuals. Among various social engineering attacks, chat-based social engineering is recognized as a critical factor in the success of cyber-attacks, particularly in small and medium-sized enterprise (SME) environments. This type of attack is attracting increasing attention due to its potential impact and ease of exploitation. According to Verizon's 2022 [2] report, the human element is a significant factor in driving 82% of cybersecurity breaches.

During a CSE attack, malicious users utilize linguistic manipulation to deceive their targets by exploiting human personality traits and technical misconfigurations. From a strategic standpoint, it is more effective [3] to isolate individual CSE attack enablers and investigate detection methods for each enabler separately. In a chat-based dialogue, such as one between an SME employee and a potential customer, interlocutors communicate through written sentences. The ability to identify one or more characteristics that can reveal the malicious intent of an interlocutor can sufficiently safeguard the SME employee against potential CSE attacks. Dialogue act (DA) is a term used to describe the function or intention of a specific utterance in a conversation, and it has already been identified [3] as one of the critical enablers of successful CSE attacks. As such, detecting dangerous DAs that may lead to a successful CSE attack is of paramount importance.

The widespread use of dialogue systems (such as Alexa and Siri) has led to a surge in research in this field. It is advantageous for us to transfer knowledge and terminology from this field to chat-based social engineering recognition tasks. The formalism used to describe these systems can facilitate the modeling of human-to-human conversations, especially CSE attacks, as both types of dialogues share common structural characteristics. Dialogue systems rely on dialogue-state tracking (DST) to monitor the state of the conversation. Similarly, in human-to-human dialogues and CSE attacks, the state of the dialogue can be tracked through the DST process, which in our case, is the state of the CSE attack. CSE attacks involve deception and manipulation in order to acquire sensitive information, such as an attacker posing as a trusted individual (e.g., a bank representative) to trick the victim into disclosing personal information. By using DST to track the dialogue state and identify when the attacker is attempting to extract sensitive information or deceive the victim, we can detect and prevent CSE attacks.

Leveraging the progress made in dialogue systems, we develop a system that performs CSE attack state tracking. The main component in the system's architecture is a BERT-based model that is trained and fine-tuned to detect the intent of an interlocutor utilizing a multi-schema ontology and related dialogue acts. The main contributions of our work are as follows:

1. A set of dialogue acts in the CSE attack context, called SG-CSE DAs.
2. A corpus, called SG-CSE Corpus, annotated with SG-CSE-DAs appropriate for CSE attack state tracking model training.
3. A multi-schema ontology for CSE attacks, called SG-CSE Ontology.
4. A BERT-based model, called SG-CSE BERT, that follows the schema-guided paradigm for CSE attack state tracking fine-tuned with SG-CSE Corpus.

2. Related Work

In DST, the main tasks are the prediction of intention, requested slots, and slot values, and although there is no similar research applying DST tasks for the purpose of recognizing CSE attacks, we present related works that deal with DST tasks in single-domain or multiple domains.

Rastogi et al. [4], introduce the Schema-Guided Dialogue dataset to face the challenges of building large-scale virtual assistants. They also present a schema-guided paradigm for task-oriented dialogue for predicting intents and slots using natural language descriptions. The Schema-Guided Dialogue corpus extends the task-oriented dialogue corpora by providing a large-scale dataset (16k+ utterances across 16 domains). Along with the novel schema-guided dialogue paradigm, a model is released that facilitates the dialogue-state tracking able to generalize to new unseen domains and services. This work, with its simplistic and clear approach, inspired our research.

Xu and Hu in [5] suggest an approach that can track unknown slot values when Natural Language Understanding (NLU) is missing, which leads to improvements in state-of-the-art results. Their technique combined with feature dropout extracts unknown slot values with great efficiency. The described enterprise-to-enterprise architecture effectively predicts unknown slot values and errors, leading to state-of-the-art results on the DSTC2 benchmark. The addition of a jointly trained classification component that is combined with the pointer network forms an efficient hybrid architecture. This work presents a different perspective as to what can be important for unknown slot prediction and how we can approach a solution.

Chao and Lane, in [6], build on BERT to learn context-aware dialogue-state tracking for scaling up to diverse domains with an unknown ontology. They propose a new end-to-end approach for scalable context-based dialogue-state tracking that uses BERT in sentence-level and pair-level representations. The proposed BERT-DST is an end-to-end dialogue-state tracking approach that directly utilizes the BERT language model for slot value identification, unlike prior methods, which require candidate generation and tagging from n-gram enumeration or slot taggers. The most important module is the BERT dialogue context encoding module, which produces contextualized representations to extract slot values from the contextual patterns. The empirical evaluation of SimM and Sim-R datasets showed that the proposed BERT-DST model was more efficient in utilizing the slot value dropout technique and encoder parameter sharing. The authors' approach to extracting slot values with an unknown domain ontology is important. Furthermore, the BERT architecture is inspiring and helped us design our model.

Deriu et al. [7], summarize the state of evaluation techniques of dialogue systems. They use a classification of dialogue systems that leads to an equally effective classification of evaluation methods. Thus, they easily comment on each evaluation method focusing on those that lead to a reduction in the involvement of human labor. In their survey paper, they compare and contrast different evaluation measures across the various classes of dialogue systems (task-oriented, conversational, and question-answering dialogue systems)—a useful review that presents the modern state of dialogue systems' evaluation techniques and helps to foresee the trends.

Xiong et al. [8], propose an approach to process multi-turn dialogues based on the combination of BERT encoding and hierarchical RNN. The authors propose a domain-independent way to represent semantic relations among words in natural language using neural context vectors. The proposed MSDU model is able to recognize intents, dialogue acts, and slots utilizing the historical information of a multi-turn dialogue. The test results using multi-turn dialogue datasets Sim-R and Sim-M showed that the MSDU model was effective and brought about a 5% improvement in FrameAcc compared with models such as MemNet and SDEN.

Zhao et al. [9], propose a schema description-driven dialog system, D3ST, which uses natural language descriptions instead of names for dialog system variables (slots and intents). They present a schema description-driven dialogue system that produces a better understanding of task descriptions and schemas, leading to better performance. Measured on the MultiWOZ, SGD, and SGD-X datasets, they demonstrate that the approach can lead to better results in all areas: quality of understanding, learning data efficiency, and robustness. Their approach differentiates between schema names and schema descriptions and recommends using natural language descriptions. This work nicely presents the flexibility that we can get using different schemas.

Ma et al. [10], propose an end-to-end machine reading comprehension for non-categorical slots and a Wide & Deep model for a categorical slot DST system, achieving state-of-the-art performance on a challenging zero-shot task. Their system, which is an end-to-end machine reading comprehension (MRC) model for the task Schema-Guided Dialogue-State Tracking, employs two novel techniques to build a model that can accurately generate slots in a user turn, with state-of-the-art performance in the DSTC 8 Track 4 dataset. This work introduces RoBERTa as a classification model for intent recognition and slot recognition.

Kim et al. [11], propose a novel paradigm to improve dialogue-state tracking that focuses on the decoder analyzing changes to memory rather than on determining the state from scratch. They use an explicitly available memory for dialogue-state tracking to improve state operations prediction and accuracy. The proposed Selectively Overwriting Memory (SOM) is a general method to improve dialogue-state tracking (DST). Initial experiments showed improved performance on open vocabulary tasks. This work presents an approach where the initial problem definition is changed to fit the proposed solution.

Kumar et al. [12], propose an architecture for robust, dialogue-state tracking using deep learning, demonstrating improved performance over the state-of-the-art. They introduce a novel model architecture for dialogue-state tracking using attention mechanisms to predict across different granularities to better handle long-range cross-domain dependencies. The proposed model employs attention mechanisms to encode information from both the recent dialogue history and from the semantic level of slots. The proposed approach is a cross-attention-based model that performs coreference resolution and slot filling in a human-robot dialogue to address complex multi-domain dialogues.

In [13], Lin et al. propose a slot-description-enhanced generative approach for zero-shot cross-domain dialogue-state tracking. They efficiently learn slot-to-value mapping and jointly learn generalizable zero-shot descriptions. They demonstrate improved performance on multimodal question answering and multi-agent dialogues. The generative framework introduced for zero-shot cross-domain dialogue-state tracking significantly improves the existing state-of-the-art for DST. Their approach helps the design and evaluation of dialogue systems for tasks that require entities to communicate about unseen domains or multiple domains.

Lin et al. [14], propose TransferQA, which transfers cross-task dialogue knowledge from general question answering to dialogue-state tracking. The TransferQA is a generative QA model that learns a transferable, domain-agnostic QA encoder for handling question-answering tasks. TransferQA is a general framework that transfers knowledge from existing machine learning models, particularly the more challenging zero-shot scenarios. Negative question sampling and context truncation techniques were introduced to construct unanswerable questions,

Li et al. [15], propose a simple syntax-aware natural language generation as a general way to perform slot inference for dialogue-state tracking, achieving a new state of the art on MultiWOZ 2.1. They present a language-independent approach for tracking slot values in task-oriented dialogues, making it possible to add a domain to an existing system (as used in the Alice personal assistant). They offer a generative, end-to-end architecture for dialogue-state tracking that only needs domain-specific substantive examples. They propose contextual language models, not built from ontologies, for determining the domains in which a multi-domain task-oriented dialogue is occurring.

3. Background

3.1. Dialogue Systems

According to Dan Jurafsky [16], human-to-human dialogues possess several properties. A crucial structural attribute is a turn, which is a singular contribution made by one speaker in the dialogue. A full dialogue comprises a sequence of turns. For instance, in Table 1, two interlocutors identified as V(ictim) and A(ttacker) exchange utterances in a small six-turn dialogue excerpted from the CSE corpus [17]. One or more utterances from the same speaker can be grouped into a single turn. Typically, an utterance from speaker V is followed by a response from speaker A, constituting an exchange. A dialogue consists of multiple exchanges.

Table 1. A sample dialogue in turns.

Turn	Utterance
V1	We don't allow Telnet, especially from the Internet; it's not secure. If you can use SSH, that'd be okay
A2	Yeah, we have SSH.
A3	So, what's the IP address?
V4	IP is [ANON]
A5	Username and password?
V6	Username is [ANON] and password is [ANON]

Nowadays, a dialogue can be conducted between humans or between a machine and a human, and the latter is named a dialogue system. Dialogue systems communicate with humans in natural language, both spoken and written, and can be classified as:

- Task-based dialogue systems where the system helps humans to complete a task.
- Conversational dialogue systems where the systems answer questions.

Task-based systems can be further classified, based on their architecture, into systems that use:

- Genial Understander System (GUS) architecture [18] which is an old simple approach to describing a dialogue structure.
- Dialogue-State (DS) architecture, in which the dialogue is modeled as a series of different states.

The primary objective of dialogue systems when interacting with a human is to elicit three key pieces of information from the user's utterance: domain classification, user intention, and requested information. Domain classification pertains to the conversation's context and can be determined through the use of a domain ontology. To ensure proper interpretation and response to user input, a formal representation of the knowledge that a conversational system must comprehend must exist. This knowledge comprises concepts and entities mentioned in the conversation, the relationships between them, and the potential dialogue states, which are encapsulated in the domain ontology. Dialogue systems rely on ontologies to link the user's input to the corresponding concepts and entities, as well as to monitor the current dialogue state. By tracking the state of the conversation and the user's objectives, the system can generate more precise and pertinent responses. The domain ontology can be defined by means of a static schema, where the dialogue system operates within a single domain, or a dynamic schema, where the dialogue system is capable of operating within multiple domains. In this context, the schema represents a framework or structure that defines the arrangement of data. Typically, a dialogue system consists of the following units:

- Natural Language Understanding (NLU) unit, which uses machine learning to interpret the user's utterance.
- Dialogue-State Tracking (DST) unit, which maintains the entire history of the dialogue.
- Dialogue Policy unit, which defines a set of rules to drive the interactions between the user and the system.
- Natural Language Generation (NLG), which generates responses in natural language.
- Text-to-Speech unit, which transforms text to speech.
- Automated Speech Recognition unit, which transforms speech into text.

3.2. Dialogue Acts

The Speech Act theory, introduced by Austin [19] and Searle [20,21] in the 1960s, has become a widely used concept in linguistics and literature studies. Today, the modern notion of speech act [22] has found applications in diverse fields such as ethics, epistemology, and clinical psychology [23]. Dialogue Acts (DAs) are a type of Speech Act that represents the communicative intention behind a speaker's utterance in a dialogue. Hence, identifying the dialogue acts of each speaker in a conversation is an essential initial step in automatically determining intention.

The Switchboard-1 corpus [24], which consists of telephone speech conversations, was one of the first corpora related to dialogue and dialogue acts. It contains approximately 2400 two-way telephone conversations involving 543 speakers. The Switchboard Dialogue Act Corpus (SwDA) [25,26] extended the Switchboard-1 corpus with tags from the SWBD-DAMSL tagset [27]. The SWBD-DAMSL tagset was created by augmenting the Discourse Annotation and Markup System of Labelling (DAMSL) tagset. Through clustering, 220 tags were reduced to 42 tags to improve the language model for the Switchboard corpus. The resulting tags include dialogue acts such as statement-non-opinion, acknowledge, statement-opinion, and agree/accept, among others. The size of the reduced set has been a matter of debate, with some studies [28–30] using a 42-label set while others have used the most commonly used four-class classification.

- Constatives: committing the interlocutor to something's being the case.
- Directives: attempts by the interlocutor to get the addressee to do something.
- Commissives: committing the interlocutor to some future course of action.
- Acknowledgments: express the interlocutor's attitude regarding the addressee with respect to some social action.

Thus, we can think of DAs as a tagset that can be used to classify utterances based on a combination of pragmatic, semantic, and syntactic criteria.

3.3. Schema-Guided Paradigm

Rastogi et al. [4] proposed a novel approach named schema-guided dialogue to facilitate dialogue-state tracking by using natural language descriptions to define a dynamic set of service schemata. A schema-guided ontology is an ontology that characterizes the domain and is confined by a specific schema or set of schemas. In this context, a schema specifies the organization of the data and the associations between the concepts and entities that are expected to be present in the dialogue. A frame is a data structure that represents the current state of the dialogue at a specific time point. It is comprised of key-value pairs, where the keys represent the entities and concepts that are relevant to the conversation, and the values represent the corresponding values of these concepts. Typically, a frame corresponds to a specific schema or set of schemas, and it can be updated as the dialogue progresses. For instance, in a restaurant booking system, a frame may contain information such as the date and time of the reservation, the number of diners, and the desired cuisine type. In the schema-guided paradigm, frames are predetermined, and the system anticipates the user's input to conform to one of the predefined frames. Although this approach can improve the robustness and accuracy of the dialogue system by providing a structured and domain-specific representation of knowledge, it also limits the flexibility of the system to handle novel and unforeseen input. A frame is typically comprised of the following components:

- Intents: represent the goal of the interlocutor; they define the task or action that the dialogue system is trying to accomplish.
- Slots: represent the information that is being requested or provided; they describe the properties of the entities or concepts that are relevant to the task.
- Slot values: represent the value of the slots; they are the actual information that has been extracted or provided during the conversation.
- Dialogue history: represents the conversation so far; it includes all the previous turns of dialogue in the current conversation.
- Constraints: represent the additional information that is useful for the task; they are used to guide the dialogue towards a successful outcome.

3.4. Dialogue-State Tracking (DST)

Dialogue-state tracking [31] is the process of maintaining an accurate representation of the current state of a conversation. This involves identifying the user's intentions and goals, as well as the entities and concepts that are mentioned in the conversation, in order to provide relevant and precise responses. To tackle the challenge of representing the dialogue state, Young et al. [32] proposed a method that leveraged dialogue acts. They employed a partially observable Markov decision process (POMDP) to build systems that could handle uncertainty, such as dialogue systems. To achieve a practical and feasible implementation of a POMDP-based dialogue system, the state can be factorized into discrete components that can be effectively represented by probability distributions over each factor [33]. These factorized distributions make it more feasible to represent the most common slot-filling applications of POMDP-based systems, where the complete dialogue state is reduced to the state of a small number of slots that require filling.

In 2020, Rastogi et al. [4] proposed the schema-guided paradigm to tackle dialogue-state tracking, which involved predicting a dynamic set of intents and slots by using their natural language descriptions as input. They also introduced a schema-guided dataset to evaluate the dialogue-state tracking tasks of dialogue systems, including domain prediction, intent prediction, and slot filling. Additionally, they developed a DST model capable of zero-shot generalization. The schema-guided design utilizes modern language models, such as BERT, to create a unified dialogue model for all APIs and services by inputting a service's schema, which enables the model to predict the dynamic set of intents and slots within the schema. This approach has gained significant attention, and annual competitions, such as the Dialogue Systems Technology Challenge (DSTC), track the progress in this field [34].

To achieve slot filling, a sequence model can be trained using different types of dialogue acts as classification labels for each individual domain. Additionally, pretrained language models, such as GPT [35], ELMo [36], BERT [37], and XLNet [38], have shown significant promise in recent years with regards to natural language processing. These models have outperformed prior algorithms in terms of generalization and zero-shot learning. Consequently, they offer an effective approach to performing zero-shot learning for language understanding. Furthermore, by leveraging pretrained language models in the schema-guided paradigm, dialogue systems can generalize to unseen service schema elements and improve their accuracy and robustness.

The primary objectives of dialogue-state tracking (DST) encompass predicting the active user intention, requested slots, and values of slots in a given conversation turn. Within the context of DST, the user intention closely relates to the service supplied by the dialogue system. The intention refers to the user input's purpose or goal, while the service represents the functionality offered by the dialogue system to achieve the user's intent. The user's intention is typically deduced from the input and ongoing conversation state and can be expressed as a label or a set of labels that indicate the user's objective. A service denotes the task or action the dialogue system intends to perform and can be defined by a group of intentions and slots. The slots indicate the requested or supplied information. For example, an intention might involve "booking a flight," while the slots could consist of "destination," "departure date," "return date," and other related information. Services may vary over time, making it critical to have a flexible and adaptable ontology that can be modified as required.

DST's capacity to handle either a closed set of slots (static ontology) or an open set of slots (dynamic ontology) is a crucial attribute. In the former, the model is capable of predicting only those predefined slot values and cannot assimilate new slot values from example data. The model generally comprises three modules: an input layer that translates each input token into an embedding vector, an encoder layer that encodes the input to a hidden state, and an output layer that predicts the slot value based on the hidden state. In the former, where the set of possible slot values is predefined, the output layer may be approached in two ways: (i) a feed-forward layer that generates all possible values associated with a specific slot; or (ii) an output layer that contrasts both the input representation and slot values and provides scores for each slot value. The scores can be normalized by applying non-linear activation functions, such as SoftMax or sigmoid, to convert them into probability distributions or individual probabilities.

In DST, the zero-shot setting enables a model to handle new intents or slots that it has not seen before without requiring additional training data. This allows the model to generalize to new contexts or situations based on the knowledge gained during training. Unlike traditional supervised learning approaches, where a DST model is trained on a specific dataset with a specific set of intents and slots, a zero-shot DST model can handle new intents and slots without prior exposure. Techniques such as transfer learning, pre-training, or meta-learning can be used to achieve this goal and learn a more general and robust model. For instance, a zero-shot DST model can utilize pre-trained language models or embeddings, which have already learned significant knowledge about language from a large corpus of text data. The model can then be fine-tuned on a smaller dataset specific to the DST task at hand.

4. The Proposed SG-CSE Attack State Tracker

4.1. Description

We propose the SG-CSE Attack State Tracker (SG-CSEAST), a system that estimates the dialogue state during a CSE attack state by predicting the intention and slot-value pairs at turn t of the dialogue. The SG-CSEAST consists of the following four units depicted in Figure 1.

- NLU: converts the utterances into a meaningful representation.
- SG-CSE BERT: takes into account the dialogue history, and outputs the estimated state of the CSE attack.
- CSE Policy: decides which mitigation action to take.
- Mitigation Action: applies the selected mitigation action.

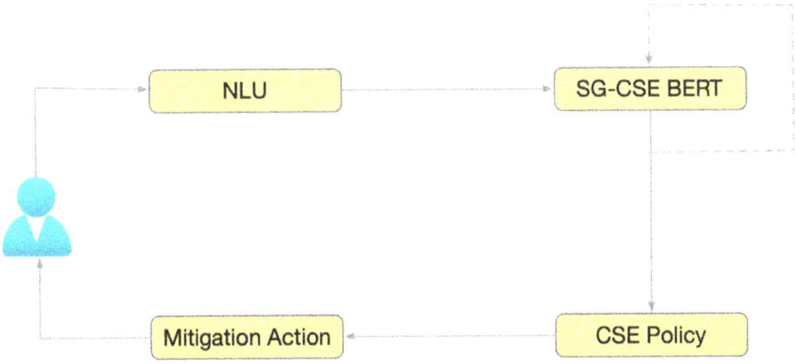

Figure 1. The SG-CSE attack State Tracker main units.

In this work, we implement the SG-CSE BERT unit, which detects information leakage and deception attempts. The information concerns SME, Information Technology (IT), and personal details, while in our context, we define deception as the attempt of a malicious user to persuade a user to use an IT resource. We follow the schema-guided paradigm utilizing Das and dialogue history. The steps toward SG-CSE BERT implementation are the following (Figure 2):

1. Using CSE Corpus [17], we extract related dialogue acts (SG-CSE Das) by mapping the utterance intention to a proposed DA.
2. We create a new corpus (SG-CSE Corpus) appropriate for DST, labeled with SG-CSE Das. This corpus is used to fine-tune and evaluate our detection model.
3. We utilize the CSE conceptual model and the CSE ontology [17] to create a schema-based ontology. Thus, different attack types are extracted.

    a. Four different CSE attack types are extracted and will be represented as services.
    b. Each service will be mapped to a specific schema.
    c. The schema will have several intents which will be mapped to the corresponding DA.

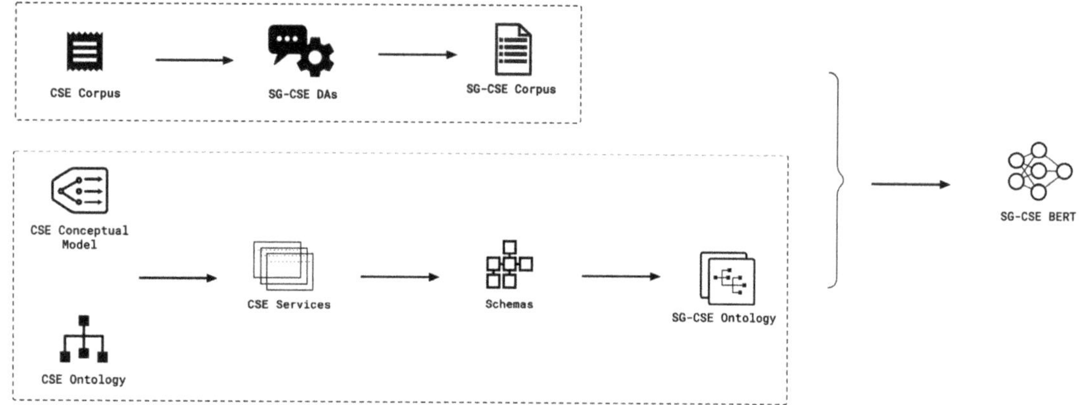

Figure 2. Implementing SG-CSE BERT.

The schema-guided paradigm makes use of a domain's ontology to create the required schema that describes each service existing in the domain. In our research, the domain is the CSE attacks, and we predict the service, user intention, and requested slot-value pairs. For example, for an utterance like: "What is the CEO's first name?" the SG-CSE BERT model should be able to build a representation like the one presented in Table 2.

Table 2. Sample representation of frame slots.

Acquire	Value
Service	CSE_SME_Info_Extraction
Intent	WH-SME_Question
Employee_name	Null
Employee_position	CEO

A CSE attack is unleashed through the exchange of utterances in turns of a dialogue. Time is counted in turns, and thus at any turn t, the CSE attack is at the state s_t and this state comprises the summary of all CSE attack history until time t. State s_t encodes the attacker's goal in the form of (slot, value) pairs. The different slot and value pairs are produced by the CSE ontology [17] and the proposed DAs that represent the in-context entities, intents, and their interconnection. The values for each slot are provided by the attacker during the CSE attack and represent her goal, e.g., during a CSE attack, at turn t the state s_t could be $s_t = \{(employee_{name}, NULL), (employee\_position, CEO)\}\}$. In such a state, the attacker's goal has been encoded for slots *employee_name, employee_position* during the dialogue.

Figure 3 depicts the dialogue system concepts and evolution that are related to our research and the concepts transferred from the dialogue systems field to the CSE attack domain for the purpose of CSE attack recognition via CSE attack-state tracking. As Figure 3 depicts, a dialogue can be realized between humans or between humans and machines. The latter is called dialogue systems and can be task-oriented or conversational. The architecture of dialogue systems can be described using frames, and among their properties, dialogue-state tracking is a property that represents the state of the dialogue at any given moment. The schema is created based on the SG-CSE domain ontology, and we can perform prediction tasks using state-of-the-art language models such as BERT.

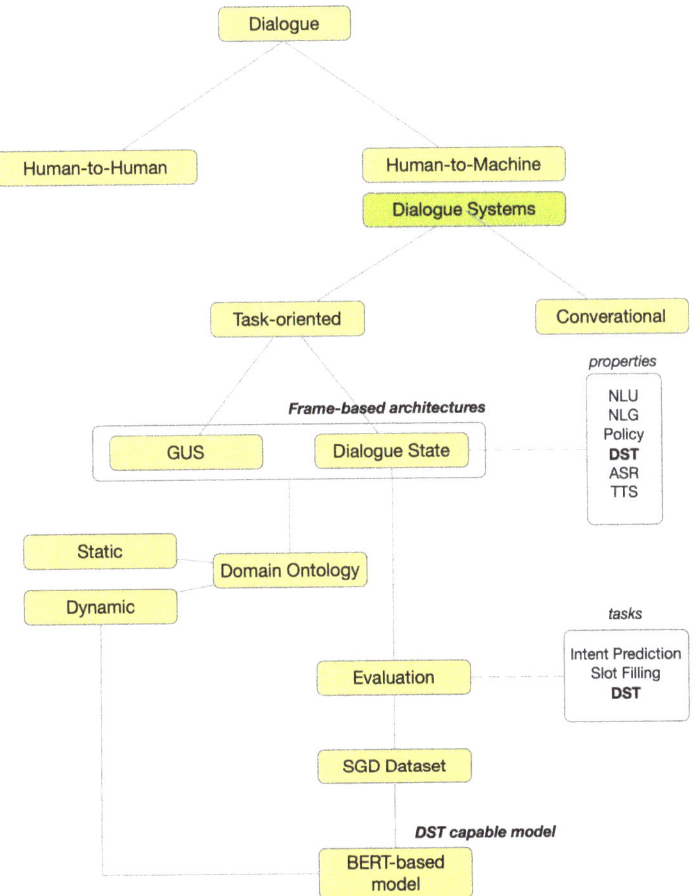

Figure 3. Dialogue System concepts related to CSE attack.

The SG-CSE BERT unit, as proposed, adheres to the schema-guided approach for zero-shot CSE attack state tracking by employing a fine-tuned BERT model. By taking various CSE attack schemas as input, the model can make predictions for a dynamic set of intents and slots. These schemas contain the description of supported intents and slots in natural language, which are then used to obtain a semantic representation of these schema elements. The use of a large pre-trained model such as BERT enables the SG-CSE BERT unit to generalize to previously unseen intents and slot-value pairs, resulting in satisfactory performance outcomes on the SG-CSE Corpus. The SG-CSE BERT unit is made up of two different components. A pre-trained BERT base cased model from Hugging Face Transformers to obtain embedded representations for schema elements, as well as to encode user utterances. The model is capable of returning both the encoding of the entire user utterance using the 'CLS' token, as well as the embedded representation of each individual token in the utterance, including intents, slots, and categorical slot values. This is achieved by utilizing the natural language descriptions provided in the schema files in the dataset. These embedded representations are pre-computed and are not fine-tuned during the optimization of model parameters in the state update module. The second component is a decoder that serves to return logits for predicted elements by conditioning on the encoded utterance. In essence, this component utilizes the encoded user utterance to make predictions regarding the user's intent and the relevant slot values. Together, these various components comprise the SG-CSE BERT unit.

4.2. The SG-CSE Dialogue Acts

Given the inherent differences between CSE attack dialogues and standard dialogues, a carefully designed set of dialogue-act labels is necessary to meet the requirements of CSE attack recognition. These dialogue acts should be able to capture the intentions of the attacker while remaining easily understandable. To create a set of appropriate dialogue acts, the utterances of both interlocutors in the CSE corpus [17] were analyzed, classified, and combined into pairs. Building on Young's paradigm outlined in [32], we propose a set of fourteen dialogue acts. Each SG-CSE DA is represented by an intent slot and has a data type and a set of values that it can take. The full list of the dialogue acts is presented in the following table (Table 3). The percentage of each dialogue act frequency is also given in the last column for the total number of utterances in the SG-CSE Corpus.

Table 3. Frequency of Dialogue Acts in SG-CSE Corpus.

Dialogue Act	Example	%
Greeting	Hello, my name is John	6
Statement	I've lost the connection	21
Uninterpreted	:-)	2
Agreement	Sure, I can	7
Question	Is this the Sales department?	11
Yes-No Question	Can you give me a copy?	9
WH-SME-Question	What is the HQ Address?	5
WH-IT-Question	What is your IP address?	8
WH-Personal Question	Which one is your personal email?	4
Rephrase	Do you have a network address?	1
Directive	Click this link, please	16
Yes Answer	No, not possible	5
Reject	I can't do this	3
Bye	Cheers!	2

The following Table 4 contains examples of dialogue acts mapped to real word utterances from the SG-CSE Corpus.

Table 4. Sample utterances and corresponding dialogue acts.

Utterance	Dialogue Act
Hello Angel, I'm experiencing a problem with my Charge 2	Greeting
I saw that the battery was leaking	Statement
Can you confirm the email associated with my [ANON] account?	Directive
Cheers!	Bye

4.3. The SG-CSE Corpus

The SG-CSE Corpus is a task-oriented dialogue corpus specifically designed for CSE attacks and is derived from the CSE corpus. Its main purpose is to serve as a training set for intent prediction, slot-value prediction, and dialogue-state tracking in the context of CSE attacks. The evaluation set of the SG-Corpus contains previously unseen services from the CSE domain, which allows us to evaluate the SG-CSE BERT model's ability to generalize in zero-shot settings. The corpus comprises various types of CSE attacks, including real-life cases and fictional scenarios. The hybrid approach used to create the SG-CSE Corpus combines characteristics of both balanced and opportunistic corpora [39,40], and it is based on the schema-guided paradigm. The SG-CSE Corpus contains 90 dialogues and is presented in Table 5.

Table 5. SG-CSE Corpus identity.

Characteristic	Value
Corpus name	SG-CSE Corpus
Collection Methods	Web scraping, pattern-matching text extraction
Corpus size	(N) 90 text dialogues/5798 sentences
Vocabulary size	(V) 4500 terms
Total no. of turns	5798
Avg. tokens per turn	8.42
No. of slots	18
No. of slot values	234
Content	chat-based dialogues
Collection date	June 2018–December 2020
Creation date	Aug. 2022

The original CSE corpus was preprocessed, and all dialogues were converted into pairs of utterances and annotated. In order to enhance the corpus, we utilized paraphrasing techniques and replaced significant verbs and nouns to generate additional dialogues. More specifically, to address the potential limitations of a small corpus, we created a list of 100 critical nouns and 100 critical verbs categorized under one of three sensitive data categories, and for each noun and verb, we generated ten new sentences where the noun was substituted with a similar word. In this regard, we utilized the Word Embeddings technique, which utilizes dense vector representations to capture the meaning of words. The embeddings are learned by moving points in the vector space based on the surrounding words of the target word. To discover synonyms, we employed a pre-trained word2vec model based on the distributional hypothesis, which posits that linguistic items with similar contextual distributions have similar meanings. By grouping words with similar contextual distributions, we created a segregation of different domain words based on their vector values. This resulted in a vector space where each unique word in the dialogues was assigned a corresponding vector, representing a vector representation of the words in the collected dialogues. The critical verbs and nouns were replaced and combined to create new phrases. The following figure (Figure 4) depicts the distribution of the number of turns in the dialogues involved.

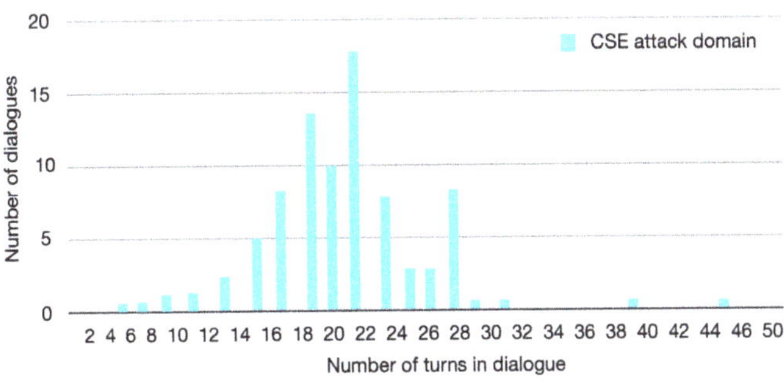

Figure 4. Distribution of turns in the SG-CSE Corpus.

Figure 5 presents the distribution of types of dialogue acts existing in the SG-CSE Corpus:

The annotation task has significantly enhanced the value and quality [41] of the SG-CSE Corpus, allowing it to be utilized for task-oriented dialogue tasks, such as DST. Additionally, natural language descriptions of the various schema elements (i.e., services, slots, and intents) are included in the corpus. To test the zero-shot generalization ability, the evaluation set includes at least five services that are not present in the training set. The SG-CSE Corpus is composed of CSE attack dialogues between two interlocutors, where each dialogue pertains to a specific CSE attack service in the form of a sequence of turns. Each turn is annotated with the active intent, dialogue state, and slot spans for the different slot values mentioned in the turn. The schema includes information such as the service name, the supported tasks (intents), and the attributes of the entities used by the service (slots).

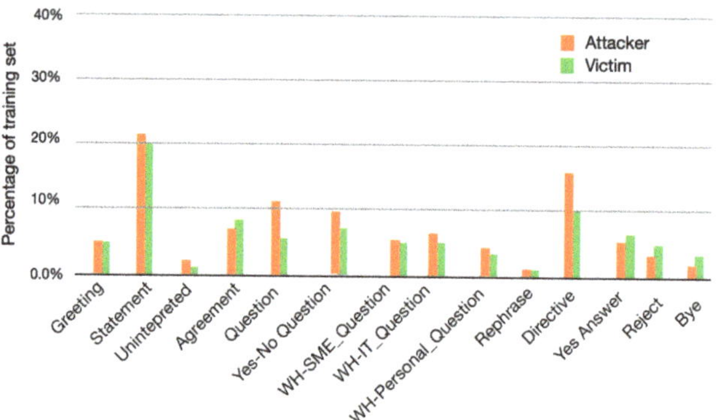

Figure 5. Distribution of types of dialogue acts in the SG-CSE Corpus.

4.4. The SG-CSE Ontology

The CSE ontology [17] is an ontology that connects social engineering concepts with cybersecurity concepts and focuses on sensitive data that could be leaked from an SME employee during a chat-based dialogue. It is an asset-oriented ontology that was derived from the corresponding concept map [17] depicted in Figure 6. The sensitive data may pertain to SME, IT, or personal details.

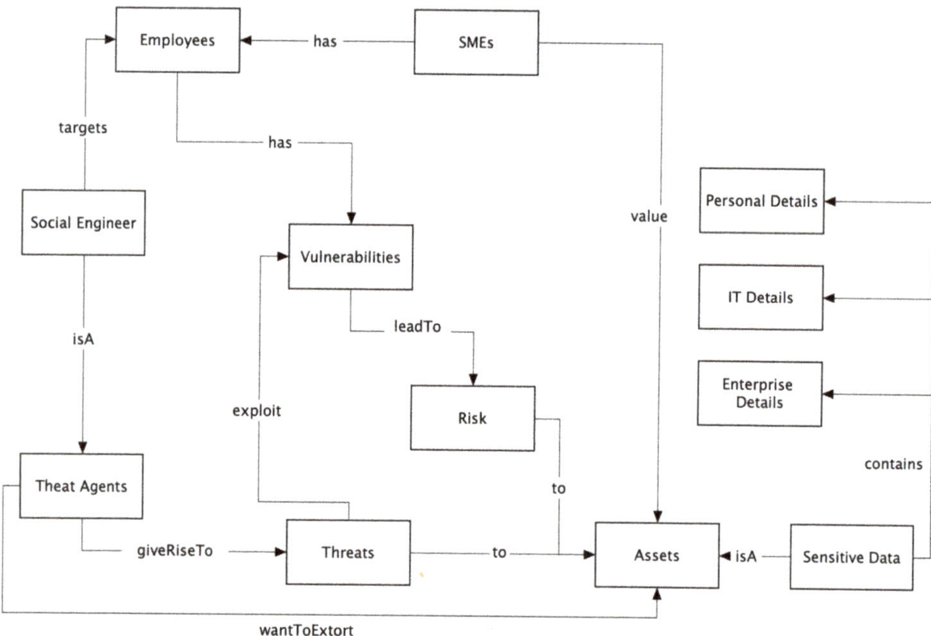

Figure 6. Concept map of the CSE attack domain.

The CSE ontology groups similar concepts in context to facilitate the hierarchical categorization of assets and is designed to be efficient for text classification tasks by limiting the depth to no more than three levels. It was created through a custom information extraction system that leveraged various text documents such as corporate IT policies, IT professionals' CVs, and ICT manuals. An excerpt of the CSE ontology created in Protégé is illustrated in Figure 7 below, along with the arc types.

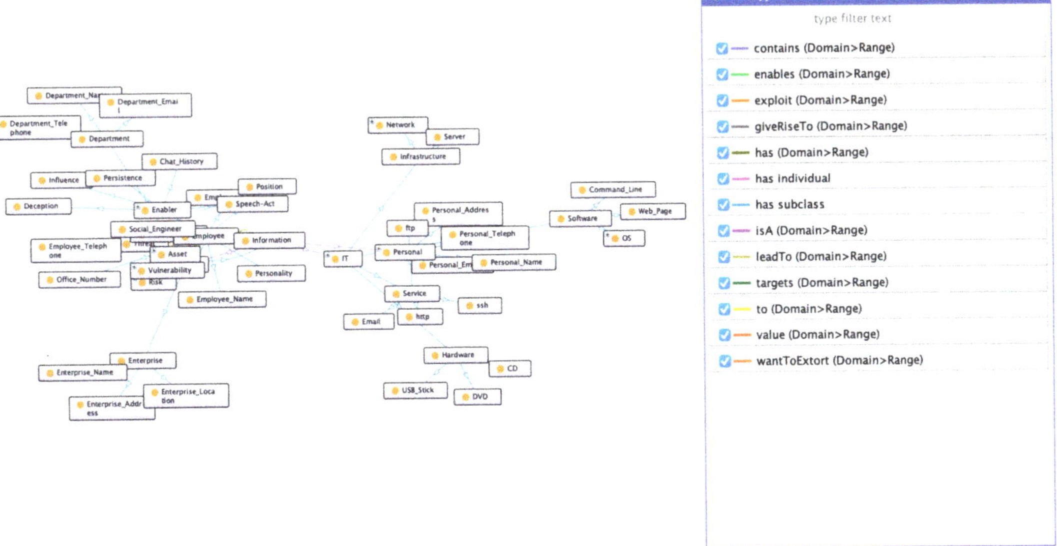

Figure 7. Excerpt of CSE ontology.

The SG-CSE ontology comprises a collection of schemas that describe the CSE attack domain based on the CSE ontology and the proposed set of fourteen SG-CSE DAs [42]. The use of a schema-based ontology enables a structured and domain-specific representation of knowledge, enhancing the robustness and accuracy of the detection task. This is achieved through the provision of predefined frames, encompassing predetermined entities and concepts, which the system can expect to encounter as input. Table 6 presents the 18 slots existing in SG-CSE in tabular form with accompanying example values.

The services present in the training set are the following four:

- CSE_SME_Info_Extraction
- CSE_IT_Info_Extraction
- CSE_Personal_Extraction
- CSE_Department_Extraction

where the slots related to each service are combinations of the aforementioned slots in Table 6.

Figure 8 depicts an excerpt of the example schema for the CSE_SME_Info_Extraction service. The dots between the boxes denote that more slots and intents exist.

4.5. The SG-CSE BERT Model

This section introduces the SG-CSE BERT model, which operates in a schema-guided setting and can condition the CSE attack service schema using the descriptions of intents and slots. To enable the model to represent unseen intents and slots, a BERT-based model pre-trained on large corpora is employed. As a result, the proposed SG-CSE is a model for zero-shot schema-guided dialogue-state tracking in CSE attack dialogues. The SG-CSE BERT model encodes all schema intents, slots, and categorical slot values into embedded representations. Since social engineering attacks can take many forms, schemas may differ in their number of intents and slots. Therefore, predictions are made over a dynamic set of intents and slots by conditioning them on the corresponding schema embedding.

Table 6. Slots in SG-CSE Corpus.

Slot	Type	Example Values
Hardware	Dictionary	CD, DVD, USB stick
Network	Numbers	192.168.13.21
Server	String	subdomain.domain.tld
Service	String	Email, FTP, ssh
Software	String	RDP, Firefox
Personal_Email	String	user@domain.tld
Personal_Name	String	George, Sandra
Personal_Telephone	Numbers	0123456789
Dept_Email	String	dept@domain.tld
Dept_Name	String	Sales, Marketing
Employee_Name	String	George, Maria
Employee_Email	String	name@domain.tld
Employee_Telephone	Numbers	0123456789
Employee_Position	String	Executive, Manager
Office_Number	Numbers	12, 23
Enterprise_Address	String	26th Av. Somewhere
Enterprise_Location	String	Athens, Thessaloniki
Enterprise_Name	String	ACME, Other

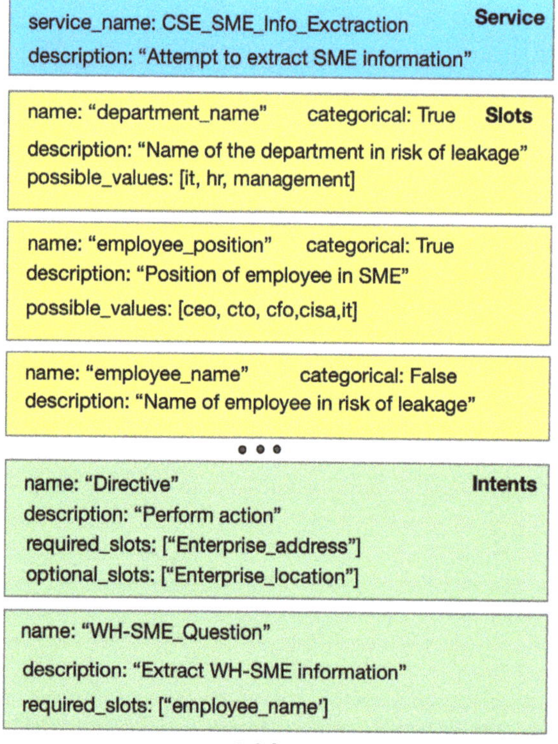

Figure 8. Example schema for CSE_SME_Info_Extraction.

The sequence pairs used for the embeddings of intent, slot, and slot value are presented in Table 7, and they are fed as input to the pre-trained BERT model (Figure 9) where

$a_1, \ldots, a_n$ and $v, \ldots, v_n$ are the two sequence tokens that are fed as a pair to SG-CSE BERT encoder. The U_{CLS} is the embedded representation of the schema and $t, \ldots, t_{n+m}$ are the token-level representations.

Table 7. Sequence pair used for embeddings.

	Sequence 1	Sequence 2
Intent	CSE attack description	Intent description
Slot	CSE attack description	Slot description
Value	Slot description	value

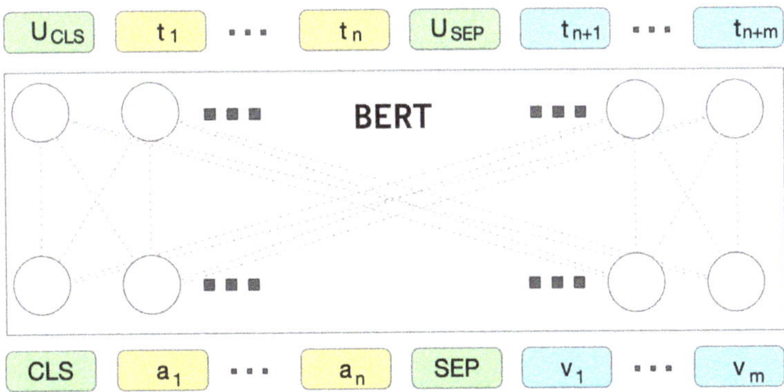

Figure 9. Pre-trained BERT model A pair of two sequences a, v is fed as input to the BERT encoder which outputs the pair's embedding U_{CLS} and the token representations $\{t_1, \ldots, t_{n+m}\}$.

Schema Embeddings: Let I, S be the intents and slots of CSE attack service and $\{i_j\}$ where $1 \leq j \leq I$ and $\{s_j\}$ where $1 \leq j \leq S$ their embeddings. The embeddings of the N non-categorical slots are denoted by $\{s_j^n\}$ where $1 \leq j \leq N \leq S$ and the embeddings for all possible values that the kth categorical slot can take are denoted by $\{v_j^k\}$ where $1 \leq j \leq V^k$ and $1 \leq k \leq C$. C is the number of categorical slots and $N + C = S$.

Utterance Embedding: A pair of two consecutive utterances between the two interlocutors is encoded and represented to embeddings as u_{CLS} and the token level representations $\{t_i\}$ where $1 \leq i \leq M$ and M the total number of tokens in the pair of utterances.

The model utilizes the schema and utterance embeddings and a set of projections [4] to proceed to predictions for active intent, requested slot, and user goal. More specifically, the **active intent,** which is the intent requested by the attacker that we are trying to recognize, takes the value 'NONE' if the model currently processes no intent. Otherwise, if i_0 is the trainable parameter in $\mathbb{R}^d$ then the intent is given by:

$$l_{int}^j = \mathcal{F}_{int}(u, i_j, 1), \ 0 \leq j \leq I$$

SoftMax function is used to normalize the logits l_{int}^j are normalized and produce a distribution over the set of intents plus the "NONE" intent. The intent with the highest probability is predicted as active.

The **requested slots,** which means the slots whose values are requested by the user in the current utterance, are given by:

$$l_{req\_slot}^j = \mathcal{F}_{req\_slot}(u, s_j, 1), \ 0 \leq j \leq S$$

The sigmoid function I is used to normalize the logits $l^j_{req\_slot}$ and get a score in the range of [0, 1]. During inference, all slots with a score > 0.6 are predicted as requested.

The **user goal** is defined as the user constraints specified over the dialogue context till the current user utterance, and it is predicted in two stages. In the first stage, a distribution of size 3 denoting the slot status taking values *none*, *harmless*, and *current* is obtained using:

$$l^j_{status} = \mathcal{F}_{status}\left(u,\, s_j,\, 3\right),\ 1 \leq j \leq S$$

If the status is predicted as *none*, its value is assumed unchanged. If the predicted status is *harmless*, then the slot gets the value *harmless*. Otherwise, the slot value is predicted using the following:

$$l^{j,k}_{value} = \mathcal{F}_{status}\left(u,\, v^k_j,\, 1\right),\ 1 \leq j \leq V^k,\ 1 \leq k \leq C$$

The SoftMax algorithm is used to map the categorical values of a variable into a distribution over the entire range of possible values. For each non-categorical slot, logits are obtained using the following:

$$l^{j,k}_{start} = \mathcal{F}_{start}\left(t_k,\, s^n_j,\, 1\right),\ 1 \leq j \leq N,\ 1 \leq k \leq M$$

$$l^{j,k}_{end} = \mathcal{F}_{end}\left(t_k,\, s^n_j,\, 1\right),\ 1 \leq j \leq N,\ 1 \leq k \leq M$$

Each of these two distributions above represents the starting and end points for the actual span of text that references the specific slot. The indices $a \leq v$ maximizing $start[a] + end[v]$ will be the boundary between spans, and the value associated with that span is assigned to that slot.

In Figure 10 above, we see the predicted CSE attack dialogue state using two turns from two different utterance pairs. In the green boxes, the active intent and slot assignments are shown, and in the orange box, we can see the related schema of the CSE attack service. The CSE attack state representation is conditioned on the CSE attack schema, which is provided as input along with the victim and attacker utterances.

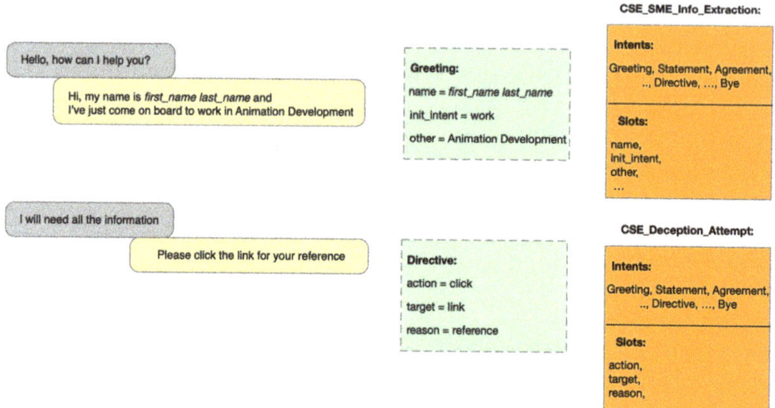

Figure 10. Two example pairs of utterances with predicted dialogue states (in dashed edges). The dialogue-state representation is conditioned on the CSE attack schema under consideration, shown in orange.

5. Results

We consider the following metrics for evaluation of the CSE attack state tracking:

- Active Intent Accuracy: The fraction of user turns for which the active intent has been correctly predicted.
- Requested Slot F1: The macro-averaged F1 score for requested slots overall eligible turns. Turns with no requested slots in ground truth and predictions are skipped.
- Average Goal Accuracy: For each turn, we predict a single value for each slot present in the dialogue state. This is the average accuracy of predicting the value of a slot correctly.
- Joint Goal Accuracy: This is the average accuracy of correctly predicting all slot assignments for a given service in a turn. Additionally, Harmonic means between seen and unseen classes.

We implemented the SG-CSE BERT model using the Hugging Face library and the BERT uncased model with 12 layers, 768 hidden dimensions, and 12 self-attention heads. To train the model, we used a batch size of 32 and a dropout rate of 0.2 for all classification heads. We also employed a linear warmup strategy with a duration of 10% of the training steps, in addition to the AdamW optimizer with a learning rate of 2×10^{-5}.

In Figure 11, the performance of the SG-CSE BERT model is depicted. SG-CSE BERT shows efficiency in Active Intent Accuracy and Requested Slots F1 and less efficiency in Average Goal Accuracy and Joint Goal Accuracy.

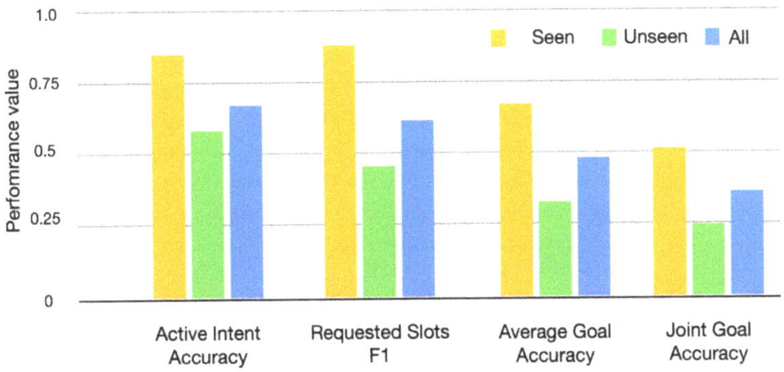

Figure 11. SG-CSE BERT model performance.

The following Table 8 presents the performance results.

Table 8. SG-CSE Attack state tracker performance.

System	Model	Parameters	Active Intent Accuracy	Req Slot F1	Acg Goal Accuracy	Joint Goal Accuracy
Seen	BERT$_{BASE}$	110 M	85.2	89.6	74.1	56.7
Unseen	BERT$_{BASE}$	110 M	53.8	48.3	31.4	24.9
All	BERT$_{BASE}$	110 M	69.5	68.9	52.7	40.8

6. Discussion

Early dialogue-state tracking (DST) datasets were developed to be specific to a particular domain due to the difficulty in building models that can effectively track dialogue states for multiple domains. However, with the recent release of multi-domain datasets and the incorporation of machine learning-based methods, it has become possible to build models that can track states for multiple domains using a single set of training data. To evaluate the ability of models to generalize in zero-shot settings, we created the SG-CSE Corpus and included evaluation sets containing unseen services. To address the issue of limited dialogue resources, data augmentation can be explored as an option. Augmenting the training dataset by adding more diverse examples can improve performance. Source-based augmentation generates sentences by changing a single variable value in a sample utterance, while target-based augmentation takes portions of sentences from different places in the training data and recombines them.

SG-CSE BERT is a model that is built around BERT, a pre-trained transformer-based model that has been trained on a large corpus of text data. BERT has demonstrated strong generalization ability across a wide range of natural language processing tasks and domains. When fine-tuned on a specific task and domain, BERT is able to learn specific patterns and features of the task and domain, which allows it to achieve good performance. However, if BERT is not fine-tuned, it may be able to detect new unseen intents, but it would not have enough information to generate the corresponding slot values. Moreover, it may not be able to detect new unseen services or new unseen domains. Large-scale neural language models trained on massive corpora of text data have achieved state-of-the-art results on a variety of traditional NLP tasks. However, although the standard pre-trained BERT is capable of generalizing, task-specific fine-tuning is essential for achieving good performance. This is confirmed by recent research, which has shown that pre-training alone may not be sufficient to achieve high accuracy in NLP tasks. For example, the performance of a pre-trained model on a downstream task may be significantly improved by fine-tuning it on a smaller in-domain dataset.

The SG-CSE Corpus is designed to test and evaluate the ability of dialogue systems to generalize in zero-shot settings. The evaluation set of the corpus contains unseen services, which is important to test the generalization ability of the model. Our evaluation set does not expose the set of all possible values for certain slots. It is impractical to have such a list for slots like IP addresses or time because they have infinitely many possible values or for slots like names or telephone numbers for which new values are periodically added. Such slots are specifically identified as non-categorical slots. In our evaluation sets, we ensured the presence of a significant number of values that were not previously seen in the training set to evaluate the performance of models on unseen values. Some slots, like hardware, software, etc., are classified as categorical, and a list of all possible values for them is provided in the schema.

SG-CSE BERT is designed to handle dialogue-state tracking for a larger number of related services within the same domain. Its schema-guided paradigm provides a structured and domain-specific representation of knowledge, which increases the model's robustness and accuracy. This, in turn, allows the system to better understand and interpret user input and track the dialogue state more accurately. The schema and frames work together to create this representation, with the schema constraining the possible states of the dialogue and the frames tracking the current state. The proposed system has shown satisfactory performance, and the experimental results demonstrate its effectiveness. Its simplistic approach also leads to more computational efficiency, which makes it a good candidate for use as a separate component in a holistic CSE attack recognition system, as proposed in previous works [3,43,44]. Additionally, SG-CSE BERT is computationally efficient and scalable to handle large schemata and dialogues. It should be noted that the performance of SG-CSE BERT is dependent on the quality and completeness of the training data. Moreover, incorporating additional sources of information, such as user context and sentiment, can enhance the system's performance in real-world scenarios. Overall, SG-CSE BERT provides a promising solution for zero-shot schema-guided dialogue-state tracking in the domain of CSE attack recognition.

A static ontology, which is a predefined and unchanging set of intents and slots, can be used for zero-shot detection in DST. However, there are some limitations to its effectiveness [45]. While a comprehensive ontology can cover a wide range of possible situations and contexts, it may not be able to encompass all possible scenarios. As a result, a model trained on such an ontology may struggle to generalize to new and unseen intents and slots. Moreover, since the ontology remains static and is not updated during the training process, it may not be able to adapt to changes in the domain over time or new types of attacks. In contrast, a dynamic ontology can be updated during the training process to adapt to new situations and contexts. This can be achieved by using unsupervised methods to extract the ontology from the data or by incorporating active learning methods that allow the system to query human experts when encountering unseen intents and slots. By using a dynamic ontology, the system can learn to recognize new intents and slots over time, improving its ability to generalize to new and unseen scenarios. Zero-shot detection in DST remains an active area of research, with new methods and approaches being developed to improve the performance of DST models in detecting unseen intents and slots. By incorporating dynamic ontologies and other techniques, future DST models may be better equipped to recognize and respond to previously unseen user input.

7. Conclusions

The aim of this study was to investigate schema-guided dialogue-state tracking in the context of CSE attacks. We created a set of fourteen dialogue acts and developed the SG-CSE Corpus using the CSE ontology and corpus. We followed the schema-guided paradigm to introduce the SG-CSE BERT, a simplistic model for zero-shot CSE attack state tracking. The performance results were promising and demonstrated the effectiveness of the approach. To provide context for the study, we discussed dialogue systems and their characteristics, which helped us define our approach. Then, we examined how concepts and terminology from task-based dialogue systems and dialogue-state tracking can be transferred to the CSE attack domain.

We focused on creating the SG-CSE DAs and SG-CSE Corpus, mapping slots and intents, and proposing the SG-CSE BERT model. The model achieved satisfactory performance results using a small model and input encoding. Although various model enhancements were attempted, no significant improvement was observed. The study suggests that data augmentation and the addition of hand-crafted features could improve the performance of the CSE attack state tracking, but further experimentation is necessary to explore these methods. The proposed model offers an advantage in the few-shot experiments, where only limited labeled data is available. Such an approach can help to create a more comprehensive and accurate understanding of CSE attacks and their underlying mechanisms, as well as to develop more effective detection and prevention strategies. For example, experts in natural language processing can help to improve the performance of dialogue-state tracking models, while experts in psychology and sociology can provide insights into the social engineering tactics used in CSE attacks and the psychological factors that make users susceptible to these attacks. Cyber-security experts can provide a deeper understanding of the technical aspects of CSE attacks and help to develop more robust and effective defense mechanisms.

In addition, it is important to emphasize the need for ethical considerations in the development of CSE attack detection systems. As these systems involve the processing of sensitive and personal information, it is crucial to ensure that they are designed and used in a way that respects privacy and protects against potential biases and discrimination. This requires a careful and transparent design process, as well as ongoing monitoring and evaluation of the system's performance and impact. Overall, a multi-disciplinary and ethical approach is essential for the development of effective and responsible CSE attack detection systems.

Future works will focus on the SG-CSE corpus augmentation techniques to present new CSE attack services. Furthermore, the proposed SG-CSE BERT will be incorporated into an ensemble system that will detect CSE attacks, where it will act as an individual component responsible for practicing dialogue-state tracking.

Author Contributions: Conceptualization, N.T., P.F. and I.M.; methodology, N.T. and I.M.; software, N.T.; evaluation, N.T. and P.F.; writing—original draft preparation, N.T. and P.F.; writing—review and editing, N.T., P.F. and I.M. All authors have read and agreed to the published version of the manuscript.

Funding: This research received no external funding.

Institutional Review Board Statement: Not applicable.

Informed Consent Statement: Not applicable.

Conflicts of Interest: The authors declare no conflict of interest.

Abbreviations

A	Attacker
ANON	Anonymized content
ASR	Automatic Speech Recognition
BERT	Bidirectional Encoder Representations from Transformers,
CEO	chief executive officer
CSE	Chat-based Social Engineering
CV	Curriculum Vitae
D3ST	Description-Driven Dialogue-State Tracking
DA	Dialogue Act
DAMSL	Dialogue Act Markup in Several Layers
DS	Dialogue State
DST	Dialogue-State Tracking
DSTC	Dialogue-State Tracking Challenge
GPT	Generative Pre-trained Transformer
GUS	Genial Architecture System
ICT	Information and Communication Technology
IT	Information Technology
MRC	Machine Reading Comprehension
MSDU	Multi-turn Spoken Dialogue Understanding
NLG	Natural Language Generation
NLU	Natural Language Understanding
POMDP	Partially Observable Markov Decision Process
RNN	Recurrent Neural Network
SG	Schema-Guided
SG-CSE Das	Schema-Guided Chat-based Social Engineering Dialogue Acts
SG-CSEAST	Schema-Guided Chat-based Social Engineering Attack State Tracker
SME	Small-Medium Enterprises
SOM	Selectively Overwriting Memory
SWBD-DAMSL	Switchboard-DAMSL
SwDA	Switchboard Dialogue Act Corpus
TTS	Text-to-Speech
V	Victim
WH	WH question (who, what, whose, which, when, where, why)

References

1. Matthews, P.H. *The Concise Oxford Dictionary of Linguistics*; OUP Oxford: Oxford, UK, 2014. [CrossRef]
2. Verizon Business. Data Breach Investigations Report'. Available online: https://www.verizon.com/business/resources/reports/dbir/ (accessed on 20 November 2022).
3. Tsinganos, N.; Sakellariou, G.; Fouliras, P.; Mavridis, I. Towards an Automated Recognition System for Chat-based Social Engineering Attacks in Enterprise Environments. In Proceedings of the 13th International Conference on Availability, Reliability and Security, Hamburg, Germany, 27–30 August 2018; p. 10. [CrossRef]
4. Rastogi, A.; Zang, X.; Sunkara, S.; Gupta, R.; Khaitan, P. Towards Scalable Multi-domain Conversational Agents: The Schema-Guided Dialogue Dataset. *arXiv* **2020**. Available online: http://arxiv.org/abs/1909.05855 (accessed on 21 October 2022).
5. Xu, P.; Hu, Q. An end-to-end approach for handling unknown slot values in dialogue state tracking. In Proceedings of the 56th Annual Meeting of the Association for Computational Linguistics (Volume 1: Long Papers), Melbourne, Australia, 15–20 July 2018; pp. 1448–1457. [CrossRef]
6. Chao, G.-L.; Lane, I. BERT-DST: Scalable End-to-End Dialogue State Tracking with Bidirectional Encoder Representations from Transformer. *arXiv* **2019**. Available online: http://arxiv.org/abs/1907.03040 (accessed on 26 October 2022).
7. Deriu, J.; Rodrigo, A.; Otegi, A.; Echegoyen, G.; Rosset, S.; Agirre, E.; Cieliebak, M. Survey on evaluation methods for dialogue systems. *Artif. Intell. Rev.* **2021**, *54*, 755–810. [CrossRef] [PubMed]

8. Xiong, W.; Ma, L.; Liao, H. An Efficient Approach based on BERT and Recurrent Neural Network for Multi-turn Spoken Dialogue Understanding. In Proceedings of the 12th International Conference on Agents and Artificial Intelligence, Valletta, Malta, 22–24 February 2020; pp. 793–800. [CrossRef]

9. Zhao, J.; Gupta, R.; Cao, Y.; Yu, D.; Wang, M.; Lee, H.; Rastogi, A.; Shafran, I.; Wu, Y. Description-Driven Task-Oriented Dialog Modeling. *arXiv* **2022**. [CrossRef]

10. Ma, Y.; Zeng, Z.; Zhu, D.; Li, X.; Yang, Y.; Yao, X.; Zhou, K.; Shen, J. An End-to-End Dialogue State Tracking System with Machine Reading Comprehension and Wide & Deep Classi-fication. *arXiv* **2020**. [CrossRef]

11. Kim, S.; Yang, S.; Kim, G.; Lee, S.-W. Efficient Dialogue State Tracking by Selectively Overwriting Memory. *arXiv* **2020**. Available online: http://arxiv.org/abs/1911.03906 (accessed on 27 October 2022).

12. Kumar, A.; Ku, P.; Goyal, A.; Metallinou, A.; Hakkani-Tur, D. MA-DST: Multi-Attention-Based Scalable Dialog State Tracking. *Proc. Conf. AAAI Artif. Intell.* **2020**, *34*, 8107–8114. [CrossRef]

13. Lin, Z.; Liu, B.; Moon, S.; Crook, P.A.; Zhou, Z.; Wang, Z.; Yu, Z.; Madotto, A.; Cho, E.; Subba, R. Leveraging Slot Descriptions for Zero-Shot Cross-Domain Dialogue StateTracking. *arXiv* **2021**. Available online: http://arxiv.org/abs/2105.04222 (accessed on 27 October 2022).

14. Lin, Z.; Liu, B.; Madotto, A.; Moon, S.; Zhou, Z.; Crook, P.A.; Wang, Z.; Yu, Z.; Cho, E.; Subba, R.; et al. Zero-Shot Dialogue State Tracking via Cross-Task Transfer. *arXiv* **2021**. Available online: http://arxiv.org/abs/2109.04655 (accessed on 27 October 2022).

15. Li, S.; Cao, J.; Sridhar, M.; Zhu, H.; Li, S.W.; Hamza, W.; McAuley, J. Zero-Shot Generalization in Dialog State Tracking through Generative Question Answering. *arXiv* **2021**. [CrossRef]

16. Jurafsky, D.; Martin, J. Speech and Language Processing. 2022. Available online: https://web.stanford.edu/~jurafsky/slp3/ (accessed on 21 October 2022).

17. Tsinganos, N.; Mavridis, I. Building and Evaluating an Annotated Corpus for Automated Recognition of Chat-Based Social Engineering Attacks. *Appl. Sci.* **2021**, *11*, 10871. [CrossRef]

18. DBobrow, G.; Kaplan, R.M.; Kay, M.; Norman, D.A.; Thompson, H.S.; Winograd, T. GUS, A frame-driven dialog system. *Artif. Intell.* **1986**, *8*, 155–173. [CrossRef]

19. Austin, J.L. *How to Do Things with Words*; Oxford University Press: Oxford, UK, 1975; Volume 88. [CrossRef]

20. Searle, J.R. *Speech Acts: An Essay in the Philosophy of Language*; Cambridge University Press: Cambridge, UK, 1969. [CrossRef]

21. Searle, J.R.; Kiefer, F.; Bierwisch, M. *Speech Act Theory and Pragmatics*; Springer: Berlin/Heidelberg, Germany, 1980; Volume 10, ISBN 978-94-009-8964-1.

22. Bach, K.; Harnish, R.M. *Linguistic Communication and Speech Acts*; MIT Press: Cambridge, MA, USA, 1979; ISBN 9780262520782.

23. Kissine, M. *From Utterances to Speech Acts*; Cambridge University Press: Cambridge, UK, 2013. [CrossRef]

24. Godfrey, J.; Holliman, E. *Switchboard-1 Release 2*; Linguistic Data Consortium: Philadelphia, PA, USA, 1993; p. 14610176 KB. [CrossRef]

25. Jurafsky, D.; Bates, R.; Coccaro, N.; Martin, R.; Meteer, M.; Ries, K.; Shriberg, E.; Stolcke, A.; Taylor, P.; Van Ess-Dykema, C. Automatic detection of discourse structure for speech recognition and understanding. In Proceedings of the 1997 IEEE Workshop on Automatic Speech Recognition and Understanding Proceedings, Santa Barbara, CA, USA, 17 December 1997; pp. 88–95. [CrossRef]

26. WS-97 Switchboard DAMSL Coders Manual. Available online: https://web.stanford.edu/~jurafsky/ws97/manual.august1.html (accessed on 19 October 2022).

27. Narayan, A.; Hedtke, J. DAMsL: A Meta-Learning Based Approach for Dialogue State Tracking. pp. 1–10. Available online: https://web.stanford.edu/class/archive/cs/cs224n/cs224n.1204/reports/custom/report34.pdf (accessed on 26 October 2022).

28. Quarteroni, S.; Ivanov, A.V.; Riccardi, G. Simultaneous dialog act segmentation and classification from human-human spoken conversations. In Proceedings of the 2011 IEEE International Conference on Acoustics, Speech and Signal Processing (ICASSP), Prague, Czech Republic, 22–27 May 2011; pp. 5596–5599. [CrossRef]

29. Liu, S.; Chen, H.; Ren, Z.; Feng, Y.; Liu, Q.; Yin, D. Knowledge Diffusion for Neural Dialogue Generation. In Proceedings of the 56th Annual Meeting of the Association for Computational Linguistics (Volume 1: Long Papers), Melbourne, Australia, 15–20 July 2018; pp. 1489–1498. [CrossRef]

30. Ortega, D.; Vu, N.T. Lexico-Acoustic Neural-Based Models for Dialog Act Classification. In Proceedings of the 2018 IEEE International Conference on Acoustics, Speech and Signal Processing (ICASSP), Calgary, AB, Canada, 15–20 April 2018; pp. 6194–6198. [CrossRef]

31. Williams, J.D.; Raux, A.; Henderson, M. The Dialog State Tracking Challenge Series: A Review. *Dialog Discourse* **2016**, *7*, 4–33. [CrossRef]

32. Young, S.; Gašić, M.; Keizer, S.; Mairesse, F.; Schatzmann, J.; Thomson, B.; Yu, K. The Hidden Information State model: A practical framework for POMDP-based spoken dialogue management. *Comput. Speech Lang.* **2010**, *24*, 150–174. [CrossRef]

33. Williams, J.D.; Young, S. Partially observable Markov decision processes for spoken dialog systems. *Comput. Speech Lang.* **2007**, *21*, 393–422. [CrossRef]

34. Rastogi, A.; Zang, X.; Sunkara, S.; Gupta, R.; Khaitan, P. Schema-Guided Dialogue State Tracking Task at DSTC8. *arXiv* **2020**. [CrossRef]

35. Radford, A.; Narasimhan, K.; Salimans, T.; Sutskever, I. Improving Language Understanding by Generative Pre-Training. pp. 1–12. Available online: https://www.cs.ubc.ca/~amuham01/LING530/papers/radford2018improving.pdf (accessed on 26 October 2022).
36. Peters, M.E.; Neumann, M.; Iyyer, M.; Gardner, M.; Clark, C.; Lee, K.; Zettlemoyer, L. Deep contextualized word representations. *arXiv* **2018**, arXiv:1802.05365.
37. Devlin, J.; Chang, M.W.; Lee, K.; Toutanova, K. Bert: Pre-training of deep bidirectional transformers for language understanding. *arXiv* **2018**, arXiv:1810.04805.
38. Yang, Z.; Dai, Z.; Yang, Y.; Carbonell, J.; Salakhutdinov, R.; Le, Q.V. XLNet: Generalized Autoregressive Pretraining for Language Understanding. *arXiv* **2020**. [CrossRef]
39. Kurdi, M.Z. *Natural Language Processing and Computational Linguistics: Speech, Morphology and Syntax*; John Wiley & Sons: Hoboken, NJ, USA, 2016.
40. Chomsky, N. *Syntactic Structures*; De Gruyter Mouton: Berlin, Germany, 2009. [CrossRef]
41. Ide, N.; Pustejovsky, J. (Eds.) *Handbook of Linguistic Annotation*, 1st ed.; Springer: Berlin/Heidelberg, Germany, 2017; ISBN 978-9402408799.
42. Duran, N.; Battle, S.; Smith, J. Sentence encoding for Dialogue Act classification. *Nat. Lang. Eng.* **2021**, 1–30. [CrossRef]
43. Tsinganos, N.; Mavridis, I.; Gritzalis, D. Utilizing Convolutional Neural Networks and Word Embeddings for Early-Stage Recognition of Persuasion in Chat-Based Social Engineering Attacks. *IEEE Access* **2022**, *10*, 108517–108529. [CrossRef]
44. Tsinganos, N.; Fouliras, P.; Mavridis, I. Applying BERT for Early-Stage Recognition of Persistence in Chat-Based Social Engineering Attacks. *Appl. Sci.* **2022**, *12*, 12353. [CrossRef]
45. Noble, B.; Maraev, V. Large-Scale Text Pre-Training Helps with Dialogue Act Recognition, but Not without Fine-Tuning. In Proceedings of the 14th International Conference on Computational Semantics (IWCS), Groningen, The Netherlands (online), 16–18 June 2021; pp. 166–172. Available online: https://aclanthology.org/2021.iwcs-1.16 (accessed on 21 October 2022).

Disclaimer/Publisher's Note: The statements, opinions and data contained in all publications are solely those of the individual author(s) and contributor(s) and not of MDPI and/or the editor(s). MDPI and/or the editor(s) disclaim responsibility for any injury to people or property resulting from any ideas, methods, instructions or products referred to in the content.

MDPI

St. Alban-Anlage 66

4052 Basel

Switzerland

www.mdpi.com

Applied Sciences Editorial Office

E-mail: applsci@mdpi.com

www.mdpi.com/journal/applsci

Disclaimer/Publisher's Note: The statements, opinions and data contained in all publications are solely those of the individual author(s) and contributor(s) and not of MDPI and/or the editor(s). MDPI and/or the editor(s) disclaim responsibility for any injury to people or property resulting from any ideas, methods, instructions or products referred to in the content.

www.ingramcontent.com/pod-product-compliance
Lightning Source LLC
LaVergne TN
LVHW071023170726
843515LV00006B/1187